OXFORD STUDIES IN PHYSICS

GENERAL EDITORS

B. BLEANEY, D. H. WILKINSON

THE MANY-BODY PROBLEM

BY

W. E. PARRY

Fellow of Oriel College, Oxford

CLARENDON PRESS · OXFORD

1973

Oxford University Press, Ely House, London W.1

GLASGOW NEW YORK TORONTO MELBOURNE WELLINGTON
CAPE TOWN IBADAN NAIROBI DAR ES SALAAM LUSAKA ADDIS ABABA
DELHI BOMBAY CALCUTTA MADRAS KARACHI LAHORE DACCA
KUALA LUMPUR SINGAPORE HONG KONG TOKYO

ISBN 0 19 851450 6

PRINTED IN GREAT BRITAIN
BY J. W. ARROWSMITH LTD, BRISTOL BS3 2NT

Many of them can be seen to be zero, and many others to be equal, but it would be convenient to be able to avoid having to think in such terms.

Second quantization allows us to do this. The idea is to express all operators in terms of a fundamental set of operators, whose matrix elements are known. This is a trick similar to the decimal system in ordinary arithmetic. To do arithmetic, we do not have to know the multiplication or addition tables for all the numbers we wish to use; we express all numbers in terms of a fundamental set, 0 to 9, and learn a set of rules which tell us how to add and multiply numbers expressed in terms of this fundamental set, using only addition and multiplication tables up to 9 by 9. In the many-body problem, we have to evaluate matrix elements of many operators: to do this we express those operators in terms of a fundamental set, called 'creation' and 'annihilation' operators, and then use a set of rules for the fundamental set. In the rest of this chapter, we define creation and annihilation operators and show how to express more complex operators in terms of them.

Exercises

1.1.1. Find the matrix element of the operator $\sum_{i=1}^{4} f(\mathbf{r}_i)$ between the state (properly normalized) in which one particle is in the αth single-particle state and three in the βth, and the state in which one is in the αth, two in the βth, and one in the γth. The particles are identical bosons, and the answer should be expressed in terms of the matrix element of f, $f_{\beta\gamma} = \int \phi_\beta^*(\mathbf{r}) f(\mathbf{r}) \phi_\gamma(\mathbf{r})\, \mathrm{d}^3\mathbf{r}$ between the single-particle states ϕ_β and ϕ_γ.

1.1.2. The Hamiltonian for the harmonic oscillator is:

$$H = \frac{p^2}{2m} + \frac{m\omega^2 x^2}{2} = -\frac{1}{2m}\frac{\partial^2}{\partial x^2} + \frac{m\omega^2 x^2}{2}.$$

Define:

$$A_+ = (p/\sqrt{m} + \mathrm{i}x\omega\sqrt{m})/\sqrt{(2\omega)}$$
$$A_- = (p/\sqrt{m} - \mathrm{i}x\omega\sqrt{m})/\sqrt{(2\omega)},$$

and show that:

(i) $H = \omega[A_+A_- + \frac{1}{2}] = \omega[A_-A_+ - \frac{1}{2}]$.
(ii) $[A_-, A_+] \equiv A_-A_+ - A_+A_- = 1$.
(iii) If $|\lambda\rangle$ is a normalized eigenvector of H with eigenvalue $\lambda\omega$, i.e.

$$H|\lambda\rangle = \omega\lambda|\lambda\rangle,$$

then

$$HA_+|\lambda\rangle = \omega(\lambda+1)A_+|\lambda\rangle$$

(i.e. $A_+|\lambda\rangle$ is an eigenvector of H with eigenvalue $\lambda+1$), and

$$HA_-|\lambda\rangle = \omega(\lambda-1)A_-|\lambda\rangle.$$

By using the fact that the eigenvalues of H must be non-negative, show that

$$\lambda = n+\tfrac{1}{2}, \qquad n = 0, 1, 2, \ldots$$

Show that

$$A_+|n+\tfrac{1}{2}\rangle = \sqrt{(n+1)}|n+\tfrac{3}{2}\rangle$$

and

$$A_-|n+\tfrac{1}{2}\rangle = \sqrt{n}|n-\tfrac{1}{2}\rangle.$$

A_+ may be called a creation operator for the excitations of the harmonic oscillator, and A_-, an annihilation operator. Note that p and x may easily be written in terms of A_+ and A_-, so that these operators will serve as the fundamental operators for this very simple problem.

1.2. Second quantization for bosons

Consider a system of identical bosons, for which is known a complete orthonormalized set of single-particle states, $\{\phi_s\}$. For example, in the case of spinless particles in a box, with periodic boundary conditions, s could stand for the momentum, $\mathbf{k}$, and $\phi_s(\mathbf{r}) = \exp(i\mathbf{k}\cdot\mathbf{r})/\sqrt{V}$, where V is the volume of the box. For particles with spin, s would have to include an indication of the spin state of the particle. The many-body wave-functions for N particles have to be symmetrical under the interchange of any two particles, and so, in order to describe a state of the system in which each particle is in one of the set $\{\phi_s\}$, we need only to state a set of integers $(n_1, n_2, \ldots n_s, \ldots)$, giving the number of particles in each single-particle state. We now interpret $|n_1, n_2, \ldots n_s, \ldots\rangle$ as a vector normalized to unity, in the Hilbert space appropriate to the system. Since the set $\{\phi_s\}$ is orthogonal and complete for the one-particle system, the set $|n_1, n_2, \ldots n_s, \ldots\rangle$ form a complete orthonormal set for the many-body system and may thus be used as a set of basis vectors. Notice that we have not specified the total number of particles, so that we may use this basis to describe a system of any number of particles; this is important, as we wish to use the grand canonical ensemble in later chapters.

One of the fundamental operators mentioned in the previous section takes us from one basis vector to another:

$$a_s^+|n_1, n_2, \ldots n_s, \ldots\rangle = \sqrt{(n_s+1)}|n_1, n_2, \ldots n_s+1, \ldots\rangle. \tag{1.2}$$

It is called a creation operator, because it puts one more particle in one of the single-particle states. The factor $\sqrt{(n_s+1)}$ is included for future convenience. By operating with products of a's on the 'vacuum' state $|0, 0, \ldots 0, 0, \ldots\rangle$ we can obtain all the basis states, and, by definition, these will be correctly symmetrized.

To find a_s, the Hermitian conjugate of a_s^+, we note that the unit operator may be written as

$$1 = \sum_{\{n'\}} |n'_1, n'_2, \dots n'_s, \dots\rangle\langle n'_1, n'_2, \dots n'_s, \dots| \tag{1.3}$$

since the set of vectors $\{|n'_1, n'_2, \dots n'_s, \dots\rangle\}$ is complete. (This may easily be verified by looking at the matrix elements of this operator between the basis vectors we have chosen: all diagonal elements are unity, and all off-diagonal elements zero.)

Thus

$$\begin{aligned} a_s|n_1, n_2, \dots n_s, \dots\rangle &= 1\, a_s|n_1, n_2, \dots n_s, \dots\rangle \\ &= \sum_{\{n'\}} |n'_1, n'_2, \dots n'_s, \dots\rangle\langle n'_1, n'_2, \dots n'_s, \dots|a_s|n_1, n_2, \dots n_s, \dots\rangle. \end{aligned} \tag{1.4}$$

But, from the definition of the Hermitian conjugate,

$$\begin{aligned} &\langle n'_1, n'_2, \dots n'_s, \dots|a_s|n_1, n_2, \dots n_s, \dots\rangle \\ &= [\langle n_1, n_2, \dots n_s, \dots|a_s^+|n'_1, n'_2, \dots n'_s, \dots\rangle]^*, \end{aligned} \tag{1.5}$$

which, from eqn (1.2), is zero unless $n'_1 = n_1, n'_2 = n_2, \dots n'_s = n_s - 1, \dots$ in which case it is $\sqrt{n_s}$. Thus

$$a_s|n_1, n_2, \dots n_s, \dots\rangle = \sqrt{n_s}|n_1, n_2, \dots n_s - 1, \dots\rangle; \tag{1.6}$$

a_s removes a particle from the sth single-particle state and is called an annihilation operator.

The fundamental operators are now defined for a system of bosons, and we may use these definitions to find all the properties we need. From eqns (1.2) and (1.6)

$$a_s^+ a_s|n_1, n_2, \dots n_s, \dots\rangle = n_s|n_1, n_2, \dots n_s, \dots\rangle. \tag{1.7}$$

The operator $a_s^+ a_s$ thus measures the number of particles in the single-particle state, s. It is now rather easy to write down the operator for the kinetic energy T, of the particles in this language. If the labels $\mathbf{k}$ stand for the single-particle wave-vector,

$$T = \sum_{\mathbf{k}} n_{\mathbf{k}}\, k^2/2m. \tag{1.8}$$

Thus, in terms of the new operators,

$$T = \sum_{\mathbf{k}} (k^2/2m) a_{\mathbf{k}}^+ a_{\mathbf{k}} \tag{1.9}$$

is the operator which, applied to any one of the basis vectors, gives the kinetic energy of the state of the many-body system described by that basis vector.

Further, in a similar fashion,

$$a_s a_s^+ |n_1, n_2, \ldots n_s, \ldots\rangle = (n_s + 1)|n_1, n_2, \ldots n_s, \ldots\rangle, \tag{1.10}$$

so that

$$(a_s a_s^+ - a_s^+ a_s)|n_1, n_2, \ldots n_s, \ldots\rangle = 1|n_1, n_2, \ldots n_s, \ldots\rangle. \tag{1.11}$$

Since this is true whatever basis vector is on the right, we may write,

$$[a_s, a_s^+] \equiv a_s a_s^+ - a_s^+ a_s = 1. \tag{1.12a}$$

Similarly, one may show that

$$[a_{s_1}, a_{s_2}] = 0 \tag{1.12b}$$

for all s_1, s_2 and that

$$[a_{s_1}, a_{s_2}^+] = 0 \tag{1.12c}$$

for all $s_1 \neq s_2$.

It is shown in Appendix A that a sum of single-particle operators $F = \sum_i f(\mathbf{r}_i, \mathbf{p}_i)$, where $\mathbf{r}_i$ is the coordinate and $\mathbf{p}_i$ the momentum operator for the ith particle, may be written in terms of the creation and annihilation operators in the form

$$F = \sum_{jk} f_{jk} a_j^+ a_k, \tag{1.13}$$

where

$$f_{jk} = \int \phi_j^*(\mathbf{r}) f(\mathbf{r}, \mathbf{p}) \phi_k(\mathbf{r})\, \mathrm{d}^3\mathbf{r}. \tag{1.14}$$

Eqn (1.9) is a special case of this result. Similarly a sum of two-particle operators $G = \sum_{s \neq t} g(\mathbf{r}_s, \mathbf{p}_s, \mathbf{r}_t, \mathbf{p}_t)$ may be written

$$G = \sum_{jklm} g_{jklm} a_j^+ a_k^+ a_m a_l, \tag{1.15}$$

where

$$g_{jklm} = \iint \phi_j^*(\mathbf{r})\phi_k^*(\mathbf{r}') g(\mathbf{r}, \mathbf{p}, \mathbf{r}', \mathbf{p}') \phi_l(\mathbf{r}) \phi_m(\mathbf{r}')\, \mathrm{d}^3\mathbf{r}\, \mathrm{d}^3\mathbf{r}'. \tag{1.16}$$

The proofs are simple, although a little tedious; one shows that the matrix elements of F and G are the same when one uses the new way of representing them as when one uses the old. The potential energy resulting from the application of an external field will be of the form of F in eqn (1.13) whereas the potential energy of particles interacting in pairs, for example, through a potential of the form given in eqn (1.1), will be of the form of G in eqn (1.15).

Exercises

1.2.1. Repeat exercise 1.1.1 using the creation- and annihilation-operator formalism.

1.2.2. The Hamiltonian for a non-interacting gas is

$$H_0 = \sum_{\mathbf{k}} (k^2/2m) a_{\mathbf{k}}^+ a_{\mathbf{k}} \quad \text{(cf. eqn (1.9)).}$$

In the Heisenberg representation, the equation of motion for an operator A without explicit time dependence is

$$\mathrm{i}\frac{\mathrm{d}A}{\mathrm{d}t} = [A, H],$$

where H is the Hamiltonian (see e.g. Gottfried (1966), p. 241). Show that

$$[a_{\mathbf{k}}^+, H_0] = -(k^2/2m) a_{\mathbf{k}}^+$$

and hence that the equation of motion for $a_{\mathbf{k}}^+$ in the Heisenberg representation is

$$\mathrm{i}\frac{\mathrm{d}a_{\mathbf{k}}^+}{\mathrm{d}t} = -(k^2/2m) a_{\mathbf{k}}^+ .$$

Write down the differential equations giving the time development of the vectors $a_{\mathbf{k}}^+|0, 0, \ldots 0, \ldots\rangle$ and $a_{\mathbf{k}}^+ a_{\mathbf{k}'}^+ a_{\mathbf{k}''}^+|0, 0, \ldots 0, \ldots\rangle$.

1.2.3. Using eqns (1.13) and (1.16) write down the Hamiltonian for a system of particles of mass m (i) in an external field, and (ii) interacting through pair-wise potentials depending only on the distance apart of the two particles in the pair.

These results may be derived in a slightly different way. In the work above, we have not specified that s should stand for any particular set of labels. Thus we may use $a^+(\mathbf{r})$ and $a(\mathbf{r})$ as operators creating and annihilating particles at the point $\mathbf{r}$, and $\rho(\mathbf{r}) = a^+(\mathbf{r})a(\mathbf{r})$ as an operator measuring the density of particles at the point $\mathbf{r}$. The potential energy of the second system is given by

$$\text{P.E.} = \tfrac{1}{2}\int\!\!\int \rho(\mathbf{r})v(|\mathbf{r}-\mathbf{r}'|)\rho(\mathbf{r}')\,\mathrm{d}^3\mathbf{r}\,\mathrm{d}^3\mathbf{r}' - \tfrac{1}{2}\int \rho(\mathbf{r})v(0)\,\mathrm{d}^3\mathbf{r}. \qquad \text{(i)}$$

But since

$$\delta(\mathbf{r}-\mathbf{r}') = \frac{1}{\sqrt{V}}\sum_{\mathbf{k}} \exp[\mathrm{i}\mathbf{k}\,.\,(\mathbf{r}-\mathbf{r}')],$$

$$a^+(\mathbf{r})|\lambda\rangle = \frac{1}{\sqrt{V}}\sum_{\mathbf{k}} \exp[\mathrm{i}\mathbf{k}\,.\,\mathbf{r}]a_{\mathbf{k}}^+|\lambda\rangle,$$

where $|\lambda\rangle$ is any vector, and V is the volume of the system. The last equation states that creating the correct superposition of plane waves is equivalent to creating a particle at a point. We may thus write

$$a^+(\mathbf{r}) = \frac{1}{\sqrt{V}} \sum_{\mathbf{k}} \exp(i\mathbf{k}\,.\,\mathbf{r}) a^+_{\mathbf{k}}$$

and substitute this into eqn (i) to obtain

$$\text{P.E.} = \tfrac{1}{2} \sum_{\mathbf{k}_1,\mathbf{k}_2,\mathbf{q}} \bar{v}(q) a^+_{\mathbf{k}_1} a^+_{\mathbf{k}_2} a_{\mathbf{k}_2-\mathbf{q}} a_{\mathbf{k}_1+\mathbf{q}},$$

where

$$\bar{v}(q) = \frac{1}{V} \int v(r) \exp(i\mathbf{q}\,.\,\mathbf{r})\, d^3\mathbf{r},$$

which agrees with the result obtained using eqns (1.15) and (1.16).

1.2.4. Consider N points of unit mass moving on the circumference of a circle, connected by strings of unit natural length with unit Young's modulus. If p_s is the momentum and q_s the displacement of the sth particle, the Hamiltonian, for small displacements, is

$$H = \tfrac{1}{2} \sum_s [p_s^2 + (q_{s+1} - q_s)^2].$$

Show that this may be written in the form

$$H = \sum_k \omega_k (b_k^+ b_k + \tfrac{1}{2})$$

where

$$\omega_k = [2(1-\cos k)]^{\frac{1}{2}},$$

$$b_k^+ = (2N\omega_k)^{-\frac{1}{2}} \sum_s (q_s \omega_k e^{iks} - ip_s e^{iks}),$$

$$b_k = (2N\omega_k)^{-\frac{1}{2}} \sum_s (q_s \omega_k e^{-iks} + ip_s e^{-iks}),$$

$$[b_k, b_k^+] = 1, \text{ etc.},$$

and

$$k = 2\pi n/N,$$

where n is integral and runs from one to N. This is a one-dimensional model of phonons in a solid. b_k^+ and b_k may be considered to create and annihilate phonons of wave number k, and H to describe a gas of non-interacting phonons. Interactions between the phonons are introduced when one considers anharmonic forces etc. The model may rather easily be extended to three dimensions (see, for example, Kittel (1963)).

If one is considering electrons in a perfect periodic lattice, with which they interact, the appropriate single-particle states for the electrons are Bloch states, with a band label as well as a wave-number label. The vibrations of the atoms of the periodic lattice are described in terms of phonons, but since these vibrations imply that the lattice will no longer be perfectly regular, the Bloch states are no longer perfect eigenstates of the Hamiltonian, even ignoring the electron–electron interaction. This effect may be described in terms of an electron–phonon interaction, which is typically of the form

$$H_{e\phi} = \sum_{\mathbf{k},\mathbf{q}} F(\mathbf{q})(b_{\mathbf{q}} - b^{+}_{-\mathbf{q}})a^{+}_{\mathbf{k}+\mathbf{q}}a_{\mathbf{k}},$$

where a^+ and a are the creation and annihilation operators for electrons (see the next section). Physically, a vibrating ion knocks an electron from one Bloch state to another. For further details see, for example, Kittel (1963).

1.2.5. In the ground state of a non-interacting gas of bosons, all the particles are in the lowest single-particle state, that with zero momentum. Since

$$a_0 a_0^+ - a_0^+ a_0 = 1,$$

and $a_0^+ a_0$ and $a_0 a_0^+$ are both approximately N, the total number of particles, Bogoliubov (1947) argued that to a good approximation the 1 in this equation could be neglected in comparison with $a_0^+ a_0$ and $a_0 a_0^+$. In that case, a_0^+ and a_0 commute with all operators, and may be treated as numbers. Call these $(N_0)^{\frac{1}{2}}$. If the gas is weakly interacting, one might hope that this approximation would still work. Using

$$N = a_0^+ a_0 + \sum_{\mathbf{k}}{}' a_{\mathbf{k}}^+ a_{\mathbf{k}} = N_0 + \sum_{\mathbf{k}}{}' a_{\mathbf{k}}^+ a_{\mathbf{k}},$$

where primes on the summation symbols indicate that the terms with zero momentum are to be omitted, show that the Hamiltonian for the interacting system may be written in the form

$$\begin{aligned} H = {} & \sum_{\mathbf{k}}{}' a_{\mathbf{k}}^+ a_{\mathbf{k}}[k^2/2m + N_0\bar{v}(k)] + \tfrac{1}{2}\sum_{\mathbf{k}}{}' N_0\bar{v}(k)[a_{\mathbf{k}}^+ a_{-\mathbf{k}}^+ + a_{\mathbf{k}}a_{-\mathbf{k}}] + \\ & + \tfrac{1}{2}N^2\bar{v}(0) + \sum_{\mathbf{kq}}{}' (N_0)^{\frac{1}{2}}\bar{v}(q)[a_{\mathbf{k}}^+ a_{\mathbf{k}-\mathbf{q}}a_{\mathbf{q}} + a_{\mathbf{k}}^+ a_{\mathbf{q}}^+ a_{\mathbf{k}+\mathbf{q}}] + \\ & + \tfrac{1}{2}\sum_{\mathbf{kk}'}{}' \bar{v}(0)a_{\mathbf{k}}^+ a_{\mathbf{k}}a_{\mathbf{k}'}^+ a_{\mathbf{k}'} + \tfrac{1}{2}\sum_{\mathbf{k}_1,\mathbf{k}_2,\mathbf{q}}{}' \bar{v}(q)a_{\mathbf{k}_1}^+ a_{\mathbf{k}_2}^+ a_{\mathbf{k}_2+\mathbf{q}}a_{\mathbf{k}_1-\mathbf{q}}. \end{aligned}$$

If the gas is weakly interacting, so that most of the particles are still in the zero-momentum state, the last three terms in this equation will be small and $N_0 \approx N$. We may in any case take the remaining terms as our zero-order approximation for the problem, writing

$$H_{\mathrm{B}} = \sum_{\mathbf{k}}{}' a_{\mathbf{k}}^+ a_{\mathbf{k}}[k^2/2m + N\bar{v}(k)] + \tfrac{1}{2}N\sum_{\mathbf{k}}{}' \bar{v}(k)[a_{\mathbf{k}}^+ a_{-\mathbf{k}}^+ + a_{\mathbf{k}}a_{-\mathbf{k}}] + \text{constant}.$$

Show that, by using the transformation

$$a_{\mathbf{k}}^{+} = s_{\mathbf{k}}\alpha_{\mathbf{k}}^{+} + t_{\mathbf{k}}\alpha_{-\mathbf{k}}$$
$$a_{\mathbf{k}} = s_{\mathbf{k}}\alpha_{\mathbf{k}} + t_{\mathbf{k}}\alpha_{-\mathbf{k}}^{+}$$

with $s_{\mathbf{k}}^2 + t_{\mathbf{k}}^2 = 1$, so that

$$[\alpha_{\mathbf{k}}, \alpha_{\mathbf{k}'}^{+}] = \delta_{\mathbf{k},\mathbf{k}'}$$
$$[\alpha_{\mathbf{k}}^{+}, \alpha_{\mathbf{k}'}^{+}] = [\alpha_{\mathbf{k}}, \alpha_{\mathbf{k}'}] = 0,$$

the Hamiltonian may be written as

$$H_{\mathrm{B}} = \text{constant} + \sum_{\mathbf{k}}{}' \omega_k \alpha_{\mathbf{k}}^{+} \alpha_{\mathbf{k}}$$

by choosing $s_{\mathbf{k}}$ and $t_{\mathbf{k}}$ suitably. This form of H_{B} describes a non-interacting gas of "quasiparticles" with energies

$$\omega_k = [(k^2/m)(N\bar{v}(k) + k^2/4m)]^{\frac{1}{2}}.$$

Note that ω_k is linear for small values of $|\mathbf{k}|$, so that the quasiparticles are phonon-like in the low-momentum region. The transformation is known as the Bogoliubov transformation, and is used in the theory of superconductivity, and in many other branches of the many-body problem.

1.3. Second quantization for fermions

The definitions for Fermi creation and annihilation operators are slightly more complicated. The wave functions of the system must now be antisymmetrical with respect to the interchange of the coordinates of any two of the particles. They can therefore be expressed in terms of a set of antisymmetric basis states, which we take as the Slater determinants of the free single-particle states. These states are to be ordered in a certain way, and this order must be kept both in the Slater determinant, so that its sign is well-defined, and in the way we write the basis vectors $|n_1, n_2, \ldots n_s, \ldots\rangle$. Notice that again, by definition, these basis states are already properly antisymmetrized, and that if we define the matrix elements of the creation operators between these states, we need no longer concern ourselves about the antisymmetry. Because of this antisymmetry, each n_s is either 0 or 1, and the creation operators must satisfy:

$$a_s^{+}|n_1, n_2, \ldots n_s = 1, \ldots\rangle = 0, \tag{1.17}$$

since a particle cannot be added to a state which is already occupied. (Note that the label s now includes the spin state of the single-particle state.) By analogy with the boson case, it would seem that one should define

$$a_s^{+}|n_1, n_2, \ldots n_s = 0, \ldots\rangle = |n_1, n_2, \ldots n_s = 1, \ldots\rangle.$$

However, this leads to somewhat inconvenient commutation relations, and very inconvenient forms for one- and two-body operators. Instead, we take

$$a_s^+|n_1, n_2, \dots n_s = 0, \dots\rangle = (-1)^{m_s}|n_1, n_2, \dots n_s = 1, \dots\rangle, \tag{1.18}$$

where

$$m_s = \sum_{r=1}^{r=s-1} n_r.$$

In other words, the sign is positive if there is an even number of particles in states with labels lower than s, and negative otherwise. The two rules (1.17) and (1.18) may be written together:

$$a_s^+|n_1, n_2, \dots n_s, \dots\rangle = (-1)^{m_s}(1-n_s)^{\frac{1}{2}}|n_1, n_2, \dots n_s+1, \dots\rangle. \tag{1.19}$$

The Hermitian conjugate of a_s^+, a_s, may be shown, by a method similar to that leading to eqn (1.6), to satisfy

$$a_s|n_1, n_2, \dots n_s, \dots\rangle = (-1)^{m_s}n_s^{\frac{1}{2}}|n_1, n_2, \dots n_s-1, \dots\rangle. \tag{1.20}$$

It is easily shown that the number operator for the sth state is still $a_s^+ a_s$, and that

$$a_s a_t^+ + a_t^+ a_s \equiv [a_s, a_t^+]_+ = \delta_{st} \tag{1.21}$$

$$[a_s^+, a_t^+]_+ = [a_s, a_t]_+ = 0.$$

Eqns (1.13) to (1.16) follow as before, by comparing matrix elements (see Appendix A).

Exercises

1.3.1. Repeat exercises 1.1.1 and 1.2.1, suitably amended, for the case of fermions.

1.3.2. Repeat exercise 1.2.2 for the case of fermions.

1.4. Other cases

Sometimes one wishes to consider the properties of a mixture of bosons and fermions, for example, a solution of He^3 in He^4 or a system of electrons and phonons. The extensions necessary to the work above are straightforward. The basis vectors are written $|n_1, n_2, \dots n_1', n_2', \dots\rangle$, where the first set of integers refers to fermion states, and the second to boson states. Creation and annihilation operators a_s^+, a_s, for the fermions, and b_t^+, b_t, for the bosons, are defined in exactly the same way as above. Each boson operator will commute with each fermion operator: otherwise, the commutation operators will be the same as before. A two-body operator $\sum_{i,j} h(\mathbf{R}_i, \mathbf{P}_i, \mathbf{r}_j, \mathbf{p}_j)$ which is a function of the coordinates and momenta both of the fermions

$\mathbf{R}_i$, $\mathbf{P}_i$, and of the bosons, $\mathbf{r}_j$, $\mathbf{p}_j$, is given in second-quantized form as

$$\sum_{klmn} h_{klmn} a_k^+ b_l^+ a_m b_n, \tag{1.22}$$

where

$$h_{klmn} = \iint \mathrm{d}^3\mathbf{R}\, \mathrm{d}^3\mathbf{r}\, \phi_k^*(\mathbf{R})\psi_l^*(\mathbf{r})h(\mathbf{R}, \mathbf{P}, \mathbf{r}, \mathbf{p})\phi_m(\mathbf{R})\psi_n(\mathbf{r}). \tag{1.23}$$

The ϕs are fermion single-particle wavefunctions, and the ψs boson single-particle wavefunctions. The potential energy of interaction between the bosons and fermions will be of this form.

So far, most of the terms appearing in the Hamiltonians for our many-body systems have had equal numbers of creation and annihilation operators. The reason for this is that usually we have interactions which do not change the total number of particles; when two particles interact, they merely change their states. This is in contrast to the situation in high-energy physics where two particles can interact to produce a third, or where a γ-ray can produce an electron-positron pair. There are occasions in many-body physics, however, when one wishes to be able to deal with interactions which do not conserve the total number of particles. There are such terms, for example, in $H_{e\phi}$ in exercise 1.2.4 and in the Hamiltonian in exercise 1.2.5 after the substitution $a_0^+, a_0 = (N_0)^{\frac{1}{2}}$. Further, since phonons in crystals can be created and destroyed at impurities, for example, we might expect the presence of terms in the phonon Hamiltonian which do not conserve the number of phonons. However, since spin cannot be created or destroyed, and since fermions carry spin, fermion creation and annihilation operators must always occur in pairs in our Hamiltonian.

The analogues of creation and annihilation operators for a system of magnetic dipoles are considered in exercise 1.4.1 below.

Exercises

1.4.1 The Heisenberg model of a ferromagnet.

Consider a system of N particles with spin $\frac{1}{2}$, localized at N lattice sites, in an external magnetic field, B', along the z-axis. We shall assume that there exists an exchange coupling, tending to align the spins, of the form, $-\sum_{i\neq j} v_{ij}\mathbf{S}_i \cdot \mathbf{S}_j$ where $\mathbf{S}_i$ is the spin operator for the ith site. v_{ij} will be taken to be positive and to depend only on the distance apart of the ith and jth sites; $v_{ij} = v(|\mathbf{r}_i - \mathbf{r}_j|)$; we may remove the restriction $i \neq j$ by defining $v_{ii} = 0$. The spin operators satisfy the usual commutation relations:

$$[S_i^x, S_j^y] = \mathrm{i}S_i^z\delta_{ij}, \quad \text{etc.}$$

If we put

$$S_i^{\pm} = S_i^x \pm \mathrm{i}S_i^y,$$

these may be written

$$[S_i^z, S_j^\pm] = \pm\delta_{ij}S_i^\pm$$
$$[S_i^+, S_j^-] = 2\,\delta_{ij}S_i^z.$$

The Hamiltonian of the system is

$$H = -B\sum_i S_i^z - \tfrac{1}{2}\sum_{ij} v_{ij}\mathbf{S}_i\,.\,\mathbf{S}_j$$
$$= -B\sum_i S_i^z - \tfrac{1}{2}\sum_{ij} v_{ij}(S_i^+S_j^- + S_i^zS_j^z),$$

where $B = \mu B'$, $\mu\mathbf{S}_i$ being the magnetic moment of the ith particle. The Hamiltonian may be put into a form which is often more useful by using the Fourier transforms of the site operators:

$$S_\mathbf{q}^+ = N^{-\frac{1}{2}}\sum_j \exp(-\mathrm{i}\mathbf{q}\cdot\mathbf{r}_j)S_j^+$$

with similar definitions for $S_\mathbf{q}^-$ and $S_\mathbf{q}^z$. Further

$$\bar{v}(\mathbf{q}) = \sum_j \exp[-\mathrm{i}\mathbf{q}\,.\,(\mathbf{r}_j - \mathbf{r}_i)]v(|\mathbf{r}_j - \mathbf{r}_i|).$$

(i) Show that

$$[S_\mathbf{q}^z, S_{\mathbf{q}'}^\pm] = \pm N^{-\frac{1}{2}}S_{\mathbf{q}+\mathbf{q}'}^\pm,$$
$$[S_\mathbf{q}^+, S_{\mathbf{q}'}^-] = 2N^{-\frac{1}{2}}S_{\mathbf{q}+\mathbf{q}'}^z,$$
$$H = -B\sqrt{N}S_0^z - \tfrac{1}{2}\sum_\mathbf{q}\bar{v}(\mathbf{q})[S_\mathbf{q}^+S_{-\mathbf{q}}^- + S_\mathbf{q}^zS_{-\mathbf{q}}^z],$$
$$\mathrm{i}\frac{\mathrm{d}}{\mathrm{d}t}S_\mathbf{q}^+ = BS_\mathbf{q}^+ + N^{-\frac{1}{2}}\sum_{\mathbf{q}'}[\bar{v}(\mathbf{q}') - \bar{v}(\mathbf{q}-\mathbf{q}')]S_{\mathbf{q}-\mathbf{q}'}^+S_{\mathbf{q}'}^z. \qquad (\alpha)$$

where sums over $\mathbf{q}$ are taken over the first Brillouin zone of the lattice.

(ii) For a ferromagnet, the ground state of the system, which we shall write $|G\rangle$, has all the spins aligned parallel to the field. Thus

$$S_i^z|G\rangle = \tfrac{1}{2}|G\rangle \quad \text{for all } i,$$

and

$$S_\mathbf{q}^z|G\rangle = \tfrac{1}{2}\delta_{\mathbf{q},\mathbf{0}}\sqrt{N}|G\rangle.$$

From eqn (α), $S_\mathbf{q}^+|G\rangle$ is thus an exact eigenstate of H with energy (cf. exercise 1.2.2)

$$\omega_\mathbf{q} = B + (\bar{v}(0) - \bar{v}(\mathbf{q}))/2.$$

$S_\mathbf{q}^+$ is said to create a spin-wave of wave-number $\mathbf{q}$. It may easily be verified that $S_\mathbf{q}^+S_{\mathbf{q}'}^+|G\rangle$ is not an exact eigenvector of H. We can interpret this by saying that the spin-waves interact.

The operators S^+, S^-, and S^z could be taken as the fundamental operators for the spin system. Unfortunately, the commutation relations they satisfy are more complicated than those for the creation and annihilation operators defined in sections 1.2 and 1.3, since the commutator of two spin operators is itself a spin operator, rather than a number (or, strictly speaking, the unit operator times a number). This leads to difficulties which we shall discuss in Chapter 6.

The extension of the results of this exercise to the case of general spin is straightforward. The spin-wave energies may also be derived by using the Holstein–Primakoff transformation (Holstein and Primakoff (1940)).

For general spin, S, define, a_j^+ and a_j:

$$S_j^+ = (2S)^{\frac{1}{2}}[1-a_j^+a_j/2S]^{\frac{1}{2}}a_j$$

$$S_j^- = (2S)^{\frac{1}{2}}a_j^+[1-a_j^+a_j/2S]^{\frac{1}{2}},$$

where the as satisfy boson commutation relations: $[a_j, a_l^+] = \delta_{jl}$, all others commuting. By calculating $(S_j^z)^2 = S(S+1)-(S_j^x)^2-(S_j^y)^2$ show that

$$S_j^z = S - a_j^+a_j \qquad (\alpha)$$

and that $S_j^\pm$ and S_j^z satisfy the correct commutation relations. Define the Fourier transforms of the as:

$$b_{\mathbf{k}} = (N)^{-\frac{1}{2}}\sum_j \exp(i\mathbf{k}\,.\,\mathbf{x}_j)a_j$$

$$b_{\mathbf{k}}^+ = (N)^{-\frac{1}{2}}\sum_j \exp(-i\mathbf{k}\,.\,\mathbf{x}_j)a_j^+.$$

Show that:

$$H = \text{const} + \sum_{\mathbf{k}} \omega_{\mathbf{k}} b_{\mathbf{k}}^+ b_{\mathbf{k}} + \text{terms of order four and higher in the } b\text{s}.$$

The second term represents spin-waves with energy $\omega_{\mathbf{k}}$ as above (where $S = \frac{1}{2}$) and the higher-order terms, the interactions between the spin-waves. Note that eqn (α) sets a condition on the a operators, since the expectation value of $S^z \geqslant -S$. This is unlikely to have a big effect for lowly excited systems, however.

Spin waves have been observed by scattering neutrons from spin systems, noting their changes in energy and momentum (Sinclair and Brockhouse (1960)) and by spin-wave resonance techniques (Kittel (1958), Seavey and Tannenwald (1958)). For further details see, for example, Phillips and Rosenberg (1966).

1.4.2. If A is an operator such that

$$[H, A] = \omega A,$$

and if $|\lambda\rangle$ is an eigenvector of H with eigenvalue λ, then $A|\lambda\rangle$ is also an eigenvector of H with energy $\omega+\lambda$. This is sometimes used as a method of determining the elementary excitations of a system.

Show that, for a system of fermions interacting through central pair forces

$$[H, a_{\mathbf{k}}^{+}] = (k^2/2m)a_{\mathbf{k}}^{+} + \sum_{\mathbf{k}',\mathbf{q}} \bar{v}(q)a_{\mathbf{k}+\mathbf{q}}^{+}a_{\mathbf{k}'}^{+}a_{\mathbf{k}'+\mathbf{q}}.$$

Thus $a_{\mathbf{k}}^{+}$ does not have the property required of A. In the random-phase approximation, however, only those terms in the sum which contain a factor $a_{\mathbf{p}}^{+}a_{\mathbf{p}}$ are retained: this factor is then replaced by its expectation value, $\langle n_{\mathbf{p}}\rangle$ in the state $|\lambda\rangle$ we are considering, for example the ground state. The equation then takes the form

$$[H, a_k^{+}] = [k^2/2m - \sum_{\mathbf{q}} (\bar{v}(q)\langle n_{\mathbf{k}+\mathbf{q}}\rangle - \bar{v}(0)\langle n_{\mathbf{q}}\rangle)]a_{\mathbf{k}}^{+}.$$

We thus have an approximation for the excitation energies of the system, which can be used in thermodynamic calculations. For further details of this method, and for higher-order correction in the case of the electron gas, see, for example, Suhl and Werthamer (1961). The random-phase approximation will be considered in more detail in Chapters 4 and 5.

Further reading

More formal treatments of the creation- and annihilation-operator formalism are given, for example, by Valatin (1961) and by Schweber (1961).

2

PERTURBATION THEORY

2.1. The perturbation expansion

FROM elementary statistical mechanics, we know that all equilibrium thermodynamic quantities of interest for a system of interacting particles can be calculated from the grand partition function:

$$Q = \sum_{\nu,N} \exp[\beta(\mu N - E_{N\nu})], \tag{2.1}$$

where β is the inverse temperature, $(\kappa \mathrm{T})^{-1}$, κ being Boltzmann's constant, μ, the chemical potential, and $E_{N\nu}$, the νth energy eigenvalue for the system with N particles. The sum is to be taken over all ν and all positive values of N, which indicates that we have chosen to work in the grand canonical ensemble. In order to avoid having always to work in the energy representation, we rewrite Q in the form

$$Q = \mathrm{Tr}\{\exp[\beta(\mu\hat{N} - H)]\} \tag{2.2}$$

where H is the Hamiltonian of the system, $\hat{N}$ the number operator, and where the trace is taken over any complete set of states for the system. In using the word complete, we must remember that the system may contain any number of particles. In the usual way, the thermodynamic potential, Ω, is defined:

$$\Omega = -\beta^{-1} \ln Q, \tag{2.3}$$

and we have

$$\Omega = -pV \tag{2.4}$$

where p is the pressure of the system and V its volume. The mean value of the number of particles is

$$\bar{N} = -\left(\frac{\partial \Omega}{\partial \mu}\right)_{T,V}, \tag{2.5}$$

and the entropy

$$S = -\left(\frac{\partial \Omega}{\partial T}\right)_{V,\mu} \tag{2.6}$$

Eqn (2.5) gives μ in terms of $\bar{N}$, so that we can find, for example, S or p in terms of T, V and $\bar{N}$ from the last three equations.

We shall thus turn our attention to the evaluation of Q. The calculation falls into two parts: as we indicated in Chapter 1, we first need to write down a perturbation expansion for Q, expressing this in terms of expectation values of products of creation and annihilation operators: we then work out a set of rules for evaluating these expectation values.

The first part of this programme is relatively simple. We split up the Hamiltonian into two parts:

$$H = H_0 + H_1, \tag{2.7}$$

where H_0 is so chosen that the properties of a system described by H_0 alone are well-known. We shall assume that our choice is such that H_0 may be written as

$$H_0 = \sum_s \varepsilon_s a_s^+ a_s, \tag{2.8}$$

that is, it describes a system of non-interacting particles (or quasi-particles such as phonons (see exercise 1.2.4)) with energies ε_s. As an example, one may think of particles in a box, in which case H_0 is usually chosen to be the kinetic energy operator, $s = \mathbf{k}$ and $\varepsilon_{\mathbf{k}} = k^2/2m$. H_1 is then the potential energy: for particles interacting through central, two-body forces, this is of the form $\frac{1}{2}\sum_{ij} v(|\mathbf{r}_i - \mathbf{r}_j|)$ which in second-quantized notation is

$$H_1 = \text{P.E.} = \frac{1}{2} \sum_{\mathbf{kk'q}} \bar{v}(q) a_{\mathbf{k}}^+ a_{\mathbf{k'}}^+ a_{\mathbf{k'+q}} a_{\mathbf{k-q}}, \tag{2.9}$$

where

$$\bar{v}(q) = V^{-1} \int \mathrm{d}^3\mathbf{r} \exp(i\mathbf{q}\,.\,\mathbf{r}) v(r).$$

For electrons in a crystal lattice, the single-particle wave-function would be a Bloch state, and would need a band label, a spin label, and a wave-number label. Alternatively, H_0 may describe non-interacting particles in an external field, the electrons in the Coulomb field of an atomic nucleus, for example, in which case the potential energy will have the more general form:

$$\text{P.E.} = \frac{1}{2} \sum_{sts't'} a_s^+ a_t^+ a_{t'} a_{s'} v_{sts't'}, \tag{2.9a}$$

where

$$v_{sts't'} = \iint \mathrm{d}^3\mathbf{r}\, \mathrm{d}^3\mathbf{r}' \phi_s^*(\mathbf{r}) \phi_t^*(\mathbf{r}') v(\mathbf{r}, \mathbf{r}') \phi_{s'}(\mathbf{r}) \phi_{t'}(\mathbf{r}')$$

(cf. eqns (1.15) and (1.16)). We shall normally use this more general form with labels s and t for states. We shall reserve the labels $\mathbf{k}$ and $\mathbf{q}$ for wave numbers. Note that the set has to be complete for the single-particle Hilbert space, so that, for particles with spin, the sum over s includes a sum over spin states for the single particle.

We define the operators $\rho^{(\beta)}$ and $S^{(\beta)}$:

$$\rho^{(\beta)} = \exp[(\mu\hat{N} - H)\beta] \tag{2.10}$$

$$= \exp[(\mu\hat{N} - H_0)\beta]S^{(\beta)}. \tag{2.11}$$

Then

$$Q = \mathrm{Tr}\{\exp[(\mu\hat{N} - H_0)\beta]S^{(\beta)}\}. \tag{2.12}$$

$\rho^{(\beta)}$ is the (unnormalized) density matrix of the system. The expectation value of any operator A in the grand canonical ensemble for a system with Hamiltonian, H, is

$$\langle A\rangle = \mathrm{Tr}\{\exp[(\mu\hat{N} - H)\beta]A\}/\mathrm{Tr}\{\exp[(\mu\hat{N} - H)\beta]\}. \tag{2.13}$$

(See, for example, ter Haar (1966)). Thus Q is related to the expectation value of $S^{(\beta)}$ in the ensemble described by the Hamiltonian H_0, which we shall call the non-interacting ensemble; we write expectation values in this ensemble $\langle\cdots\rangle_0$. Then

$$Q = \mathrm{Tr}\{\exp[(\mu\hat{N} - H_0)\beta]S^{(\beta)}\}\ \mathrm{Tr}\{\exp[(\mu\hat{N} - H_0)\beta]\}/\mathrm{Tr}\{\exp[(\mu\hat{N} - H_0)\beta]\} \tag{2.14}$$

$$= Q_0\langle S^{\beta}\rangle_0, \tag{2.15}$$

where

$$Q_0 = \mathrm{Tr}\{\exp[(\mu\hat{N} - H_0)\beta]\} \equiv \mathrm{Tr}\{\rho_0^{(\beta)}\}$$

is the grand partition function for the non-interacting ensemble.

To obtain a perturbation expansion of S, we differentiate eqns (2.10) and (2.11) with respect to β:

$$\frac{\partial\rho^{(\beta)}}{\partial\beta} = (\mu\hat{N} - H)\rho^{(\beta)}$$

$$= (\mu\hat{N} - H_0)\exp[(\mu\hat{N} - H_0)\beta]S^{(\beta)} + \exp[(\mu\hat{N} - H_0)\beta]\frac{\partial S^{(\beta)}}{\partial\beta}. \tag{2.16}$$

If we operate on the left of this equation with $\exp[-(\mu\hat{N} - H_0)\beta]$ we obtain

$$\exp[-(\mu\hat{N} - H_0)\beta](\mu\hat{N} - H_0 - H_1)\exp[(\mu\hat{N} - H_0)\beta]S^{(\beta)}$$

$$= \exp[-(\mu\hat{N} - H_0)\beta](\mu\hat{N} - H_0)\exp[(\mu\hat{N} - H_0)\beta]S^{(\beta)} + \frac{\partial S^{(\beta)}}{\partial\beta}. \tag{2.17}$$

We may write this:

$$\frac{\partial S^{(\beta)}}{\partial\beta} = -H_1(\beta)S^{(\beta)}. \tag{2.18}$$

where $H_1(\tau)$ is defined by

$$H_1(\tau) = \exp[-(\mu\hat{N} - H_0)\tau]H_1 \exp[(\mu\hat{N} - H_0)\tau]. \tag{2.19}$$

This trick is very similar to the transformation to the interaction representation in the treatment of the time-dependent Schrödinger equation. The equation (2.18) which results from it is particularly suitable for solution by iteration. We first need a boundary condition for S, which is obtained by putting $\beta = 0$ in eqns (2.10) and (2.11), giving $S^{(0)} = \mathit{1}$. Integrating eqn (2.18) we obtain

$$S^{(\beta)} = \mathit{1} - \int_0^\beta H_1(\tau_1)S^{(\tau_1)}\,d\tau_1. \tag{2.20}$$

This equation itself can be used to give $S^{(\tau_1)}$, which we can substitute into the right-hand side to give

$$S^{(\beta)} = \mathit{1} - \int_0^\beta H_1(\tau_1)\,d\tau_1 + \int_0^\beta d\tau_1 H_1(\tau_1)\int_0^{\tau_1} d\tau_2 H_1(\tau_2)S^{(\tau_2)}. \tag{2.21}$$

Repeating the process again and again, we build up the iterative solution:

$$\begin{aligned} S^{(\beta)} = \mathit{1} &- \int_0^\beta H_1(\tau_1)\,d\tau_1 + \int_0^\beta d\tau_1 H_1(\tau_1)\int_0^{\tau_1} d\tau_2 H_1(\tau_2) - \cdots + \\ &+(-1)^n \int_0^\beta d\tau_1 \int_0^{\tau_1} d\tau_2 \ldots \int_0^{\tau_{n-1}} d\tau_n H_1(\tau_1)H_1(\tau_2)\ldots H_1(\tau_n) + \ldots \end{aligned} \tag{2.22}$$

The first part of our programme is already achieved, since, using eqns (2.15) and (2.22), we can express Q as a power series in H_1, each term in which is the expectation value in the non-interacting ensemble of a product of operators, which can be expressed in terms of creation and annihilation operators. Our next task, and one that takes considerably longer, is to learn how most conveniently to evaluate these expectation values.

It is inconvenient, in the nth order term in eqn (2.22), to have the upper limits of the integrals dependent on other integration variables. If the H_1s were ordinary functions rather than operators, we could easily avoid this difficulty. Consider an integral similar to the second-order term,

$$I = \int_0^\beta d\tau_1 \int_0^{\tau_1} d\tau_2\, f(\tau_1)f(\tau_2)$$

except that f is such that $f(\tau_1)f(\tau_2) = f(\tau_2)f(\tau_1)$. The area of integration is

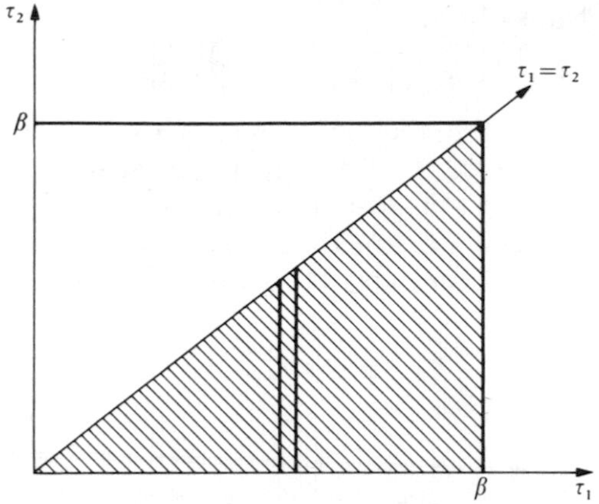

FIG. 2.1. The area of integration is shaded.

shown in Fig. 2.1. The integral over the upper triangle is

$$\int_0^\beta d\tau_2 \int_0^{\tau_2} d\tau_1\, f(\tau_1)f(\tau_2),$$

which, by interchanging τ_1 and τ_2, we can write

$$\int_0^\beta d\tau_1 \int_0^{\tau_1} d\tau_2\, f(\tau_2)f(\tau_1),$$

which is the same as our original integral I since the order of the fs is unimportant. Hence,

$$I = \tfrac{1}{2}\int_0^\beta d\tau_1 \int_0^\beta d\tau_2\, f(\tau_1)f(\tau_2). \tag{2.23}$$

This argument may easily be generalized. If we wish to evaluate the integral

$$I_n = \int_0^\beta d\tau_1 \int_0^{\tau_1} d\tau_2 \ldots \int_0^{\tau_{n-1}} d\tau_n f(\tau_1)f(\tau_2)\ldots f(\tau_n),$$

we notice that the limits on the integrals are such that

$$\tau_1 \geq \tau_2 \geq \tau_3 \ldots \geq \tau_{n-1} \geq \tau_n, \tag{2.24}$$

and that if we do the integral over the range with any two of the τs interchanged, we still have I_n. There are $n!$ possible interchanges, and so

$$I_n = (n!)^{-1} \int_0^\beta d\tau_1 \int_0^\beta d\tau_2 \dots \int_0^\beta d\tau_n f(\tau_1) f(\tau_2) \dots f(\tau_n).$$

Unfortunately, the argument breaks down when we are integrating operators, because the order of the operators now matters and eqn (2.23) is no longer necessarily true. We can, at a cost, make the same idea work, however, by noticing that in our original integral the operators are so ordered that the τ-labels get bigger as we move from right to left in the product. Thus if we define a τ-ordering operator, P, which arranges all operators it acts on in such a way that their τ-labels always increase from left to right, we have

$$\int_0^\beta d\tau_1 \int_0^{\tau_1} d\tau_2 \dots \int_0^{\tau_{n-1}} d\tau_n H_1(\tau_1) H_1(\tau_2) \dots H_1(\tau_n)$$

$$= (n!)^{-1} \int_0^\beta d\tau_1 \int_0^\beta d\tau_2 \dots \int_0^\beta d\tau_n P\{H_1(\tau_1) H_1(\tau_2) \dots H_1(\tau_n)\} \qquad (2.25)$$

since now each interchange amongst the τ-labels in eqn (2.25) will produce an exact replica of the original integral.

There is one slight difficulty associated with this definition in that it does not tell us what to do when two or more of the τs are equal. There are two ways in which this can occur. Firstly, in the n-dimensional 'volume' we are considering in I_n there will be 'surfaces' where $\tau_i = \tau_j$ (in two dimensions, there is a line where $\tau_1 = \tau_2$): but if we first do the integral over τ_i there will be only one point in the range, when $\tau_i = \tau_j$, at which we are uncertain about the product. Provided that we know that the integrand is finite, its value at this one point can make no difference to the total integral; in mathematical terms, the measure of the set where the integral is undefined is zero. Thus as far as the first type of equality of the τs is concerned, it does not matter how we define our P-product for equal τs.

We obtain a definite answer when we consider the second possible way. We wish to express the H_1s in terms of creation and annihilation operators. We can easily attach τ-labels to each of these operators since

$$\begin{aligned}
&\exp[-\tau(\mu\hat{N} - H_0)] ABCD \exp[\tau(\mu\hat{N} - H_0)] \\
&\quad = \exp[-\tau(\mu\hat{N} - H_0)] A \exp[\tau(\mu\hat{N} - H_0)] \exp[-\tau(\mu\hat{N} - H_0)] B \dots \times \\
&\quad\quad \times D \exp[\tau(\mu\hat{N} - H_0)] \\
&\quad = A(\tau) B(\tau) C(\tau) D(\tau).
\end{aligned}$$

However, we now have several operators with the same τ-label. For this type of τ-equality, we must clearly leave the τ-operators under the P-ordering sign in the order in which they appear in the original $H_1(\tau)$.

The operator P introduced above turns out not to be the most convenient one. We introduce a slightly more complicated ordering operator, $\mathscr{T}$, the only justification for this step being that it makes things very much simpler later on:

$$\mathscr{T}\{...\} \equiv (-1)^A P\{...\}, \tag{2.26}$$

where A is the number of interchanges of fermion creation or annihilation operators needed to achieve the correct order. Thus if the as are fermion operators and the bs boson operators, choosing simple values for the τs,

$$P\{a^+(7)a(4)a(6)a^+(8)b(1)b^+(5)a(3)b(9)a^+(2)\}$$
$$= b(9)a^+(8)a^+(7)a(6)b^+(5)a(4)a(3)a^+(2)b(1)$$
$$\mathscr{T}\{a^+(7)a(4)a(6)a^+(8)b(1)b^+(5)a(3)b(9)a^+(2)\}$$
$$= (-1)^{3+1}b(9)a^+(8)a^+(7)a(6)b^+(5)a(4)a(3)a^+(2)b(1)$$

since we need no fermion-operator interchanges to bring $b(9)$ to the left, three to bring $a^+(8)$ to its required position, one for $a(6)$, and none for the other operators. It is easy to see from this example that the number A is not unique; but two As for the same ordering operation must differ at most by an even integer. As we remarked in the previous chapter, H_1 must contain an even number of fermion operators, so that A will always be even for products such as those occurring in eqn (2.25). Thus

$$\int_0^\beta d\tau_1 \int_0^{\tau_1} d\tau_2 \ldots \int_0^{\tau_{n-1}} d\tau_n H_1(\tau_1)H_1(\tau_2)\ldots H_1(\tau_n)$$
$$= (n!)^{-1}\int_0^\beta d\tau_1 \int_0^\beta d\tau_2 \ldots \int_0^\beta d\tau_n \mathscr{T}\{H_1(\tau_1)H_1(\tau_2)\ldots H_1(\tau_n)\}. \tag{2.27}$$

A shorthand way of writing eqn (2.22) is

$$S^{(\beta)} = \mathscr{T}\left\{\exp\left[-\int_0^\beta H_1(\tau)\,d\tau\right]\right\}. \tag{2.28}$$

It must be emphasized that this is useful only as a quick way of writing the integral: we cannot first do the integral inside the exponential, then take the exponential, and finally wonder what to do about the $\mathscr{T}$-ordering operator. The only way to sort out the meaning of eqn (2.28) is to write down the series for the exponential: the meaning of the $\mathscr{T}$-ordering operator is then quite clear, but unfortunately we are then back with the series given by eqn (2.22).

2.2. Wick's theorem

If we rewrite the H_1s in eqn (2.27) in terms of creation and annihilation operators, we see, combining eqns (2.15) and (2.22), that we need to evaluate expectation values like

$$\mathrm{Tr}\{\rho_0^{(\beta)}\mathscr{T}(abcd\,...\,ef)\}/\mathrm{Tr}\{\rho_0^{(\beta)}\}$$

where each letter stands for a creation or annihilation operator and has a τ-label. We may take the trace in the occupation number representation for the single-particle states to which the creation and annihilation operators refer. Thus we shall be evaluating matrix elements like

$$\langle n_1, n_2, \ldots n_i, \ldots|\exp\left[\sum_s \beta(\mu-\varepsilon_s)a_s^+ a_s\right]abcd\,...\,ef|n_1, n_2, \ldots n_i, \ldots\rangle$$

which, because the occupation numbers of the state on each side of this matrix element are the same, will be zero unless the number of creation operators in the product $abcd\,...\,ef$ is equal to the number of annihilation operators for each type of particle we have in the system. (Note that $\exp[\sum_s \beta(\mu-\varepsilon_s)a_s^+ a_s]$ leaves the number of particles the same.)

We first consider a simpler problem by omitting all τ-labels and also the $\mathscr{T}$-ordering operator. We shall consider:

$$I_{2n} = \mathrm{Tr}\{\rho_0 abcd\,...\,ef\}.$$

Wick's theorem gives us the expectation value of a product of $2n$ operators, such as this, in terms of products of expectation values of two operators. It is most easily proved by induction (Gaudin 1960): We express I_{2n} in terms of traces of products of $2(n-1)$ operators by bringing the operator a to the end of the product by commuting, or anticommuting it, with each of the other operators in turn. We then use the invariance of the trace under cyclic permutations to put a in front of ρ_0, and finally we commute a with ρ_0. Thus

$$I_{2n} = \varepsilon\mathrm{Tr}\{\rho_0 bacd\,...\,ef\} + \mathrm{Tr}\{\rho_0[a, b]_{-\varepsilon}cd\,...\,ef\}. \tag{2.29}$$

(If a and b are two boson operators or one boson and one fermion operator, $[a, b]_{-\varepsilon}$ stands for the commutator of a with b; for two fermion operators it stands for the anticommutator. Thus ε will be $+1$ for the first case and -1 for the second. We make this choice so that the second term in eqn (2.29) contains the product of ρ_0 with $2(n-1)$ creation and annihilation operators). Proceeding in the same way, we obtain

$$\begin{aligned}\mathrm{Tr}\{\rho_0 abcd\,...\,ef\} &= \mathrm{Tr}\{\rho_0[a, b]_{-\varepsilon}cd\,...\,ef\} + \varepsilon\,\mathrm{Tr}\{\rho_0 b[a, c]_{-\varepsilon}d\,...\,ef\} + \ldots \\ &\quad + \mathrm{Tr}\{\rho_0 bcd\,...\,e[a, f]_{-\varepsilon}\} + \varepsilon\,\mathrm{Tr}\{\rho_0 bcd\,...\,efa\},\end{aligned} \tag{2.30}$$

where, in working out the sign of the last terms, we have used the fact that there is an even number of fermion operators in the product.

Using the invariance of the trace under cyclic permutation† we may write the last term in the form

$$\varepsilon\,\mathrm{Tr}\{a\rho_0 bcd \dots ef\}.$$

In Appendix B it is proved that:

$$\begin{aligned} a\rho_0 &= \rho_0 a \exp[\beta(\varepsilon_s - \mu)] \quad \text{if } a = a_s^+ \\ &= \rho_0 a \exp[-\beta(\varepsilon_s - \mu)] \quad \text{if } a = a_s. \end{aligned}$$

We write this symbolically:

$$a\rho_0 = \rho_0 a \exp(\pm\beta E). \tag{2.31}$$

We then have, for the last term,

$$\varepsilon\,\mathrm{Tr}\{a\rho_0 bcd \dots ef\} = \varepsilon \exp(\pm\beta E)\,\mathrm{Tr}\{\rho_0 abcd \dots ef\}.$$

Thus, from eqn (2.30),

$$\begin{aligned} &\mathrm{Tr}\{\rho_0 abcd \dots ef\}[1 - \varepsilon \exp(\pm\beta E)] \\ &\quad = \mathrm{Tr}\{\rho_0 [a, b]_{-\varepsilon} cd \dots ef\} + \varepsilon\,\mathrm{Tr}\{\rho_0 b[a, c]_{-\varepsilon} d \dots ef\} + \cdots \\ &\qquad + \mathrm{Tr}\{\rho_0 bcd \dots e[a, f]_{-\varepsilon}\}. \end{aligned}$$

If we write

$$[a, b]' \equiv [a, b]_{-\varepsilon}/[1 - \varepsilon \exp(\pm\beta E)], \tag{2.32}$$

and call it a restricted paired contraction, we have

$$\begin{aligned} \mathrm{Tr}\{\rho_0 abcd \dots ef\} &= \mathrm{Tr}\{\rho_0 cd \dots ef\}[a, b]' + \varepsilon\,\mathrm{Tr}\{\rho_0 bd \dots ef\}[a, c]' + \cdots \\ &\quad + \mathrm{Tr}\{\rho_0 bcd \dots e\}[a, f]', \end{aligned} \tag{2.33}$$

where $[a, b]'$ has been taken outside the trace, since it is either zero or proportional to the identity operator, in which case it may be treated as a c-number. We may therefore express the original expectation value of $2n$ operators in terms of expectation values of $2(n-1)$ operators.

The following theorem may now be proved by induction: The expectation value in the non-interacting ensemble of the product of n creation and n annihilation operators $\mathrm{Tr}\{\rho_0 abcd \dots ef\}/\mathrm{Tr}\{\rho_0\}$ is the sum of all the possible products of n restricted paired contractions amongst the $2n$ operators, multiplied by $(-1)^B$, where B is the number of interchanges of fermion

† Notice that it is important at this stage that we should be working in the grand canonical ensemble. $\mathrm{Tr}\{ABC\} = \mathrm{Tr}\{CAB\}$ only if the operators A, B, and C, operating on a vector in our space, produce a vector in that space. If we were working in the canonical ensemble, our space would be the Hilbert space appropriate to N particles, with N fixed. The operator a^+, acting on a vector in this space, produces a vector in the Hilbert space for $N+1$ particles, and so the trace is no longer invariant. There is no such difficulty in the grand canonical ensemble, since the space we are working in is then the product space of the spaces for $1, 2, \dots N, N+1, \dots$ particles.

operators needed to bring the contracted operators together; the order of the two operators within the contraction is the same as the order in the original product. Thus, for example, if a, b, c, d are all fermion operators

$$\langle abcd\rangle_0 = [a, b]'[c, d]' + (-1)[a, c]'[b, d]' + (-1)^2[a, d]'[b, c]'.$$

To prove the theorem, we note that the work above shows that if it is true for $2(n-1)$ operators, it is true for $2n$ operators. But for two operators, the theorem states:

(i) $\mathrm{Tr}\{\rho_0 a_s a_t\}/\mathrm{Tr}\{\rho_0\} = [a_s, a_t]' = 0$

since the commutator (or anticommutator) vanishes,

(ii) $\mathrm{Tr}\{\rho_0 a_s^+ a_t^+\}/\mathrm{Tr}\{\rho_0\} = 0,$

(iii) $\mathrm{Tr}\{\rho_0 a_s^+ a_t\}/\mathrm{Tr}\{\rho_0\} = [a_s^+, a_t]' = \dfrac{[a_s^+, a_t]}{1-\varepsilon\exp[\beta(\varepsilon_s-\mu)]}$

$$= \delta_{s,t}/\{\exp[\beta(\varepsilon_s-\mu)]-\varepsilon\},$$

(iv) $\mathrm{Tr}\{\rho_0 a_s a_t^+\}/\mathrm{Tr}\{\rho_0\} = \delta_{s,t}(1+\varepsilon/\{\exp[\beta(\varepsilon_s-\mu)]-\varepsilon\}).$

The results (i) and (ii) are correct since, as we noticed above, the trace should be zero if the number of creation operators is not equal to the number of annihilation operators. For (iii), note that $\mathrm{Tr}\{\rho_0 a_s^+ a_t\}/\mathrm{Tr}\{\rho_0\}$ is the expectation value in the non-interacting ensemble of $a_s^+ a_t$, which is zero unless $s = t$, in which case it is $\langle n_s\rangle_0 = \{\exp[\beta(\varepsilon_s-\mu)]-\varepsilon\}^{-1}$ from elementary statistical mechanics. Similarly, result (iv) may be seen to be true by using $a_s a_t^+ = \delta_{s,t} + \varepsilon a_t^+ a_s$. Thus the theorem is true for any two creation or annihilation operators, and our proof by induction is complete.

We now have to see how the τ-labels and the $\mathscr{T}$-ordering operator of eqn (2.27) affect this result. Consider the second-order term in $\langle S^{(\beta)}\rangle_0$,

$$\tfrac{1}{2}\int_0^\beta \mathrm{d}\tau_1 \int_0^\beta \mathrm{d}\tau_2\, \mathrm{Tr}\{\rho_0 \mathscr{T}[H_1(\tau_1)H_1(\tau_2)]\}/\mathrm{Tr}\{\rho_0\},$$

in the case when H_1 is of the form of eqn (2.9a) and when we are dealing with bosons. The integrand of this term will then be of the form†

$$\mathrm{Tr}\{\rho_0 \mathscr{T}[\bar{a}_1(\tau_1)\bar{a}_2(\tau_1)a_3(\tau_1)a_4(\tau_1)\bar{a}_5(\tau_2)\bar{a}_6(\tau_2)a_7(\tau_2)a_8(\tau_2)]\}/\mathrm{Tr}\{\rho_0\}.$$

Using the results Appendix B (iv) and (v), this may be written, for $\tau_1 > \tau_2$:

$$\frac{\mathrm{Tr}\{\rho_0 a_1^+ a_2^+ a_3 a_4 a_5^+ a_6^+ a_7 a_8\}\exp[\tau_1(\varepsilon_1+\varepsilon_2-\varepsilon_3-\varepsilon_4)+\tau_2(\varepsilon_5+\varepsilon_6-\varepsilon_7-\varepsilon_8)]}{\mathrm{Tr}\{\rho_0\}}$$

† We write $\bar{a}_1(\tau)$ for $\exp[(H_0-\mu\hat{N})\tau]a_1^+ \exp[(H_0-\mu\hat{N})\tau]$ rather than $a_1^+(\tau)$, because the latter might be mistaken for $[a_1(\tau)]^+$ which is $\exp[-(H_0-\mu\hat{N})\tau]a_1^+ \exp[(H_0-\mu\hat{N})\tau]$.

to which our theorem above may be applied, to give a sum of terms like $[a_1^+, a_7]'[a_2^+, a_3]'[a_5^+, a_8]'[a_4, a_6^+]' \exp[\tau_1 ...]$. For $\tau_1 < \tau_2$ we have,

$$\frac{\text{Tr}\{\rho_0 a_5^+ a_6^+ a_7 a_8 a_1^+ a_2^+ a_3 a_4\} \exp[\tau_1(\varepsilon_1+\varepsilon_2-\varepsilon_3-\varepsilon_4)+\tau_2(\varepsilon_5+\varepsilon_6-\varepsilon_7-\varepsilon_8)]}{\text{Tr}\{\rho_0\}},$$

and the term corresponding to the one above will now be

$$[a_7, a_1^+]'[a_2^+, a_3]'[a_5^+, a_8]'[a_6^+, a_4]' \exp[\tau_1 ...],$$

that is, it is the same apart from the order of the operators within the contractions. For fermion operators, we shall have a factor of $(-1)^C$ as well as the contractions. C is the number (in this case, even) of interchanges of fermion operators needed to bring the contracted operators together, whether these interchanges occur in achieving the correct τ-ordering, or in bringing together the contracted operators after the τ-ordering. (It is here that the amended definition of ordering, from the P-product to the $\mathscr{T}$-product, is a great help.)

We may combine the two results above by defining the contraction of two operators, missing out the word restricted, as:†

$$\begin{aligned}\{a_1(\tau_1), \bar{a}_2(\tau_2)\}' &= [a_1(\tau_1), \bar{a}_2(\tau_2)]' \qquad \tau_1 > \tau_2 \\ &= \varepsilon[\bar{a}_2(\tau_2), a_1(\tau_1)]' \qquad \tau_2 \geqslant \tau_1.\end{aligned} \tag{2.34}$$

Since, from above, $[a_1(\tau_1), \bar{a}_2(\tau_2)]' = \langle a(\tau_1)\bar{a}_2(\tau_2)\rangle_0$, we may rewrite this:

$$\{a_1(\tau_1), \bar{a}_2(\tau_2)\}' = \langle \mathscr{T}\{a_1(\tau_1)\bar{a}_2(\tau_2)\}\rangle_0. \tag{2.34a}$$

Similarly

$$\{\bar{a}_2(\tau_2), a_1(\tau_1)\}' = \langle \mathscr{T}\{\bar{a}_2(\tau_2)a_1(\tau_1)\}\rangle_0.$$

Then

$$\langle \mathscr{T}\{\bar{a}_1(\tau_1)\bar{a}_2(\tau_1)a_3(\tau_1)a_4(\tau_1)\bar{a}_5(\tau_2)\bar{a}_6(\tau_2)a_7(\tau_2)a_8(\tau_2)\rangle_0$$

is given by the sum of all terms like

$$(-1)^C\{\bar{a}_1(\tau_1), a_7(\tau_2)\}'\{\bar{a}_2(\tau_1), a_3(\tau_1)\}'\{\bar{a}_5(\tau_2), a_8(\tau_2)\}'\{a_4(\tau_1), \bar{a}_6(\tau_2)\}'$$

whether $\tau_1 >$ or $\leqslant \tau_2$.

Generalizing these ideas in an obvious way for the $\mathscr{T}$-ordered product of $2n$ operators, we have the full Wick's theorem:

The expectation value in the non-interacting ensemble of the $\mathscr{T}$-ordered product of n creation an n annihilation operators is the sum of all possible products of n contractions amongst the $2n$ operators, multiplied by a factor of $(-1)^C$, where C is number of interchanges of fermion operators needed

† This notation is unusual, but it is convenient for our present purposes, and it will soon be abandoned.

to bring together the contracted operators in the order in which they appear in the contraction.

Since, by definition,

$$\langle \mathscr{T}\{\bar{a}_2(\tau_2)a_1(\tau_1)\}\rangle_0 = \varepsilon\langle \mathscr{T}\{a_1(\tau_1)\bar{a}_2(\tau_2)\}\rangle_0, \tag{2.35}$$

the order in which the operators appear within a contraction no longer matters. If we change the order of the operator within a contraction, we change by one the number of interchanges of operators, and the extra factor of ε which we thus pick up from Wick's theorem cancels the factor of ε in the eqn (2.35).

It is convenient to bring the contracted operators together in such a way that the annihilation operator lies to the left of the creation operator. Then the only contraction we shall need to use is $\{a_s(\tau_1), \bar{a}_t(\tau_2)\}'$ which we shall call, for reasons which will appear later, the free-particle propagator, or Green function.† We shall write

$$\begin{aligned} g_{st}(\tau_1, \tau_2) &= \langle \mathscr{T}\{a_s(\tau_1)\bar{a}_t(\tau_2)\}\rangle_0 \\ &= \delta_{st}(\varepsilon\langle n_s\rangle_0 + 1)\exp[(\tau_2-\tau_1)(\varepsilon_s-\mu)] \qquad \tau_1 > \tau_2 \\ &= \delta_{st}\varepsilon\langle n_s\rangle_0 \exp[(\tau_2-\tau_1)(\varepsilon_s-\mu)] \qquad \tau_1 \leqslant \tau_2, \end{aligned} \tag{2.36}$$

and define

$$g_{ss}(\tau_1, \tau_2) \equiv g_s(\tau_1, \tau_2).$$

We note that g is a function only of $\tau_1 - \tau_2$, and that, if we write it as $g_s(\tau_1-\tau_2)$, we shall use it in the range $-\beta \leqslant \tau_1-\tau_2 \leqslant \beta$, since the τs in eqn 2.28 lie in the range $0 \leqslant \tau \leqslant \beta$. Further, when $\tau_1-\tau_2 \leqslant 0$, $\beta+\tau_1-\tau_2 \geqslant 0$ so that, in this case:

$$\begin{aligned} g_s(\tau_1-\tau_2) &= \varepsilon\langle n_s\rangle_0 \exp[(\tau_2-\tau_1)(\varepsilon_s-\mu)], \\ g_s(\beta+\tau_1-\tau_2) &= (\varepsilon\langle n_s\rangle_0+1)\exp[(\tau_2-\tau_1-\beta)(\varepsilon_s-\mu)] \\ &= \langle n_s\rangle_0 \exp[(\tau_2-\tau_1)(\varepsilon_s-\mu)] \end{aligned}$$

(using $\langle n_s\rangle_0 = \{\exp[\beta(\varepsilon_s-\mu)]-\varepsilon\}^{-1}$). Therefore

$$g_s(\tau_1-\tau_2+\beta) = \varepsilon g_s(\tau_1-\tau_2) \qquad \tau_1-\tau_2 \leqslant 0. \tag{2.37}$$

Similarly,

$$g_s(\tau_1-\tau_2-\beta) = \varepsilon g_s(\tau_1-\tau_2) \qquad \tau_1-\tau_2 > 0. \tag{2.38}$$

† An argument on the merits of calling them Green functions or Green's functions has been waged for some time. The author tends to agree with ter Haar, a strong advocate for the former, by analogy with Bessel functions, or Legendre functions, but has to admit that in speech he usually calls them Green's functions, that combination coming more readily to his lips.

Exercises

2.2.1. Use Wick's theorem to evaluate for the non-interacting grand ensemble the fluctuations in the number of particles in a given state,

$$[\langle n_s n_s \rangle_0 - (\langle n_s \rangle_0)^2]^{1/2}$$

and in the total number of particles $[\langle \hat{N}\hat{N}\rangle_0 - (\langle \hat{N}\rangle_0)^2]^{1/2}$. Notice that the latter is of the order of $\bar{N}^{1/2}$, in most cases, but that for the boson gas below the Bose–Einstein transition temperature, it is of the order $\bar{N}$. In the first case $[\langle \hat{N}\hat{N}\rangle_0 - (\langle \hat{N}\rangle_0)^2]^{1/2}/\bar{N} = \mathcal{O}(1/\sqrt{\bar{N}})$ and the fluctuations in the total number of particles is very small.

2.2.2. From eqn (2.15), (2.22), and (2.27), the first- and second-order terms in the expansion of Q are:

$$Q_0\left[\left\langle -\int_0^\beta \mathcal{T}\{H_1(\tau)\,\mathrm{d}\tau\}\right\rangle_0\right]$$

and

$$Q_0\left[\left\langle \int_0^\beta\int_0^\beta \mathrm{d}\tau_1\,\mathrm{d}\tau_2 \mathcal{T}\{H_1(\tau_1)H_1(\tau_2)\}\right\rangle_0\right].$$

If H_1 takes the form in eqn (2.9a), there are two sets of non-zero contractions in the first term and twenty-four in the second. Write out some (or if you are energetic all) of these.

How many sets are there in nth order?

2.2.3. For phonons, or particles whose number is not fixed, we must put $\mu = 0$ in eqn (2.1). Show that the Green functions are then:

$$\begin{aligned} d_s(\tau_1-\tau_2) &\equiv \langle \mathcal{T}\{b_s(\tau_1)\bar{b}_s(\tau_2)\}\rangle_0 \\ &= \langle n_s\rangle_0 \exp[(\tau_2-\tau_1)\omega_s] \qquad \tau_1 \leqslant \tau_2 \\ &= [\langle n_s\rangle_0+1]\exp[(\tau_2-\tau_1)\omega_s] \qquad \tau_1 > \tau_2 \end{aligned}$$

where

$$\langle n_s\rangle_0 = \{\exp(\beta\omega_s)-1\}^{-1}.$$

Since, in the electron–phonon interaction, b and b^+ occur only in the combination $(b_{\mathbf{q}} - b^+_{-\mathbf{q}})$ (see example 1.2.4) it is sometimes more convenient to use

$$D_{\mathbf{q}}(\tau_1-\tau_2) \equiv \langle \mathcal{T}\{[b_{\mathbf{q}}(\tau_1)-\bar{b}_{-\mathbf{q}}(\tau_1)][\bar{b}_{\mathbf{q}}(\tau_2)-b_{-\mathbf{q}}(\tau_2)]\}\rangle.$$

2.3. The diagrammatic representation of the perturbation series

Wick's theorem has taken us a long way towards our goal. The chief difficulty we foresaw in Chapter 1 was the evaluation of complicated matrix elements which would appear in the perturbation series. This has now been avoided. Unfortunately, from a product of n creation and n annihilation operators, we can form $n!$ contractions, and so if H_1 contains four operators, there will be $(2n)!$ terms in the nth order of perturbation theory. Moreover, the sign of each term, for fermion operators, will be rather tedious to determine. As an aid to remembering all the contractions, and also in determining the sign to be taken with each product, it is convenient to develop a way of representing each term by a diagram. In each $H_1(\tau)$ we have, in the most frequent case, four operators, and a matrix element of the potential,

$$\tfrac{1}{2}v_{s_1s_2s_3s_4}\bar{a}_{s_1}(\tau)\bar{a}_{s_2}(\tau)a_{s_4}(\tau)a_{s_3}(\tau).$$

We shall represent this as a horizontal dotted line for $\frac{1}{2}v_{s_1s_2s_3s_4}$ with two full lines leaving it, for the creation operators, and two full lines entering, for the annihilation operators:

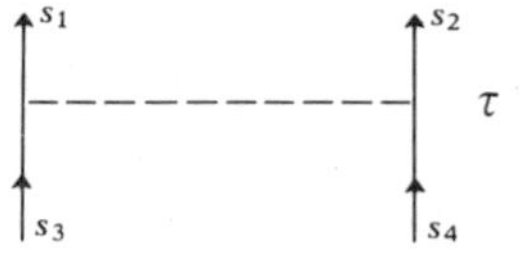

We shall call this a vertex. We note that the line leaving the left-hand side of the vertex represents the first operator in H_1, so that, at this stage, the side of a vertex to which a full line is attached is important. We have added a τ-label to tell us the τ-label on the operators. To indicate that two operators are contracted together, we join their two lines together. Thus, in first order, we have to evaluate

$$\int_0^\beta d\tau \left\langle \mathcal{T}\left\{ \sum_{s_1s_2s_3s_4} \tfrac{1}{2}v_{s_1s_2s_3s_4}\bar{a}_{s_1}(\tau)\bar{a}_{s_2}(\tau)a_{s_4}(\tau)a_{s_3}(\tau)\right\}\right\rangle_0$$

$$= \int_0^\beta d\tau \sum_{s_1s_2s_3s_4} \tfrac{1}{2}v_{s_1s_2s_3s_4}[\{a_{s_3}(\tau),\bar{a}_{s_1}(\tau)\}'\{a_{s_4}(\tau),\bar{a}_{s_2}(\tau)\}'+$$

$$+\varepsilon\{a_{s_3}(\tau),\bar{a}_{s_2}(\tau)\}'\{a_{s_4}(\tau),\bar{a}_{s_1}(\tau)\}']$$

$$= \int_0^\beta d\tau \sum_{s_1s_2s_3s_4} \tfrac{1}{2}v_{s_1s_2s_3s_4}[g_{s_3s_1}(0)g_{s_4s_2}(0)+\varepsilon g_{s_3s_2}(0)g_{s_4s_1}(0)].$$

These two terms are represented:

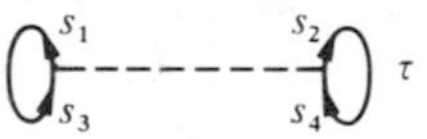

(a)

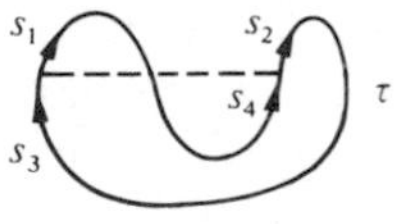

(b)

Notice that, since the only non-zero contractions are between an annihilation operator and a creation operator, the two arrows on a full line must point in the same direction: we shall often use this fact, and put only one arrow on each line. There is no particular merit in having all the arrows at a vertex pointing upwards, provided we keep to the convention that ingoing arrows represent annihilation operators and outgoing, creation operators; thus (b) may be drawn more elegantly:

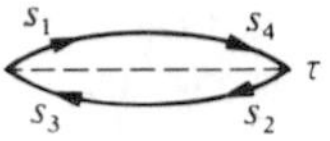

All possible contractions are obtained by joining the lines at the vertices in all possible ways, the arrows on a line always pointing in the same direction. Thus, in second order, we have the terms shown in Fig. 2.2. (Compare with the results of exercise 2.2.2.) The diagrams (i), (v), and (xiii), for example, represent respectively the terms:

$$\text{(i)} = \sum_{\substack{s_1s_2s_3s_4\\ s_1's_2's_3's_4'}} \tfrac{1}{2}v_{s_1s_2s_3s_4}\tfrac{1}{2}v_{s_1's_2's_3's_4'}\tfrac{1}{2}\int_0^\beta\!\!\int_0^\beta \mathrm{d}\tau_1\,\mathrm{d}\tau_2\, g_{s_3s_1}(0)g_{s_4s_2}(0)g_{s_3's_1'}(0)g_{s_4's_2'}(0), \tag{2.39}$$

$$\text{(v)} = \sum_{\substack{s_1s_2s_3s_4\\ s_1's_2's_3's_4'}} \tfrac{1}{2}v_{s_1s_2s_3s_4}\tfrac{1}{2}v_{s_1's_2's_3's_4'}\tfrac{1}{2}\int_0^\beta\!\!\int_0^\beta \mathrm{d}\tau_1\,\mathrm{d}\tau_2\, g_{s_3's_1}(\tau_2-\tau_1)g_{s_3s_1'}(\tau_1-\tau_2)$$
$$\times g_{s_4's_2}(\tau_2-\tau_1)g_{s_4s_2'}(\tau_1-\tau_2), \tag{2.40}$$

$$\text{(xiii)} = \varepsilon\sum_{\substack{s_1s_2s_3s_4\\ s_1's_2's_3's_4'}} \tfrac{1}{2}v_{s_1s_2s_3s_4}\,\tfrac{1}{2}v_{s_1's_2's_3's_4'}\,\tfrac{1}{2}\int_0^\beta\!\!\int_0^\beta \mathrm{d}\tau_1\,\mathrm{d}\tau_2\, g_{s_4s_1}(0)g_{s_4's_1'}(0)$$
$$\times g_{s_3s_2'}(\tau_1-\tau_2)g_{s_3's_2}(\tau_2-\tau_1). \tag{2.41}$$

The first two factors of $\frac{1}{2}$ come with the v matrix elements (see eqn (2.9)); the third comes from the factor $1/n!$ in the nth order term in the perturbation series for S (see eqn (2.25)). The factor of ε in eqn (2.41) is the factor $(-1)^C$ in Wick's theorem. Considerable simplification of these expressions may be made by noticing that $g_{st}(\tau_1-\tau_2)$ contains a factor δ_{st}. Further, in many cases, the ss stand for momentum labels and $v_{\mathbf{k}_1\mathbf{k}_2\mathbf{k}_3\mathbf{k}_4} = \bar{v}(|\mathbf{k}_1-\mathbf{k}_3|\,\delta_{\mathbf{k}_1+\mathbf{k}_2,\mathbf{k}_3+\mathbf{k}_4}$ (see eqn (2.9)).

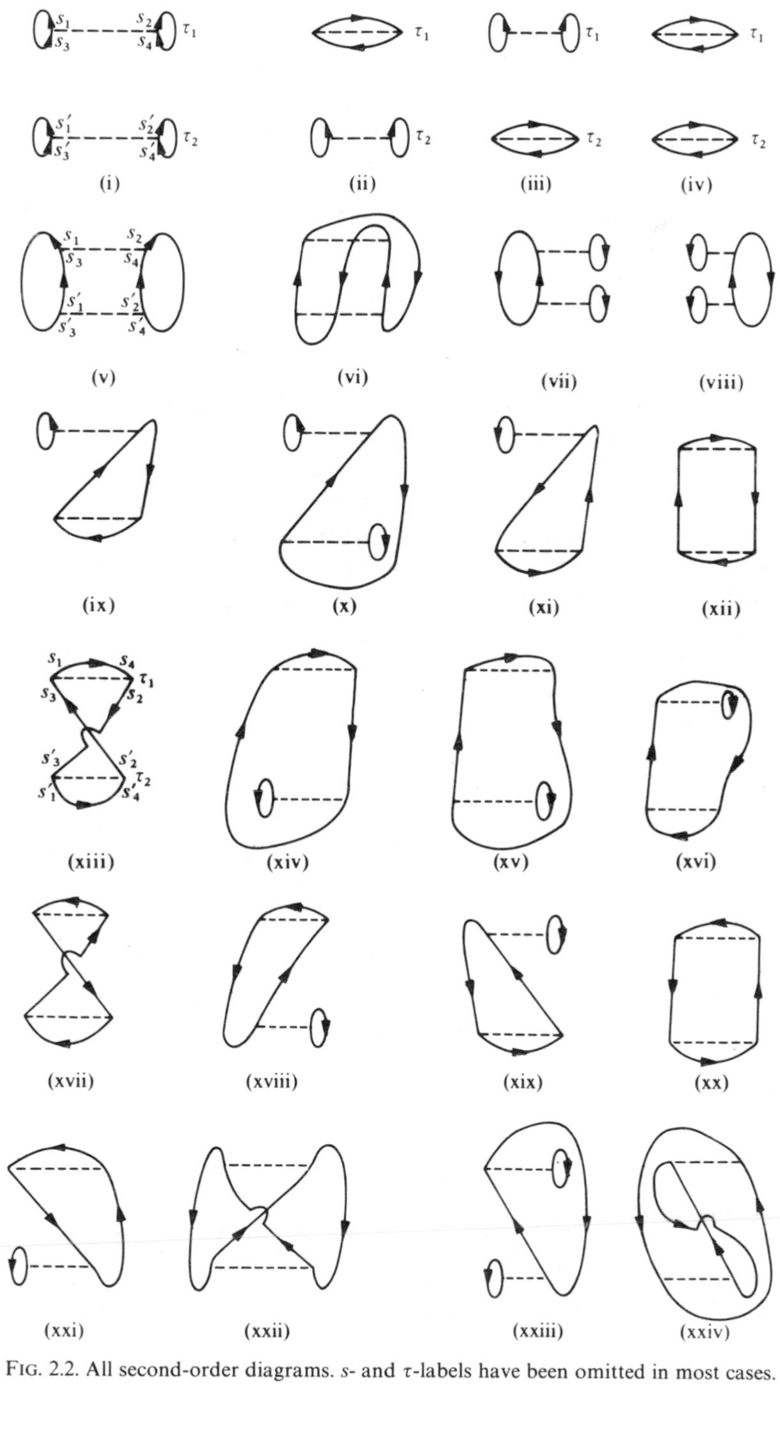

FIG. 2.2. All second-order diagrams. s- and τ-labels have been omitted in most cases.

At first sight, the factor $(-1)^C$ seems to be very difficult to obtain from the form of the diagram: but a little work leads to the very convenient theorem that $(-1)^C$ is $(-1)^D$, where D is the number of fermion loops of full lines, in the diagram. That is, if all the operators in our present example are for fermions, for diagram (i), $D = 4$, for (v), $D = 2$, and for (xiii) $D = 1$.

To see why this is so, we confine our considerations for the next few lines to the *sign* of the term only. As far as this is concerned, we may rewrite H_1 as

$$\frac{1}{2}\sum_{s_1s_2s_3s_4} v_{s_1s_2s_3s_4} a^{+}_{s_1} a^{+}_{s_2} a_{s_4} a_{s_3} \to \frac{1}{2}\sum v_{s_1s_2s_3s_4} a^{+}_{s_1} a_{s_3} a^{+}_{s_2} a_{s_4},$$

since to achieve this order we need to make two interchanges of fermion operators, thus leaving the sign unchanged. We now have the operators at one end of a vertex together in our algebraic expression, and we shall call these a vertex pair. From each loop, we arbitrarily choose one vertex pair—call it number one. We move the creation operator (call it number two) contracted with the annihilation operator of pair number one to the right of that annihilation operator. We cannot say in general how many interchanges of fermion operators this involves, but if we carry along with this creation operator the annihilation operator, number two, in its vertex pair (which, by our first move, will be on the immediate right of creation operator number two), we must make an even number of interchanges, since we shall move two operators the same number of times. We then look for the creation operator contracted with annihilation operator number two, and move this, together with its vertex pair, to its correct position. We continue this operation until we have moved the annihilation operator (number s) to be contracted with creation operator number one. Our operators will now look like:

(uncontracted operators) $a^{+}a$ (vertex pair number one) [a contracted together with] $a^{+}a$ (number two) [a contracted together with] $a^{+}a \ldots a^{+}a$ (number three) ... $a^{+}a$ (number s) (uncontracted operators)

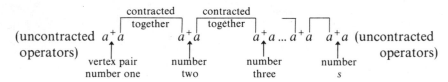

We have so far made an even number of interchanges. To put together all contracted operators in the correct order we need to take the final annihilation operator, labelled s, and put it to the left of creation operator number one. This will involve an odd number of interchanges. But the lines representing the contractions of these operators together make up a closed fermion loop: consequently, if we are considering a fermion loop, we must associate a minus sign with it. The same argument may now be applied to the other loops, and the theorem follows. The ease with which the sign of a term can now be calculated is one great advantage of the diagrammatic method; counting the number of interchanges needed to bring together the contracted operators is a laborious business, but we can now determine the sign by glancing at the form of the diagram.

An inspection of Fig. 2.2 nevertheless tempts one to despair. If there are twenty-four second-order contributions the prospect of going to much higher orders is not a happy one. However, we shall now spend some time grouping diagrams together so that we are left with only two second-order diagrams.

Exercises

2.3.1. Write down the contributions of some of the diagrams in Fig. 2.2, and compare these with the answers to exercise 2.2.2.

2.3.2. Call the contribution of all first-order diagrams C_1. Show that the contribution of the first four diagrams of Fig. 2.2 is equal to $(C_1)^2/2$.

2.3.3. Draw some third-order diagrams and write down the contribution they represent to the perturbation series. Show that there is a subset of these diagrams the contributions of which are $(C_1)^3/3$. Guess at a general theorem.

2.3.4. Set up a diagrammatic expansion for the electron–phonon problem with interaction as in exercise 1.2.4 (i.e. no electron–electron and no phonon–phonon interaction). Using a wavy line to represent a free-phonon propagator D show that the only second-order diagram is:

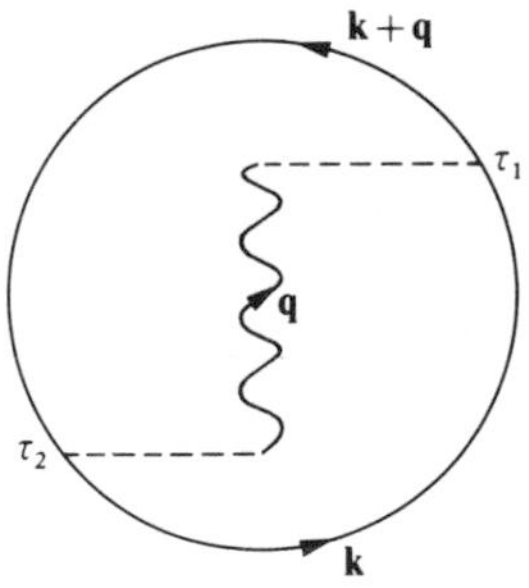

(Since D contains both $\langle \mathscr{T}\{b_q(\tau_1)\bar{b}_q(\tau_2)\}\rangle_0$ and $\langle \mathscr{T}\{\bar{b}_{-q}(\tau_1)b_{-q}(\tau_2)\}\rangle_0$ the arrow on the wavy line must be taken to indicate only the direction in which the momentum is carried.) The diagram

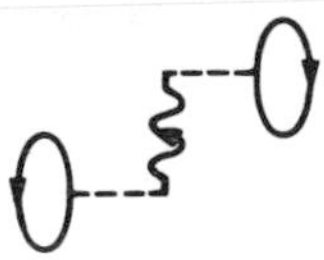

should not be included. Why not?

2.4. Symmetrization of vertices

It is easy to see that the contributions of diagrams (vii) and (viii) of Fig. 2.2 are the same: they are mirror reflections of each other about a vertical axis and, from eqn (2.9a), $v_{s_1s_2s_3s_4} = v_{s_2s_1s_4s_3}$. The same is true of the pairs (ix) and (xix), (xiv) and (xviii), and (xv) and (xxi). (The diagrams do not always appear as mirror images, but can be redrawn so that they are.) Diagrams (i), (ii), (iii), (iv), (v), (vi), (xxii) and (xxiv) are their own mirror images. Similarly, the contributions of (vii) and (xxiii) will be the same; (xxiii) is the diagram (vii) with the bottom vertex twisted round. Similar pairs are (v) and (xxiv), and (vii) and (x): for the latter the top vertex is twisted. If there are n vertices, we shall obtain $(2)^n$ diagrams from taking one and making all possible twists of the vertices. The contributions of these diagrams will all be equal. Unfortunately, as we have seen above, not all such twists lead to different diagrams.

There is a different type of similarity between diagrams (v) and (vi): they would be the same if we were to fail to distinguish between the two lines leaving a vertex, and between the two lines entering. We have already grouped together the vertices

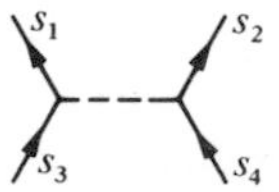

and

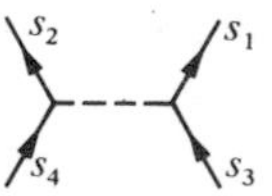

;

we now plan to take with these

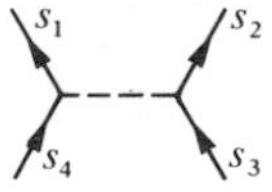

and

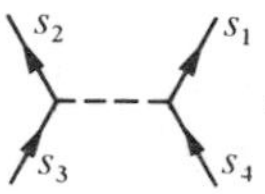

.

We shall represent the sum of all four vertices by a square; thus

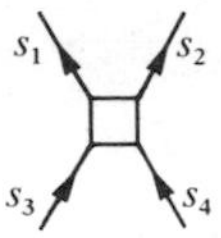

includes

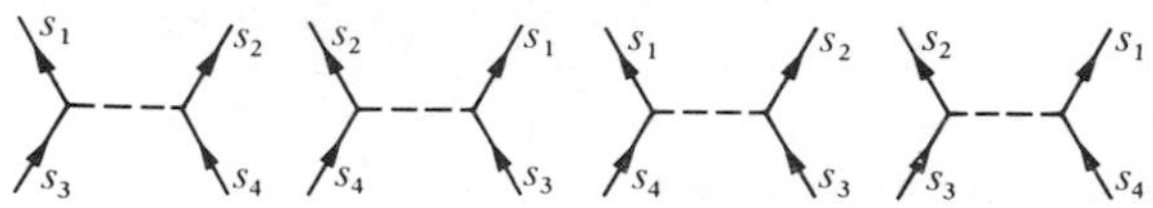

.

By allowing this type of grouping, we take together (v), (vi), (xxii), and (xxiv) of Fig. 2.2 under the general description—'Each of the τ_1-creation operators is contracted with one of the τ_2-annihilation operators; each of the τ_2-creation operators is contracted with one of the τ_1-annihilation operators', and we represent it:

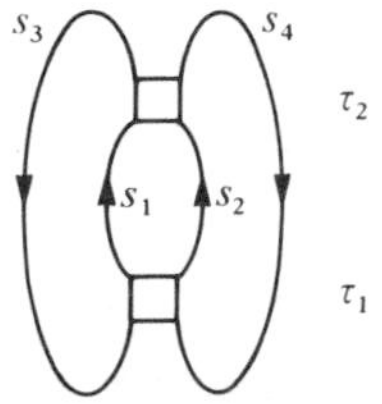

FIG. 2.3.

Diagrams (i) to (iv) could be drawn:

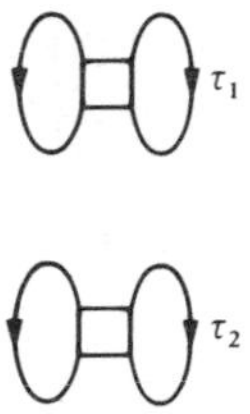

FIG. 2.4.

under the general description: 'Each τ_1-creation operator is contracted with a τ_1-annihilation operator; each τ_2-creation operator is contracted with a τ_2-annihilation operator.' All the remaining diagrams of Fig. 2.2 can be included in:

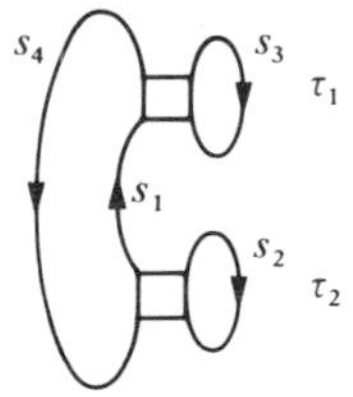

FIG. 2.5.

described: 'One of the τ_1-creation operators is contracted with one of the τ_2-annihilation operators; one of the τ_2-creation operators is contracted with one of the τ_1-annihilation operators; the other τ_1-creation operator is contracted with a τ_1-annihilation operator, and the other τ_2-creation operator is contracted with a τ_2-annihilation operator.'

However, in our present formulation of the problem, the terms we have now grouped together do not always give identical contributions since the vertex

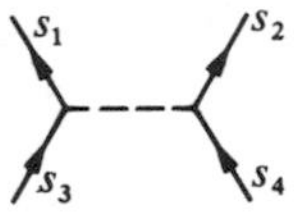

has associated with it the matrix element $v_{s_1s_2s_3s_4}$ and

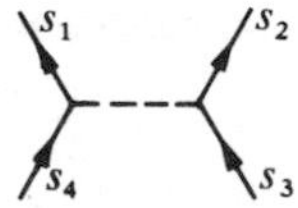

has $v_{s_1s_2s_4s_3}$ and these are not in general the same. In the fermion case there is also a difficulty about signs: Fig. 2.3 has two closed loops, but some of the diagrams in Fig. 2.2 which we hope to include in it have only one. In order to deal with these difficulties we go back to eqn (2.9a) and rewrite H_1:

$$H_1 = \tfrac{1}{2} \sum_{s_1s_2s_3s_4} v_{s_1s_2s_3s_4} a^+_{s_1} a^+_{s_2} a_{s_4} a_{s_3}$$

$$= \tfrac{1}{4} \sum_{s_1s_2s_3s_4} (v_{s_1s_2s_3s_4} a^+_{s_1} a^+_{s_2} a_{s_4} a_{s_3} + v_{s_1s_2s_4s_3} a^+_{s_1} a^+_{s_2} a_{s_3} a_{s_4}),$$

since s_3 and s_4 are dummy indices. Thus

$$H_1 = \tfrac{1}{4} \sum_{s_1s_2s_3s_4} (v_{s_1s_2s_3s_4} + \varepsilon v_{s_1s_2s_4s_3}) a^+_{s_1} a^+_{s_2} a_{s_4} a_{s_3},$$

by commuting or anticommuting the annihilation operators in the final term. If we put $v_{s_1s_2s_3s_4} + \varepsilon v_{s_1s_2s_3s_4} = \Gamma^0_{s_1s_2s_3s_4}$, we have

$$H_1 = \tfrac{1}{4} \sum_{s_1s_2s_3s_4} \Gamma^0_{s_1s_2s_3s_4} a^+_{s_1} a^+_{s_2} a_{s_4} a_{s_3} \tag{2.42}$$

and we could repeat all that we have previously said in this chapter in terms of Γ^0 instead of v. Γ^0 is called a symmetrized vertex function. The difficulty is now passed. For bosons

$$\Gamma^0_{s_1s_2s_3s_4} = \Gamma^0_{s_1s_2s_4s_3} = \Gamma^0_{s_2s_1s_3s_4} = \Gamma^0_{s_2s_1s_4s_3}.$$

Thus, provided we replace v by Γ^0, we may group together the diagrams in the manner discussed above, and all the diagrams grouped together have the same contribution. We then have an additional factor of four for each 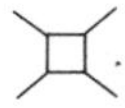 vertex, to account for the fact that it includes four arrangements of the vertex labels. This happily cancels the factor of $\frac{1}{4}$ outside the sum in H_1 in eqn (2.42).

Similar considerations apply for fermions. Here, however,

$$\Gamma^0_{s_1s_2s_3s_4} = -\Gamma^0_{s_1s_2s_4s_3} = -\Gamma^0_{s_2s_1s_3s_4} = \Gamma^0_{s_2s_1s_4s_3}.$$

Fortunately, every time we interchange the two creation operators of a vertex, or the two annihilation operators, we change the number of closed loops by one. Thus the sign of the whole contribution is still given correctly by $(-1)^D$, where D is the number of closed loops, since each such interchange also carries with it a (-1) from Γ^0. (Note that in counting closed loops, continuous lines must be treated as staying at the same side of a vertex as they pass through: thus  has two closed loops, not one .

The final difficulty we have to deal with in this matter, already mentioned above, is that not all the interchanges we have been discussing lead to different diagrams. For example, since each 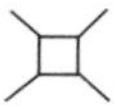 vertex corresponds to four >- - -< vertices, the diagram in Fig. 2.3 should have sixteen diagrams corresponding to it in Fig. 2.2. As we have seen, it has, in fact, only four diagrams. This is further illustrated in Fig. 2.6 and 2.7 where we have isolated three or two vertices forming parts of a diagram. All the terms included in Fig. 2.6 lead to different diagrams in the >- - -< representation; in Fig. 2.7 they do not. This difficulty is going to occur if and only if two lines leaving a vertex together also come together at a vertex. Such a pair of lines is called an equivalent pair. Thus if there are n_e pairs of equivalent lines in a diagram, one has to divide by $(2)^{n_e}$ to allow for this effect.

Exercise

2.4.1. Show that, by taking into consideration the equivalent lines, the contributions of the diagrams of Figs. 2.3, 2.4, and 2.5 reproduce exactly those of Fig. 2.2.

2.5. Disconnected diagrams

The first four diagrams of Fig. 2.2 differ from the others in an obvious respect: they separate into two parts with no connections between the two. We shall describe them as disconnected, and the others as connected. In higher orders, we may have several disconnected parts in a diagram; for example, the seventh-order diagram of Fig. 2.8, has four disconnected parts. One might suspect that the contributions of such diagrams could be

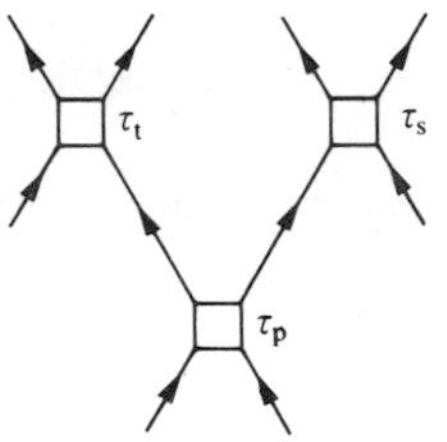

(One τ_p-creation operator is contracted with a τ_s-annihilation operator, and the other with a τ_t-annihilation operator.)

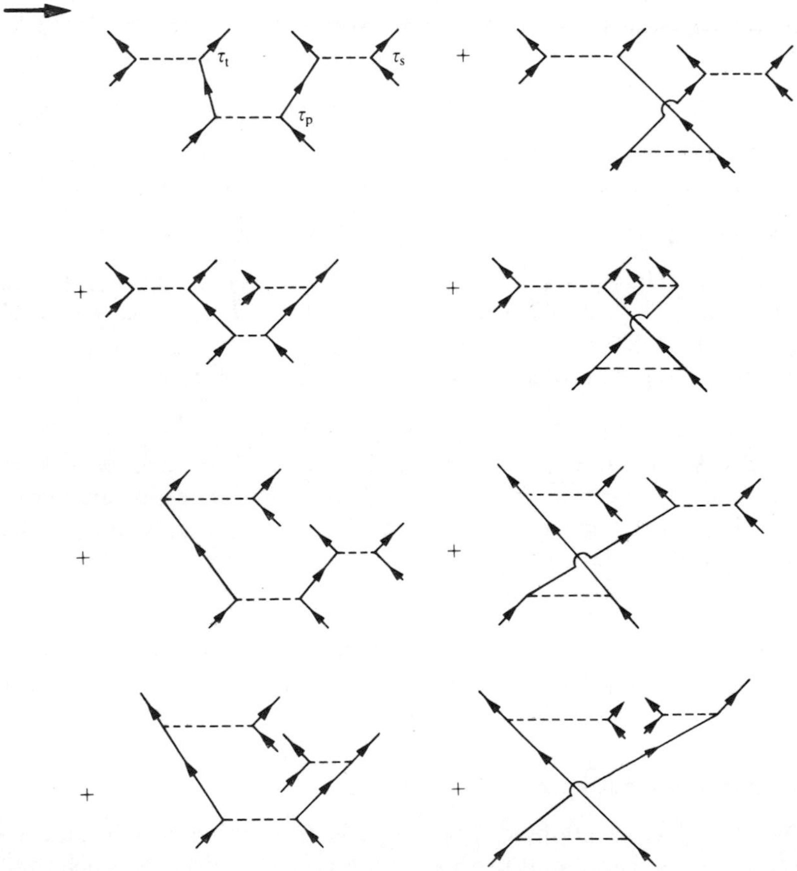

+ 56 other diagrams obtained by interchanging the τ_p-annihilation operators, the τ_s- and the τ_t-creation operators.

FIG. 2.6. The diagrams shown are obtained from the first by interchanging the τ_p-creation operators, the τ_s- and the τ_t-annihilation operators.

(Each τ_p-creation operator is contracted with a τ_s-annihilation operator.)

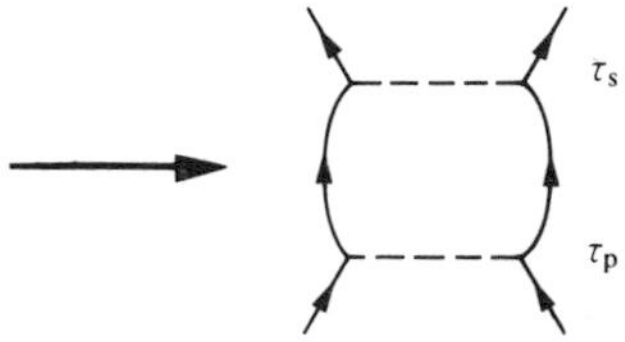

+ 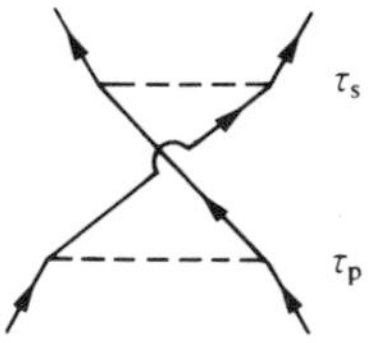

(Interchange τ_s-annihilation operators; τ_p-creation operators left unchanged.)

+ 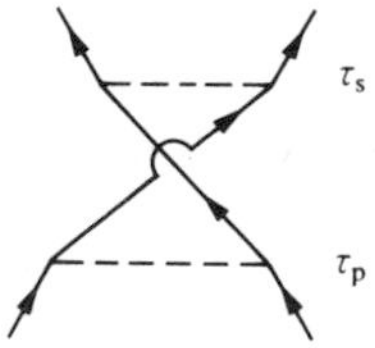

(Interchange τ_p-creation operators; τ_s-annihilation operators left unchanged.)

+ 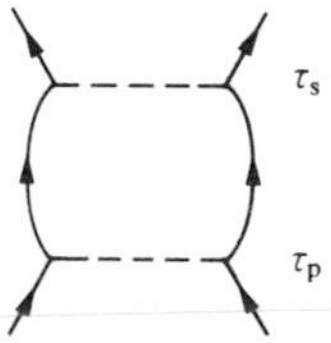

(Interchange τ_p-creation operators and τ_s-annihilation operators.)

+ 12 other diagrams obtained by interchanging the τ_s-annihilation operators and the τ_p-creation operators.

FIG. 2.7. The first and last diagrams (in the non-symmetrized form) represent the same set of contractions, as do the second and third.

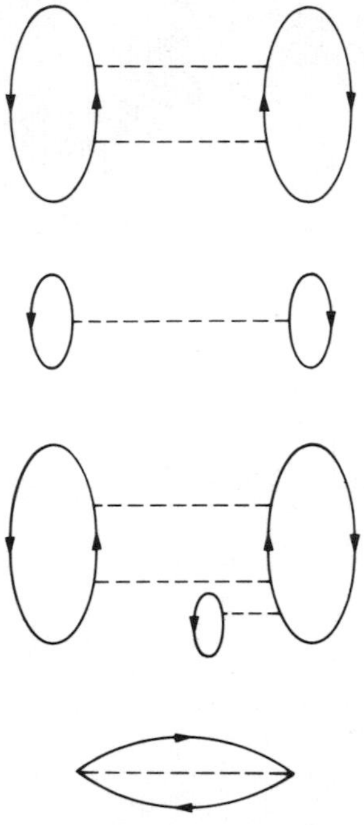

FIG. 2.8.

expressed in terms of the lower-order diagrams of which they are composed, and the results of exercises 2.3.2 and 2.3.3 support this conjecture. It looks as if the contribution of the diagram in Fig. 2.8 is a product of two first-order contributions, one second-order, and one third-order. To see that this is the case, consider a disconnected diagram with vertices at $\tau_1, \tau_2, \ldots$ in one connected part, $\tau'_1, \tau'_2, \ldots$ in a second connected part, etc. The contribution of the whole diagram will be

$$D = \int_0^\beta \cdots \int_0^\beta \mathrm{d}\tau_1\, \mathrm{d}\tau_2 \ldots \mathrm{d}\tau'_1\, \mathrm{d}\tau'_2 \ldots \mathrm{d}\tau''_1\, \mathrm{d}\tau''_2 \ldots f_1(\tau_1, \tau_2, \ldots) \times$$
$$\times f_2(\tau'_1, \tau'_2, \ldots) f_3(\tau''_1, \tau''_2, \ldots),$$

where $f_1(\tau_1, \tau_2, \ldots)$ is a function of $\tau_1, \tau_2, \ldots$ only, $f_2(\tau'_1, \tau'_2, \ldots)$ of $\tau'_1, \tau'_2, \ldots$ only, etc. Thus

$$D = \left[\int_0^\beta \cdots \int_0^\beta \mathrm{d}\tau_1\, \mathrm{d}\tau_2 f_1(\tau_1, \tau_2, \ldots)\right] \times \left[\int_0^\beta \cdots \int_0^\beta \mathrm{d}\tau'_1\, \mathrm{d}\tau'_2 f_2(\tau'_1, \tau'_2, \ldots)\right] \ldots . \tag{2.43}$$

That is, D is the product of the contributions of the lower-order diagrams of which the disconnected diagram is composed, apart, possibly, from numerical factors which we shall have to consider later. This property of disconnected diagrams is in itself very convenient, but there is a remarkable theorem depending on it which removes the need to consider disconnected diagrams any further. So far, we have

$$Q = Q_0 \quad (1 + \text{contribution of all diagrams, connected and disconnected}).$$

The theorem, sometimes called the Linked Cluster Theorem, says that this series can be rearranged into the form:

$$Q = Q_0\left(1 + \sum_{n=1} [\mathscr{C}(\beta)]^n/n!\right), \tag{2.44}$$

where $\mathscr{C}(\beta)$ is the sum of the contributions of all connected diagrams.

The proof of the theorem, consisting mostly of combinatorics, is given in Appendix C, which we suggest should be read after section 2.6. To see that the theorem is a possible one, we write

$$\mathscr{C}(\beta) = \mathscr{C}_a(\beta) + \mathscr{C}_b(\beta) + \ldots$$

where $\mathscr{C}_i(\beta)$ is the contribution of a particular connected diagram. From $[\mathscr{C}(\beta)]^2$, in eqn (2.44) we shall obtain terms like $\mathscr{C}_a(\beta)\mathscr{C}_b(\beta)$ which, according to eqn (2.43) we should expect to have from disconnected diagrams with two parts. Similarly, $[\mathscr{C}(\beta)]^n$ will contain terms we should expect from disconnected diagrams with n parts. The proof of the theorem will thus have only to show that the numerical factors of the disconnected parts are given correctly by eqn (2.44).

Eqn (2.44) may be written

$$Q = Q_0 \exp \mathscr{C}(\beta), \tag{2.44a}$$

and, from eqn (2.3),

$$\begin{aligned}\Omega &= -\beta^{-1} \ln Q \\ &= \Omega_0 - \mathscr{C}(\beta)/\beta,\end{aligned} \tag{2.45}$$

Thus, if we now consider the expansion for Ω rather than for Q, we need only include connected diagrams. Thus in the expansion for Ω in terms of symmetrized vertices, we have only two diagrams in second order.

There is a further reason for expecting the existence of a theorem something like the linked cluster theorem, which may be understood by considering the volume dependence of the contributions of connected and disconnected diagrams. For particles in a box, interacting through central two-body forces, each matrix element at a vertex is proportional to $(\text{Volume})^{-1}$, and carries with it a δ-function on the four momentum labels entering and leaving. (See eqn (2.9)). In a connected nth order diagram, there are n matrix elements, and $2n$ propagators. Each propagator has a momentum label (or two momentum labels which must be equal (eqn 2.36), and these are to be summed over, subject to the conditions imposed by the δ-functions at the vertices. However, if we satisfy these conditions at $n-1$ of the vertices the condition at the nth vertex is automatically satisfied. We thus have $2n-(n-1) = n+1$ independent sums over momentum labels. For large systems, we may replace the sum by an integral, in the usual way, using a density of states which is proportional to the volume. Thus the total contribution of a connected diagram is proportional to the volume. Similarly, the contribution to Q of a diagram which has s disconnected parts is proportional to the sth power of the volume. (s of the momentum restrictions are now satisfied automatically). Thus, our expression for Q contains all powers of the volume. On the other hand, Ω which is $-\kappa T \ln Q$, is an extensive thermodynamic function, and should thus be proportional to the volume. Thus, when we take the logarithm of our series for Q, we must get rid of all higher powers of the volume. At this stage, one can guess the form of the linked cluster theorem.

Exercises

2.5.1 Consider the series of disconnected diagrams shown in the figure. (We use unsymmetrized vertices for simplicity.)

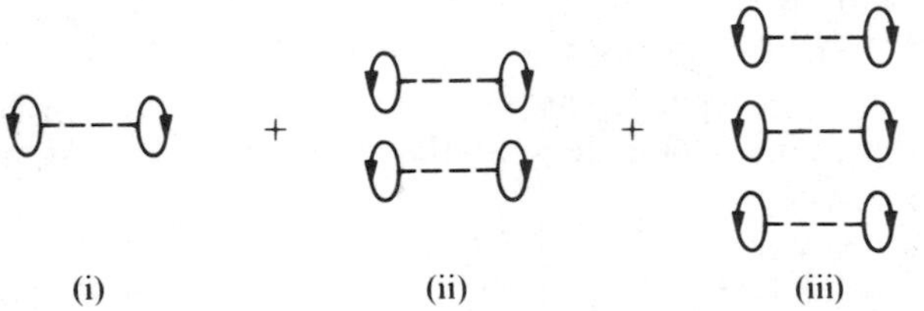

(i) (ii) (iii)

Call the contribution of (i), r. Show that the contributions of (ii) and (iii) are $r^2/2!$ and $r^3/3!$ respectively (see the expansion for S, eqns (2.22) and (2.27)) and that the sum of the infinite series is $\exp(r)-1$.

2.5.2. What is the contribution of

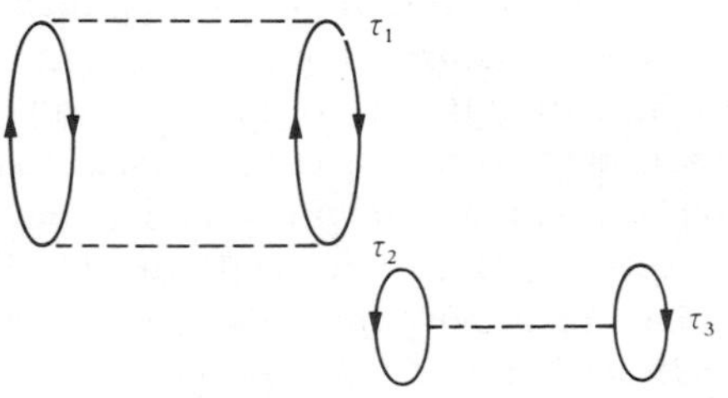

in third order? Compare this with the product of

in second order, and

in first order.

2.5.3. For the electron–phonon interaction of exercise (3.4), draw the first few terms of an infinite series which will sum in a manner similar to that in exercise 2.5.1.

2.6. τ-labels

Our next reduction in the number of diagrams comes from a consideration of the τ-labels of vertices. So far, the two diagrams in Fig. 2.9 are to be counted as different, as they represent different sets of contractions. The τs have all to be integrated over the same range, however, and are thus merely dummy indices. The contributions of the two diagrams will thus be identical, and we need no longer distinguish between them, if we multiply the contribution by a suitable factor. At first sight, it looks as if there are $n!$ identical contributions coming from the $n!$ ways of interchanging the τ-labels amongst themselves. Not all these interchanges lead to different sets of contractions, however, as can be seen in the case of the second-order diagrams in Fig. 2.10; both represent the same set of contractions.

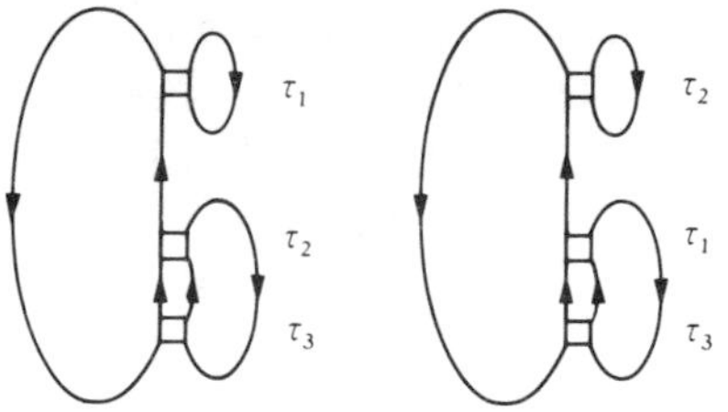

FIG. 2.9.

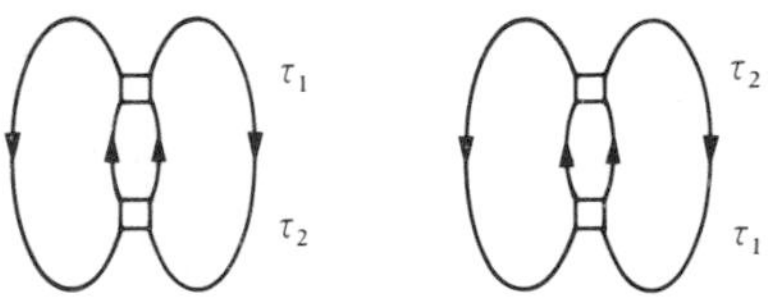

FIG. 2.10.

Thus if we agree to represent by one diagram all the diagrams that can be obtained by interchanging τ-labels, we must multiply the contribution of this one diagram by $n!/p$, where n is the order of the diagram, and p is a (very inconvenient) factor, called the coordination number of the diagram, which gives the number of permutations of the τ-labels amongst themselves that lead to identical sets of contractions. The $n!$ is convenient, in that it cancels the $1/n!$ which occurs in the expansion of $S^{(\beta)}$ (eqn (2.22) and (2.27)). The coordination number is rather a nuisance, although in realistic calculations one is usually dealing with low-order diagrams, in which case it is easy to see what p is from the form of the diagram, or one is summing the contributions of a series of diagrams of a particular type, perhaps to all orders, so that the coordination number of the particular nth order diagrams one is considering is a simple function of n. (See, for example, (iii) of Fig. 2.11). We shall see in the next chapter how it is possible in some calculations to avoid altogether the use of the coordination number. A few examples are given in Fig. 2.11.

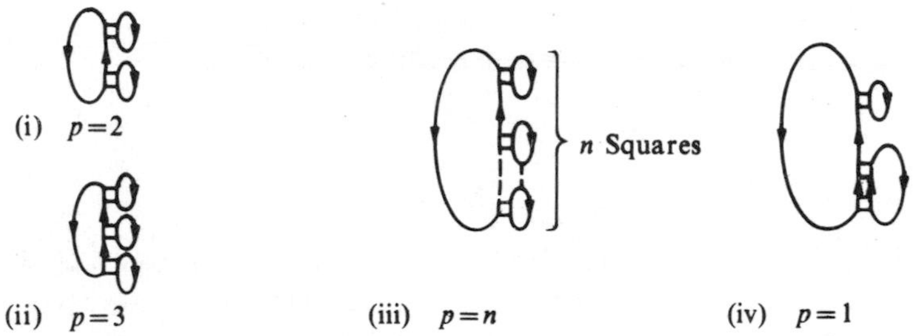

FIG. 2.11. Examples of coordination numbers.

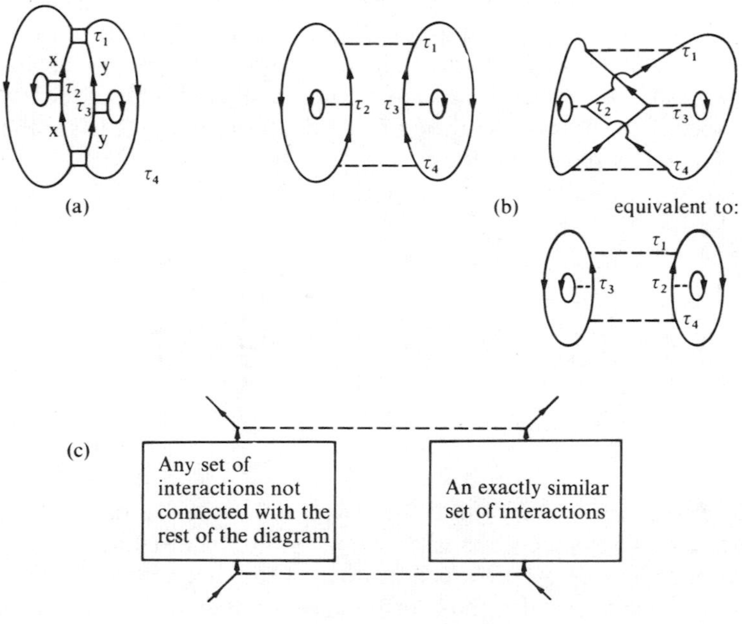

FIG. 2.12.

There is a certain amount of cross-play between this reduction and that of section 2.4. Consider, for example, the diagram shown in Fig. 2.12(a). Before the τ-label interchange, the two non-symmetrized diagrams (from the set of 256 which the symmetrized diagram includes) shown in (b) are different since, although they may be redrawn to look very similar, the τ-labelling is different. However, as soon as the one set of τ-labels is to be taken to include all permutations of the labels amongst themselves, the two diagrams in Fig. 2.12(b) become the same. This may either be taken into account through p, or, more conveniently, we may extend the definition of equivalent lines of section 4 to that of equivalent arms, defined pictorially in Fig. 2.12(c), of which, of course, the pair $(x, x)(y, y)$ in Fig. 2.12(a) form an example.

Although the changes discussed in this section clearly reduce the number of diagrams we need to consider in high orders, we are still left with two second-order diagrams.

Exercise

2.6.1. Check the coordination numbers of the diagrams in Fig. 2.11, and draw some more diagrams and evaluate their coordination numbers.

2.7. Rules for calculating the perturbation expansion for Ω in the τ-representation

We now summarize the work of the last six sections in a set of rules for evaluating the nth order term in the expansion of Ω:

Draw n squares each with two lines entering and two leaving. Each square has a τ-label and each line a state label. Join up these lines in all possible different ways with the restrictions that the diagrams so obtained must be connected, and that the arrows on any line must point in the same direction. The word 'different' must be interpreted in the light of what has been done above in our efforts to reduce the number of diagrams. Each diagram represents a contribution to Ω which is made up of the following factors:

(i) $\Gamma^0_{s_1 s_2 s_3 s_4}$ for each vertex square with lines s_1 and s_2 leaving and s_3 and s_4 entering,

(ii) $g_s(\tau_i - \tau_j)$ for each full line with label s from τ_j to τ_i,

(iii) $(-1)^n$, from the expansion of S (eqn (2.22)),

(iv) $(-1)^D$, where D is the number of closed fermion loops in the diagram (section 2.3),

(v) $1/p$, where p is the coordination number of the diagram (section 2.6),

(vi) $1/(2)^{n_e}$, where n_e is the number of equivalent arms in the diagram (sections 2.4 and 2.6), and

(vii) $-1/\beta$, from the definition of Ω.

Finally, sum over all the state (usually momentum and spin) labels of the lines, and integrate over all the τs from 0 to β.

The integrand in the contribution of each diagram depends only on the differences between the τs. The final integral will thus give a factor of β. Thus we may fix one vertex, at $\tau = 0$, say, perform only $n-1$ integrals, and omit the β^{-1} of the last factor.

Here it should be emphasized that if one is ever in doubt about factors to be included, one should return to Wick's theorem. We need all possible contractions between the operators in any order in perturbation theory. The awkward factors have resulted from rearrangements of these contractions, and by retracing the steps of these rearrangements it is possible to write down coordination numbers, numbers of pairs of equivalent lines, and so on, in a straightforward way.

As an example in the use of these rules, we consider the second-order diagrams, Figs. 2.3 and 2.5:

Rule	Fig. 2.3.	Fig. 2.5.
(i)	$\Gamma^0_{s_1s_2s_3s_4}\Gamma^0_{s_3s_4s_1s_2}$	$\Gamma^0_{s_1s_2s_4s_2}\Gamma^0_{s_4s_3s_1s_3}$
(ii)	$g_{s_1}(\tau_1-\tau_2)g_{s_2}(\tau_1-\tau_2)g_{s_3}(\tau_2-\tau_1)g_{s_4}(\tau_2-\tau_1)$	$g_{s_1}(\tau_1-\tau_2)g_{s_2}(0)g_{s_3}(0)g_{s_4}(\tau_2-\tau_1)$
(iii)	$(-1)^2$	$(-1)^2$
(iv)	1	$\begin{cases}(-1) \text{ for fermions,} \\ +1 \text{ for bosons}\end{cases}$
(v)	$\frac{1}{2}$	$\frac{1}{2}$
(vi)	$\frac{1}{4}$	1
(vii)	$-1/\beta$	$-1/\beta$

The integrals for the second-order terms in Ω can now be written down in a straightforward way.

Exercise

2.7.1. Write down the zero- and first-order terms in the perturbation series for Ω for a gas of bosons interacting through pair forces, and the resulting equations for $\bar{N}$ and for pV. The equation for $\bar{N}$ cannot be solved for μ (this cannot be done even in the non-interacting case). At high temperatures, $\exp[-\beta(\varepsilon_{\mathbf{k}}-\mu)] \ll 1$. In this case, and when the first order term in Ω is very much less than the zeroth-order term, show that

$$pV/\kappa T = \bar{N} + \frac{A}{V}$$

and find the form of A. Compare this with, for example, the van der Waal's equation. Why is this treatment unsatisfactory for a realistic dilute gas?

2.7.2. Example of the summation of an infinite subset of terms in the perturbation expansion.

One of the simplest infinite subsets of diagrams is shown in Fig. 2.13. It contains all the diagrams concerning the scattering of one particle by another, when the two particles are allowed to interact with each other any number of times, but when neither interacts with the rest of the system. The contributions

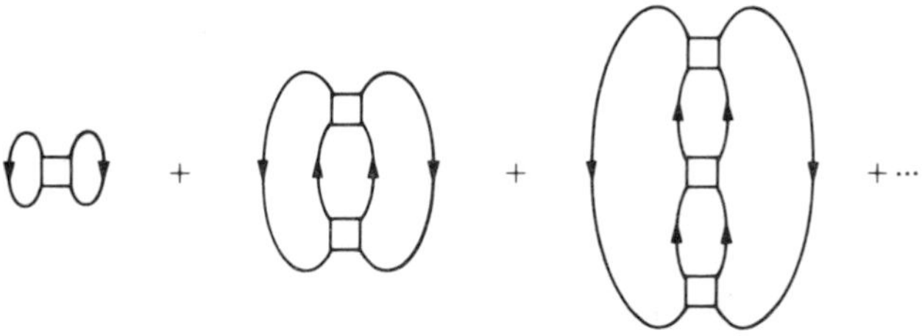

FIG. 2.13.

are affected by the rest of the system, however, through the statistical factors contained in the Green functions. The coordination number of the nth-order diagram is n, since cyclic permutations of the vertices leave the contractions invariant. We may rid ourselves of this slightly awkward factor by multiplying each vertex by λ; if we then evaluate $\lambda\,\partial\Omega/\partial\lambda$ instead of Ω, each nth-order diagram must be multiplied by n, which cancels the $1/n$ from the coordination number. The number of pairs of equivalent lines in the nth-order diagrams is n.

We consider the factors contributing to $\lambda\,\partial\Omega/\partial\lambda$ coming from a part of the diagrams:

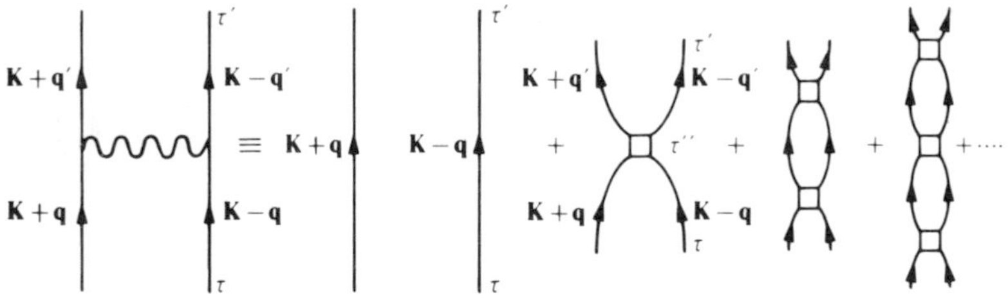

FIG. 2.14.

and call this $L(\mathbf{K}, \mathbf{q}, \mathbf{q}', \tau', \tau)$. (We are working in the momentum representation.) The sum of the diagrams in Fig. 2.13 is then given by

$$\beta^{-1} \sum_{\mathbf{K},\mathbf{q},\mathbf{q}'} \Gamma^0(\mathbf{K}+\mathbf{q}, \mathbf{K}-\mathbf{q}, \mathbf{K}+\mathbf{q}', \mathbf{K}-\mathbf{q}')L(\mathbf{K}, \mathbf{q}, \mathbf{q}', \tau, \tau)$$

as shown in Fig. 2.15.

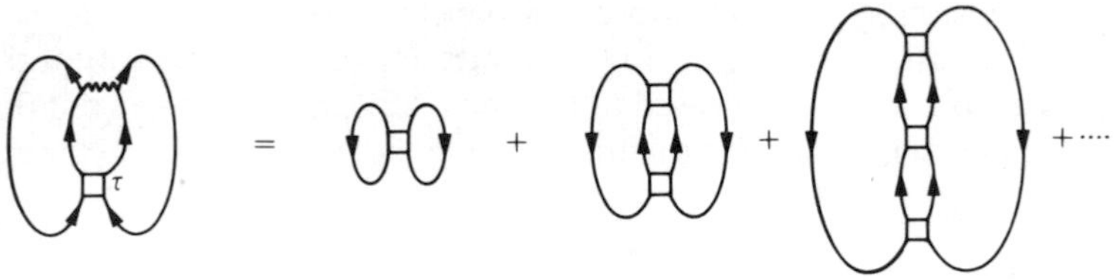

FIG. 2.15.

The equation for L is shown in diagrammatic form in Fig. 2.16. Algebraically:†

$$L(\mathbf{K}, \mathbf{q}, \mathbf{q}', \tau', \tau) = \tfrac{1}{2}\delta_{\mathbf{q},\mathbf{q}'} g_{\mathbf{K}+\mathbf{q}}(\tau'-\tau) g_{\mathbf{K}-\mathbf{q}}(\tau'-\tau)$$

$$-\tfrac{1}{2}\sum_{\mathbf{q}''}\int_0^\beta \mathrm{d}\tau''\, g_{\mathbf{K}+\mathbf{q}'}(\tau'-\tau'') g_{\mathbf{K}-\mathbf{q}'}(\tau'-\tau'') \times$$

$$\times \Gamma^0(\mathbf{K}+\mathbf{q}', \mathbf{K}-\mathbf{q}', \mathbf{K}+\mathbf{q}'', \mathbf{K}-\mathbf{q}'') \times L(\mathbf{K}, \mathbf{q}, \mathbf{q}'', \tau'', \tau). \quad (2.46)$$

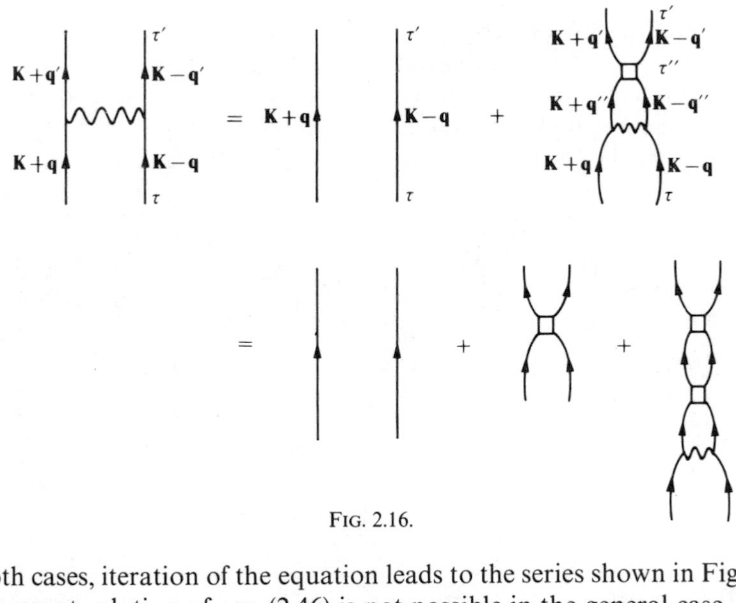

FIG. 2.16.

In both cases, iteration of the equation leads to the series shown in Fig. 2.14.

The exact solution of eqn (2.46) is not possible in the general case. It has been solved by Thouless (1960) for fermions for the case in which Γ^0 is 'separable' that is, is of the form

$$\Gamma^0(\mathbf{K}+\mathbf{q}', \mathbf{K}-\mathbf{q}', \mathbf{K}+\mathbf{q}, \mathbf{K}-\mathbf{q}) = \gamma_{\mathbf{q}'}\,\gamma_{\mathbf{q}}$$

† We have formally counted the two 'external' lines in Fig. 2.15 as equivalent, since they are to be joined at a vertex, as in Fig. 2.14.

and the solution is of interest in the theory of superconductivity. We refer the reader to Thouless' paper for further details.

2.7.3. Write down the rules for Ω for the electron–phonon Hamiltonian of exercise (2.3.4) and work out the contribution of some simple diagrams.

2.8. The ω-representation

The integrals which have to be performed after one has written down the seven factors of section 2.7 are tiresome to evaluate, as the range of each has usually to be split into two parts; the integrand contains factors $g(\tau_1-\tau_2)$ which change their form as we cross the line $\tau_1 = \tau_2$. We therefore make a transformation so that these integrals can be done at once. This carries with it certain disadvantages, but these are usually outweighed by the advantages, so that most many-body calculations are done in the new representation.

We first notice that $g(\tau_1, \tau_2)$ is a function only of $\tau_1-\tau_2$ (see eqn (2.36)), and that $\tau_1-\tau_2$ has a range of $-\beta$ to β. Within that range we may expand g in a Fourier series:

$$g_s(\tau_1-\tau_2) \equiv g_s(\tau) = -\beta^{-1}\sum_l \exp(-\mathrm{i}\pi l\tau/\beta)\bar{g}_l, \qquad l \text{ integral}$$

$$\equiv -\beta^{-1}\sum_l \exp(-\mathrm{i}\omega_l\tau)\bar{g}(s, \mathrm{i}\omega_l) \tag{2.47}$$

where the $\bar{g}(s, \mathrm{i}\omega_l)$ are Fourier coefficients yet to be determined, and $\omega_l = \pi l/\beta$.† Eqn (2.37) tells us that only the coefficients with l even are different from zero for bosons, and only the odd-l coefficients for fermions. That is, we may redefine ω_l: $\omega_l = 2\pi l/\beta$ for bosons; $\omega_l = (2l+1)\pi/\beta$ for fermions. The coefficients are easily determined by using the normal inversion formulae for Fourier series (see, for example, Courant and Hilbert (1953)) and eqn (2.36):

$$\bar{g}(s, \mathrm{i}\omega_l) = -\tfrac{1}{2}\int_{-\beta}^{\beta} \mathrm{d}\tau\, g_s(\tau)\exp(\mathrm{i}\omega_l\tau)$$

$$= -\tfrac{1}{2}\varepsilon\langle n_s\rangle_0\int_{-\beta}^{0} \mathrm{d}\tau\exp[-(\varepsilon_s-\mu)\tau+\mathrm{i}\omega_l\tau] -$$

$$-\tfrac{1}{2}[\varepsilon\langle n_s\rangle_0+1]\int_{0}^{\beta} \mathrm{d}\tau\exp[-(\varepsilon_s-\mu)\tau+\mathrm{i}\omega_l\tau]$$

$$= [\mathrm{i}\omega_l-\varepsilon_s+\mu]^{-1}, \tag{2.48}$$

where we have used $\langle n_s\rangle_0 = \{\exp[\beta(\varepsilon_s-\mu)]-\varepsilon\}^{-1}$.

† The rather strange notation for the Fourier coefficients proves to be convenient.

We now need to sort out the τ-dependence of a diagram with a vertex labelled τ. For such a diagram, part of which is shown in Fig. 2.17, this dependence will come entirely from the factors

$$g_{s_1}(\tau-\tau_a)g_{s_2}(\tau-\tau_b)g_{s_3}(\tau_c-\tau)g_{s_4}(\tau_d-\tau).$$

We express each of these in terms of its Fourier expansion, writing

$$g_{s_i}(\tau') = -\beta^{-1}\sum_{l_i} \bar{g}(s_i, i\omega_{l_i})\exp(-i\omega_{l_i}\tau')$$

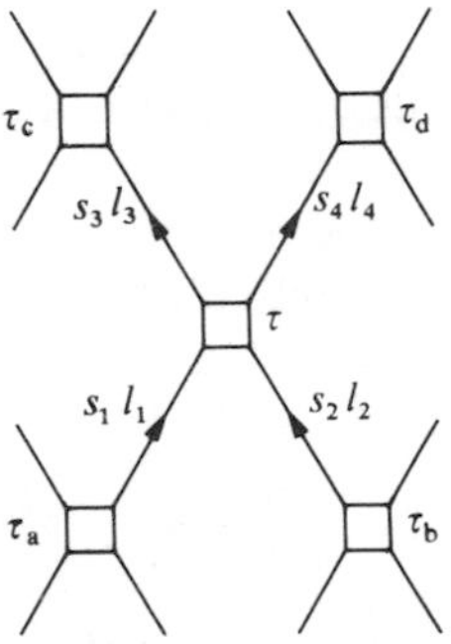

FIG. 2.17.

and indicate in the diagram the summation index we have used for each Green function by a label at the side of the line representing it. We find that the τ-dependence of the diagram is of the form

$$\exp[-i(\omega_{l_1}+\omega_{l_2}-\omega_{l_3}-\omega_{l_4})\tau]$$

and the τ-integration we need to do is very easy, giving β for $\omega_{l_1}+\omega_{l_2}-$ $-\omega_{l_3}-\omega_{l_4} = 0$, or $l_1+l_2-l_3-l_4 = 0$, and zero otherwise. This can be done for each vertex. There is thus a 'conservation of ω_l law' for the sum of the l-labels entering and leaving each vertex. At the last vertex, however, l will be conserved automatically, as a result of its conservation at the other vertices.

We have thus got rid of all the τ-integrals at the cost of introducing a new label for each line. These labels also have to be summed over, subject to a conservation law at each vertex. This is very similar to the transformation to the energy representation in conventional field theory, but now our 'energies' are discrete rather than continuous variables.

We now write out the rules for calculating the nth order term in the expansion of Ω in the ω-representation:

Draw all different nth-order connected diagrams as above, omitting the τ-labels on the vertices, and giving each line an l-label as well as a state label.

The factors making up the contribution of a particular diagram to Ω are:

(i) $\Gamma^0_{s_1s_2s_3s_4} \times \beta\delta_{l_1+l_2,l_3+l_4}$ for each vertex square with lines (s_1, l_1), (s_2, l_2) leaving and (s_3, l_3), (s_4, l_4) entering. (The factor $\beta\delta$ comes from the transformation to the ω-representation.)

(ii) $-\beta^{-1}[i\omega_l - \varepsilon_s + \mu]^{-1}$ for each full line with labels s and l (see eqns (2.47) and (2.48). See also (ii)′ below)

(iii) $(-1)^n$

(iv) $(-1)^D$, where D is the number of closed fermion loops in the diagram.

(v) $1/2^{n_e}$, where n_e is the number of equivalent arms in the diagram.

(vi) $1/p$ where p is the coordination number of the diagram.

(vii) $-\beta^{-1}$.

Finally, sum over the s and l labels of the full lines.

In this representation, the second-order diagrams appear as in Fig. 2.18, and have the contributions:

(a)
$$\sum_{\substack{s_1s_2s_3s_4\\ l_1l_2l_3l_4}} \Gamma^0_{s_1s_2s_3s_4}\Gamma^0_{s_3s_4s_1s_2}\beta^2\delta_{l_1+l_2,l_3+l_4}(-\beta)^{-4}\times$$
$$\times[i\omega_{l_1}-\varepsilon_{s_1}+\mu]^{-1}[i\omega_{l_2}-\varepsilon_{s_2}+\mu]^{-1}[i\omega_{l_3}-\varepsilon_{s_3}+\mu]^{-1}\times$$
$$\times[i\omega_{l_4}-\varepsilon_{s_4}+\mu]^{-1}(-1)^2\times(\varepsilon)^2\times\tfrac{1}{4}\times\tfrac{1}{2}\times(-\beta)^{-1}, \tag{2.49}$$

(b)
$$\sum_{\substack{s_1s_2s_3s_4\\ l_1l_2l_3l_4}} \Gamma^0_{s_1s_2s_3s_2}\Gamma^0_{s_3s_4s_1s_4}\beta^2\delta_{l_1,l_3}(-\beta)^{-4}\times$$
$$\times[i\omega_{l_1}-\varepsilon_{s_1}+\mu]^{-1}[i\omega_{l_2}-\varepsilon_{s_2}+\mu]^{-1}[i\omega_{l_3}-\varepsilon_{s_3}+\mu]^{-1}\times$$
$$\times[i\omega_{l_4}-\varepsilon_{s_4}+\mu]^{-1}(-1)^2(\varepsilon)^3\tfrac{1}{2}\,.\,(-\beta)^{-1}. \tag{2.50}$$

A trick which is helpful in evaluating the sums is described in Appendix D.

We might expect some difficulty to arise in the case of the diagram of Fig. 2.18(b), since in the τ-representation we should have to be careful to write $\varepsilon\langle a^+_{s_4}a_{s_4}\rangle_0$ and not $\langle a_{s_4}a^+_{s_4}\rangle_0$ for the s_4 line which starts and ends at the same τ. (See the discussion on p. 21.) This means that $g_s(0)$ is to be taken as

$$\underset{\substack{\tau\to 0\\ \tau<0}}{Lt}\; g_s(\tau).$$

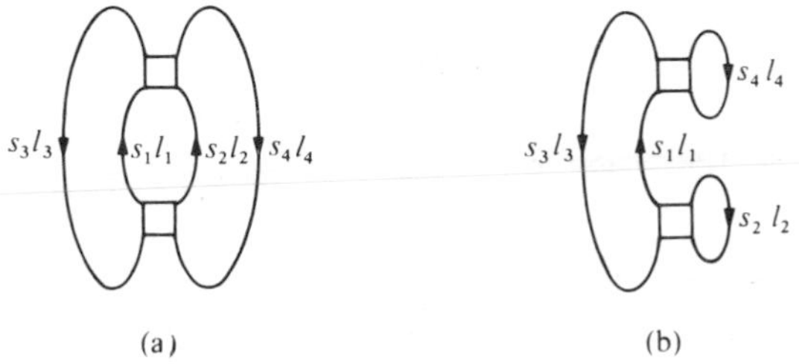

FIG. 2.18.

Thus, from eqn (2.47),

$$g_s(0) = -\underset{\substack{\tau\to 0\\ \tau<0}}{\mathrm{Lt}}\ (\beta)^{-1}\sum_l \bar{g}(s, \mathrm{i}\omega_l)\,\mathrm{e}^{-\mathrm{i}\omega_l\tau}$$

$$= -\underset{\substack{\tau\to 0\\ \tau<0}}{\mathrm{Lt}}\ (\beta)^{-1}\sum_l \frac{\exp(-\mathrm{i}\omega_l\tau)}{\mathrm{i}\omega_l-\varepsilon_s+\mu}$$

$$= \varepsilon\langle n_s\rangle_0 \text{ as required}$$

where the technique described in Appendix D has been used to evaluate the sum.

(ii)′ Thus lines entering and leaving the same vertex should have a factor

$$-\underset{\substack{\tau\to 0\\ \tau<0}}{\mathrm{Lt}}\ \beta^{-1}[\mathrm{i}\omega_l-\varepsilon_s+\mu]^{-1}\exp(-\mathrm{i}\omega_l\tau).$$

Exercise

2.8.1. Write down the first-order terms for Ω in the ω-representation, and, by using the trick for evaluating the sums described in Appendix D, show that the contribution is the same as that obtained in exercise 2.7.1.

2.8.2. Using the techniques described in Appendix D, evaluate the sums in eqns (2.49) and (2.50), and compare your answer with the results obtained in the τ-representation.

In the examples we have considered, eqns (2.49) and (2.50) and exercises 2.8.1 and 2.8.2 the advantages of the ω-representation are by no means apparent, since the sums one has to perform appear to be just as tiresome as the integrals in the τ-representation. The advantages will be more apparent when we consider the interacting Green functions in the next chapter.

2.8.3. Show that in the ω-representation, the free-phonon Green functions are:

$$\bar{d}_q(\mathrm{i}\omega_n) = [\mathrm{i}\omega_n-\omega_q]^{-1}, \qquad \omega_n = 2\pi n/\beta,$$

$$\bar{D}_q(\mathrm{i}\omega_n) = 2\omega_q[(\mathrm{i}\omega_n)^2-(\omega_q)^2]^{-1}.$$

(See exercises 2.2.3, 2.3.4 and 2.7.3, and note that $\omega_q = \omega_{-q}$). Write down the rules for Ω for the electron–phonon interaction in the ω-representation and evaluate the contribution of some simple diagrams.

2.9. Other formulations of the rules

As one might expect, many other formulations of the rules for Ω are possible, and we shall mention a few of them in this section. Firstly, it is sometimes more convenient not to use the fully symmetrized vertex of section 2.4. We can either go back to the first representation of the perturbation series, in which we distinguish between the two sides of a vertex, or we can use the symmetry of the potential energy matrix element

$$v_{s_1 s_2 s_3 s_4} = v_{s_2 s_1 s_4 s_3}$$

and fail to distinguish between the two sides of a vertex, thus removing the factor of $\frac{1}{2}$ which appears in front of $v_{s_1 s_2 s_3 s_4}$, but complicating the form of the rules, since we then have to be careful about equivalent lines. We shall sometimes use this formulation in what follows.

A rather more drastic reformulation of the rules has been given by Bloch and De Dominicis (1958, 1959). We return to eqn (2.22), and do not use eqn (2.25). As before, we use Wick's theorem to develop a diagrammatic expansion of the perturbation series, but a diagram now includes only one order of the vertices, and our final integrations must be over the range

$$\beta > \tau_1 > \tau_2 \ldots > \tau_n > 0,$$

as we have not used eqn (2.25). Thus the diagrams in Fig. 2.19 are now to be counted as different. By taking together all possible τ-orderings of the unlinked parts of diagrams, it is possible to prove a linked-cluster theorem as above. (All possible orderings must be taken in order that the unlinked parts should factorize.)

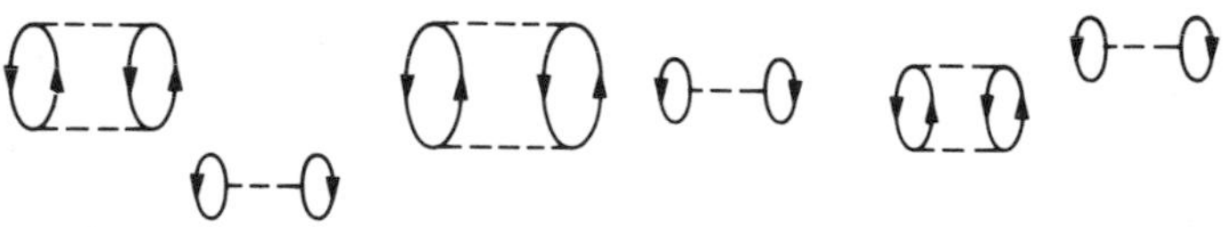

(a) (b) (c)

FIG. 2.19. Three diagrams which would be regarded as the same in the original formulation of the rules, but as different in the formulation described in section 2.9.

To work out the τ_i-dependence of a diagram, we notice that from eqns (iv) and (v) of Appendix B

$$\exp[(H_0 - \mu\hat{N})\tau_n]H_1 \exp[-(H_0 - \mu\hat{N})\tau_n] \exp[(H_0 - \mu\hat{N})\tau_{n-1}]H_1 \times$$
$$\times \exp[-(H_0 - \mu\hat{N})\tau_{n-1}] \ldots \exp[(H_0 - \mu\hat{N})\tau_i]H_1 \exp[-(H_0 - \mu\hat{N})\tau_i] \ldots$$
$$= \exp(\Delta E_n \tau_n)H_1 \exp(\Delta E_{n-1}\tau_{n-1})H_1 \ldots \exp(\Delta E_i \tau_i)H_1 \ldots \quad (2.51)$$

where ΔE_i is the sum of $\varepsilon_s - \mu$ for the lines leaving the ith vertex minus the same sum for the lines entering. Eqn (2.51) also gives the τ_i-dependence of the ensemble average of the product. One may easily deduce the new rules for the contribution of an nth order diagram in the τ-representation:

(i) A factor $\frac{1}{2}v_{s_1s_2s_3s_4}$ for each vertex with lines labelled s_1, s_2 leaving and s_3, s_4 entering. (The factor $\frac{1}{2}$ indicates that we are working in an unsymmetrized formulation; symmetrization can be carried out in exactly the same way as in section 2.4.)

(ii) For each interaction with label τ_i, a factor $\exp(\Delta E_i \tau_i)$ with ΔE_i as defined above.

(iii) For each ascending line with label s a factor f_s^+, and for each descending line, or line leaving and entering the same vertex, a factor f_s^-, where:

$$f_s^+ = 1+\varepsilon f_s^- = \{1-\varepsilon \exp[-\beta(\varepsilon_s-\mu)]\}^{-1},$$

$$f_s^- = \{\exp \beta(\varepsilon_s-\mu)-\varepsilon\}^{-1}.$$

(These factors come from $\langle a_s a_s^+ \rangle_0$ and $\langle a_s^+ a_s \rangle_0$ in Wick's theorem.)

(iv) A factor $(-1)^{h+l}$, where h is the number of descending Fermion lines and l is the number of closed Fermion loops.

(v) A factor $-(\beta)^{-1}(-1)^n$.

Finally, integrate over all the τs over the range

$$\beta > \tau_1 > \tau_2 \ldots > \tau_n > 0.$$

A τ-independent form of these rules may be obtained by making the transformation:

$$\begin{aligned} \alpha_n &= \tau_n & \tau_n &= \alpha_n \\ \alpha_{n-1} &= \tau_{n-1}-\tau_n & \tau_{n-1} &= \alpha_{n-1}+\alpha_n \\ \alpha_{n-2} &= \tau_{n-2}-\tau_{n-1} & \tau_{n-2} &= \alpha_{n-2}+\alpha_{n-1}+\alpha_n \\ &\cdots\cdots & &\cdots\cdots \\ \alpha_1 &= \tau_1-\tau_2 & \tau_1 &= \alpha_1+\alpha_2+\cdots+\alpha_n. \end{aligned}$$

The τ-integrals then take the form

$$I = \int \ldots \int \mathrm{d}\alpha_1 \ldots \mathrm{d}\alpha_n \exp[\alpha_1 \Delta E_1 + \alpha_2(\Delta E_1+\Delta E_2)+ \cdots +$$

$$+\alpha_n(\Delta E_1+\Delta E_2+\cdots+\Delta E_n)]$$

with the restrictions on the range of integration $\sum_{i=1}^n \alpha_i < \beta$; $\alpha_i > 0$. We write:

$$\mathscr{E}_j = -\sum_{i=1}^{j} \Delta E_i,$$

$$\mathscr{E}_n = -\sum_{i=1}^{n} \Delta E_i = 0, \quad \text{for an } n\text{th-order diagram},$$

and call $\mathscr{E}_j$ the energy of excitation in the interval τ_j to τ_{j+1}. It is the energy of all the ascending lines minus the energy of all the descending lines in that

$$\Delta E_1 = \varepsilon_{s_1}+\varepsilon_{s_2}-\varepsilon_{s_3}-\varepsilon_{s_4}$$
$$\Delta E_2 = \varepsilon_{s_3}+\varepsilon_{s_4}-\varepsilon_{s_5}-\varepsilon_{s_6}$$
$$\Delta E_3 = \varepsilon_{s_5}+\varepsilon_{s_6}-\varepsilon_{s_4}-\varepsilon_{s_3}$$

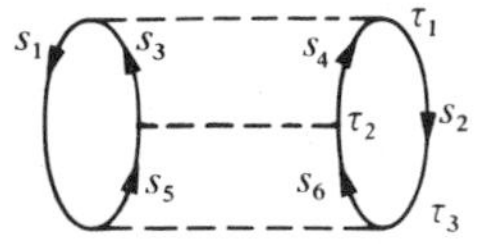

$$\mathscr{E}_1 = \varepsilon_{s_3}+\varepsilon_{s_4}-\varepsilon_{s_1}-\varepsilon_{s_2}$$
$$\mathscr{E}_2 = \varepsilon_{s_5}+\varepsilon_{s_6}-\varepsilon_{s_1}-\varepsilon_{s_2}$$
$$\mathscr{E}_3 = 0$$

FIG. 2.20.

interval (see Fig. 2.20). One then has

$$I = \int \ldots \int \mathrm{d}\alpha_1 \ldots \mathrm{d}\alpha_n \exp\left[-\sum_{i=1}^{n} \alpha_i \mathscr{E}_i\right].$$

To get rid of the restriction $\sum_{i=1}^{n} \alpha_i < \beta$ we use

$$\frac{1}{2\pi \mathrm{i}} \int_{-a+\mathrm{i}\infty}^{-a-\mathrm{i}\infty} \mathrm{d}z \exp(-zx)/z = \begin{array}{ll} 1 & x > 0 \\ 0 & x < 0 \end{array},$$

where a is a positive constant. Thus, if $b > -\mathscr{E}_i$ for all i,

$$I = \frac{1}{2\pi \mathrm{i}} \int_{-b+\mathrm{i}\infty}^{-b-\mathrm{i}\infty} \mathrm{d}z \exp(-z\beta)/z \int \ldots \int_{\alpha_i > 0} \mathrm{d}\alpha_1 \ldots \mathrm{d}\alpha_n \prod_{i=1}^{n} \exp[\alpha_i(z-\mathscr{E}_i)]$$

$$= \frac{(-1)^n}{2\pi \mathrm{i}} \int_{-b+\mathrm{i}\infty}^{-b-\mathrm{i}\infty} \mathrm{d}z\, \mathrm{e}^{-z\beta}/z^2 \prod_{i=1}^{n-1} (z-\mathscr{E}_i)^{-1}.$$

b may be taken to be as large as we like and the contour may be closed by the infinite semicircle on the right of the complex z-plane. Since there are poles in the integrand only along the real axis, the contour may be deformed so that the integral is taken along either side of the real axis as in Fig. (2.21).

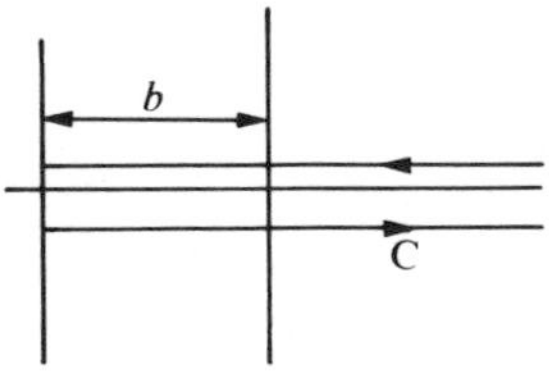

FIG. 2.21.

The rules in this formulation are then the same as those given above except for rules (ii) and (v) which are replaced by:

(ii) for each interval between successive vertices, a factor $(z-\mathscr{E}_i)^{-1}$ where $\mathscr{E}_i$ is the energy of excitation in that interval,

(v) $-(2\pi \mathrm{i}\beta)^{-1} \exp(-\beta z)/z^2$.

The contribution to Ω is then obtained by integrating over z along the contour C shown in Fig. 2.21.

These rules may themselves be put in several alternative forms by using the cyclic invariance of the trace in eqn (2.22). We refer the interested reader to the original papers for further details. (Bloch and De Dominicis 1958, 1959.)

Bibliography

The work in this chapter follows most closely Luttinger and Ward (1960). Similar work is described by Matsubara (1955), Thouless (1957), Montroll and Ward (1958), Bloch and De Dominicis (1958), (1959).

3

GREEN FUNCTIONS AND CORRELATION FUNCTIONS

3.1. The full Green function

CONSIDER those parts of a diagram which have one line entering and one leaving:

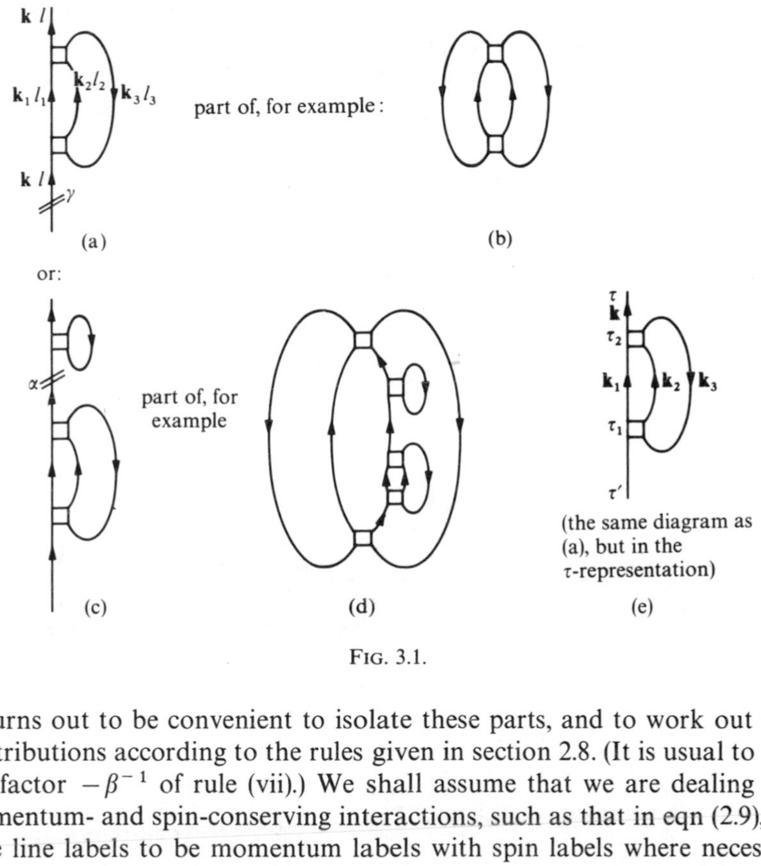

FIG. 3.1.

It turns out to be convenient to isolate these parts, and to work out their contributions according to the rules given in section 2.8. (It is usual to omit the factor $-\beta^{-1}$ of rule (vii).) We shall assume that we are dealing with momentum- and spin-conserving interactions, such as that in eqn (2.9), and take line labels to be momentum labels with spin labels where necessary. We shall use the symbol $\mathbf{k}$ to indicate the momentum and spin of the state. In the ω-representation, this contribution will depend on the l and $\mathbf{k}$-labels of the ingoing line, which, since the total l and $\mathbf{k}$ are conserved at each vertex, will be the same as the l and $\mathbf{k}$ of the outgoing line. (This is illustrated in

Fig. 3.1(a).) In the τ-representation, the ingoing and outgoing lines will still have $\mathbf{k}$-labels; the point of entry of the ingoing line, and of exit of the outgoing line will have τ-labels (τ' and τ of Fig. 3.1(e)). Intermediate τ-labels (τ_1 and τ_2 of Fig. 3.1(e)) will be integrated over.

These diagrams, with one ingoing and one outgoing line, can be divided into two classes: members of the first can be separated into two disconnected parts by cutting a single full line, and members of the second cannot be so divided. We call the first type 'improper' diagrams, and the second 'proper'. Thus Fig. 3.1(a) is a proper diagram, and Fig. 3.1(c) is improper, since it can be separated into two parts by cutting the full line at α. (In this classification, we exclude trivial separations such as that achieved by a cut at γ in Fig. 3.1(a).)

It is convenient to introduce a symbol $-\beta\Sigma(\mathbf{k}, i\omega_l)$ for the sum of the contributions of all proper diagrams, not including the factors in each diagram for the ingoing and outgoing lines. We call Σ the proper self-energy, or mass operator. The simplest diagrams for $-\beta\Sigma(\mathbf{k}, i\omega_l)$ and their contributions are:

(a)

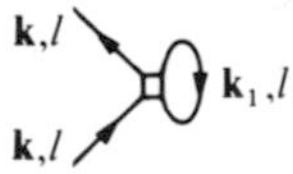

$$-\beta\Sigma^{(a)}(\mathbf{k}, i\omega_l) = \mathop{\mathrm{Lt}}_{\substack{\tau\to 0\\ \tau<0}} \sum_{\mathbf{k}_1, l_1} (-1)^2 \beta\Gamma^0_{\mathbf{kk}_1\mathbf{kk}_1}\varepsilon\beta^{-1} \times \times \frac{\exp(-i\omega_{l_1}\tau)}{i\omega_{l_1}-(\varepsilon_{\mathbf{k}_1}-\mu)},$$

(b)

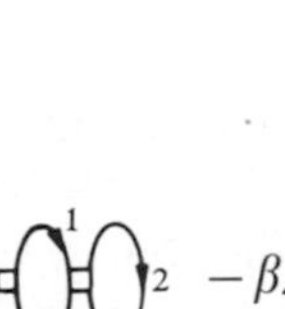

$$-\beta\Sigma^{(b)}(\mathbf{k}, i\omega_l) = \sum_{\substack{\mathbf{k}_1\mathbf{k}_2\mathbf{k}_3\\ l_1 l_2 l_3}} \Gamma^0_{\mathbf{k}_1\mathbf{k}_2\mathbf{kk}_3}\Gamma^0_{\mathbf{kk}_3\mathbf{k}_1\mathbf{k}_2}\varepsilon\tfrac{1}{2}\beta^{-3}(-1)^5 \times \times \beta^2\delta_{l_1+l_2, l+l_3}[i\omega_{l_1}-(\varepsilon_{\mathbf{k}_1}-\mu)]^{-1} \times \times [i\omega_{l_2}-(\varepsilon_{\mathbf{k}_2}-\mu)]^{-1}[i\omega_{l_3}-(\varepsilon_{\mathbf{k}_3}-\mu)]^{-1},$$

(c)

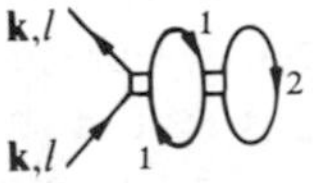

$$-\beta\Sigma^{(c)}(\mathbf{k}, i\omega_l) = \mathop{\mathrm{Lt}}_{\substack{\tau\to 0\\ \tau<0}} \sum_{\substack{\mathbf{k}_1\mathbf{k}_2\\ l_1 l_2}} \beta^2\Gamma^0_{\mathbf{kk}_1\mathbf{kk}_1}\Gamma^0_{\mathbf{k}_1\mathbf{k}_2\mathbf{k}_1\mathbf{k}_2}(\beta)^{-3} \times \times (-1)^5[i\omega_{l_1}-(\varepsilon_{\mathbf{k}_1}-\mu)]^{-2} \times \times [i\omega_{l_2}-(\varepsilon_{\mathbf{k}_2}-\mu)]^{-1} \times \times \exp(-i\omega_{l_2}\tau). \qquad (3.1)$$

A factor of $\frac{1}{2}$ has been included in (b) for the two equivalent lines, but the coordination number has been taken as unity in all cases for reasons to be discussed later.

In the τ-representation, Σ will be a function of the momentum label of the ingoing and outgoing lines, and is also to be taken to be a function of the labels, τ_1 and τ_2, of the initial and final interactions in the diagrams for Σ. (Any other τ-labels within the diagram for Σ are to be integrated over the range 0 to β.) Thus:

(a′) 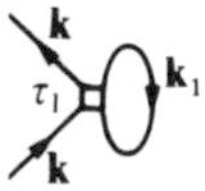

$$\Sigma^{(a)}(\mathbf{k}, \tau_1, \tau_2) = \delta(\tau_1-\tau_2)\sum_{\mathbf{k}_1}(-1)\Gamma^0_{\mathbf{kk}_1\mathbf{kk}_1}\varepsilon g_{\mathbf{k}_1}(0),$$

(b′) 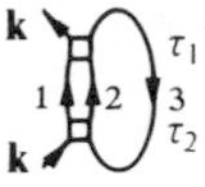

$$\Sigma^{(b)}(\mathbf{k}, \tau_1, \tau_2) = \sum_{\mathbf{k}_1\mathbf{k}_2\mathbf{k}_3}\Gamma^0_{\mathbf{k}_1\mathbf{k}_2\mathbf{kk}_3}\Gamma^0_{\mathbf{kk}_3\mathbf{k}_1\mathbf{k}_2}\varepsilon\tfrac{1}{2}\times$$
$$\times g_{\mathbf{k}_1}(\tau_1-\tau_2)g_{\mathbf{k}_2}(\tau_1-\tau_2)g_{\mathbf{k}_3}(\tau_2-\tau_1),$$

(c′) 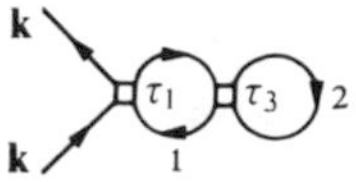

$$\Sigma^{(c)}(\mathbf{k}, \tau_1, \tau_2) = \delta(\tau_1-\tau_2)\sum_{\mathbf{k}_1\mathbf{k}_2}\Gamma^0_{\mathbf{kk}_1\mathbf{kk}_1}\Gamma^0_{\mathbf{k}_1\mathbf{k}_2\mathbf{k}_1\mathbf{k}_2}\times$$
$$\times\int_0^\beta \mathrm{d}\tau_3\, g_{\mathbf{k}_1}(\tau_3-\tau_1)g_{\mathbf{k}_1}(\tau_1-\tau_3)g_{\mathbf{k}_2}(0).$$

Let us call $-\beta^{-1}\bar{G}(\mathbf{k}, \mathrm{i}\omega_l)$ the sum of the contributions of all diagrams, proper and improper, with one line, labelled $\mathbf{k}$, l, entering and one leaving: in $\bar{G}$ we shall include the contributions of the ingoing and outgoing lines. If $-\beta\Sigma$ is represented diagrammatically by a circle and $-\beta^{-1}\bar{G}$ by a heavy line, we have:

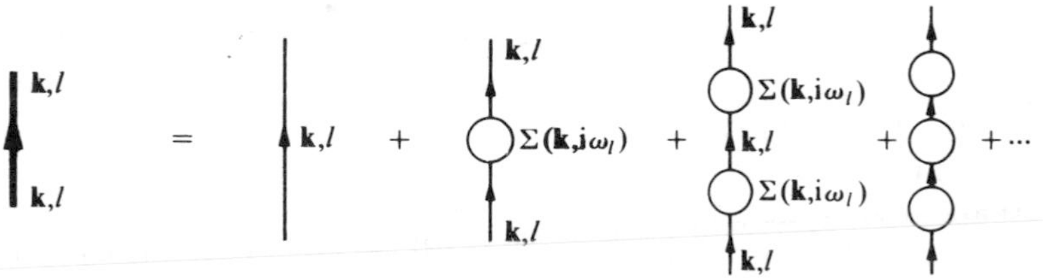

(3.2a)

$$-\beta^{-1}\bar{G}(\mathbf{k}, \mathrm{i}\omega_l) = -\beta^{-1}[\bar{g}(\mathbf{k}, i\omega_l)+\bar{g}(\mathbf{k}, \mathrm{i}\omega_l)\Sigma(\mathbf{k}, \mathrm{i}\omega_l)\bar{g}(\mathbf{k}, \mathrm{i}\omega_l)+$$
$$+\bar{g}\Sigma\bar{g}\Sigma\bar{g}+\ldots] \qquad (3.2b)$$

since any improper diagram can be split into two, or three, or four ... parts by cutting single lines. This series may be summed to give

$$\bar{G}(\mathbf{k}, \mathrm{i}\omega_l) = \frac{\bar{g}}{1-\Sigma\bar{g}} = \frac{1}{\bar{g}^{-1}-\Sigma}$$

$$= \frac{1}{\mathrm{i}\omega_l-(\varepsilon_{\mathbf{k}}-\mu)-\Sigma}. \tag{3.3}$$

This equation may also be obtained by noticing that eqn (3.2) is equivalent to:

(3.2c)

$$\bar{G} = \bar{g}+\bar{g}\Sigma\bar{G}$$

as may easily be seen by iteration. Eqn (3.2c) may then be solved for $\bar{G}$ to give eqn (3.3).

The equation for G in the τ-representation may also be written down

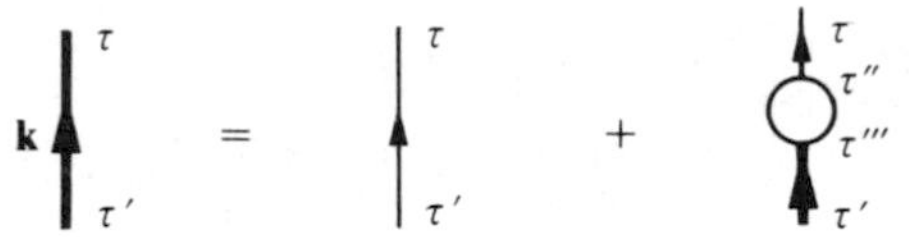

(3.4)

$$G(\mathbf{k}, \tau, \tau') = g_{\mathbf{k}}(\tau-\tau')+\int_0^\beta\int_0^\beta \mathrm{d}\tau''\,\mathrm{d}\tau'''\, g_{\mathbf{k}}(\tau-\tau'')\Sigma(\mathbf{k}, \tau'', \tau''')G(\mathbf{k}, \tau''', \tau').$$

This, however, is an integral equation, and its solution cannot immediately be written down. To solve it, the standard procedure would be to take a Fourier transform, and this at once leads us to the ω-representation. Here is one important advantage of this representation.

We may now perhaps see the reasons behind our nomenclature. g is the non-interacting form of G: thus we call G the full Green function, or the interacting Green function. Further, $\Sigma(\mathbf{k}, \mathrm{i}\omega_l)$ appears in $\bar{G}$ in a very similar position to $\varepsilon_{\mathbf{k}}$ and in a hand-waving way may be thought of as contributing to the energy of the particle of momentum **k**; hence the term "self energy". A too-literal interpretation of the idea can easily lead to difficulties, however.

We shall now spend some time discussing the properties of $\bar{G}$. The reason for this will be apparent at a glance at the figure on p. vi from which it is seen that Green functions play a most important role in many-body calculations. One of our purposes in the next few sections will be to trace the connection between the various definitions and uses of Green functions in the literature.

The rules for calculating $-\beta\bar{G}(\mathbf{k}, \mathrm{i}\omega_n)$ are very similar to those for Ω. In the ω-representation they are:

Draw all different connected diagrams with one line entering and one line leaving, both labelled, $\mathbf{k}, l$. Each internal line has a momentum–spin label and an l-label. The factors making up the contribution of the diagram to $-\beta^{-1}\bar{G}$ are:

(i) $(-1)\Gamma^0_{\mathbf{k}_1\mathbf{k}_2\mathbf{k}_3\mathbf{k}_4} \times \beta\delta_{l_1+l_2, l_3+l_4}$ for each vertex square with lines $\mathbf{k}_1, l_1$ and $\mathbf{k}_2, l_2$ leaving, and $\mathbf{k}_3, l_3$ and $\mathbf{k}_4, l_4$ entering. (This includes rules (i) and (iii) of section 2.8.)

(ii) $-\beta^{-1}[\mathrm{i}\omega_{l_1} - \varepsilon_{\mathbf{k}_1} + \mu]^{-1}$ for each full line with labels $\mathbf{k}_1, l_1$.

(ii)′ Lines entering and leaving the same vertex should have a factor

$$-\operatorname*{Lt}_{\substack{\tau\to 0\\ \tau<0}} \beta^{-1}[\mathrm{i}\omega_{l_1} - \varepsilon_{\mathbf{k}_1} + \mu]^{-1} \exp(-\mathrm{i}\omega_{l_1}\tau).$$

(iii) $1/2^{n_e}$, where n_e is the number of equivalent arms.

(iv) $(-1)^D$, where D is the number of closed fermion loops.

Finally, sum over all internal $\mathbf{k}$ and l labels.

In view of eqn (3.3), it is nearly always better to evaluate $\Sigma(\mathbf{k}, \mathrm{i}\omega_n)$. The rules for $-\beta\Sigma(\mathbf{k}, \mathrm{i}\omega_n)$ are the same as those for $\bar{G}$ except for (ii) which should read $-\beta^{-1}[\mathrm{i}\omega_{l_1} - \varepsilon_{\mathbf{k}_1} + \mu]^{-1}$ for each internal full line with labels $\mathbf{k}_1, l_1$.

In the τ-representation:

Draw all different diagrams with one line entering at τ' and one leaving at τ, both labelled $\mathbf{k}$. Each vertex has a τ-label, and each internal line a $\mathbf{k}$ label. The factors making up the contribution to $G(\mathbf{k}, \tau, \tau')$ are:

(i) $-\Gamma^0_{\mathbf{k}_1\mathbf{k}_2\mathbf{k}_3\mathbf{k}_4}$ for each vertex square with lines $\mathbf{k}_1, \mathbf{k}_2$ leaving, and $\mathbf{k}_3, \mathbf{k}_4$ entering.

(ii) $g_{\mathbf{k}}(\tau_1, \tau_2)$ for each full line with label $\mathbf{k}$ from τ_2 to τ_1.

(iii) $(-1)^D$, where D is the number of closed fermion loops.

(iv) $1/2^{n_e}$, where n_e is the number of equivalent arms.

Finally, sum over all the state labels of the internal lines, and integrate, from 0 to β, over the τ-labels of the vertices.

The absence of the inconvenient $1/p$ factor is to be noted. The coordination number of each diagram for G is unity on account of the presence of the ingoing and outgoing lines; in the τ-representation, we see that interchanging the internal τ-labels cannot now lead to the same set of contractions. (See, for example, Fig. 3.2.)

The absence of the $1/p$ factor means that the contribution of a part of a diagram for G of a particular form is the same wherever it occurs. For example, the contributions of the ringed portions of Fig. 3.3(b) and (c) are the same. In diagrams for Ω, however, this is not necessarily the case, as the coordination number may depend on the way in which the isolated part is connected with the rest of the diagram, as in Fig. 3.3(d) and (e). Since, so far, G has been defined only as part of a diagram for Ω, this may be thought to be

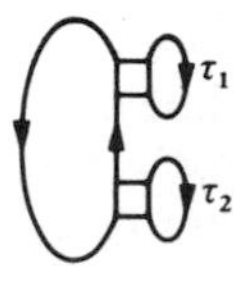

$p = 2$, since interchanging τ_1 and τ_2 leads to the same set of contractions.

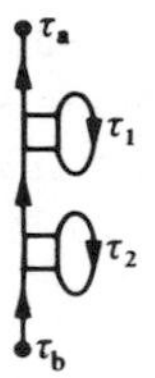

$p = 1$, since interchanging τ_1 and τ_2 leads to a different set of contractions.

FIG. 3.2.

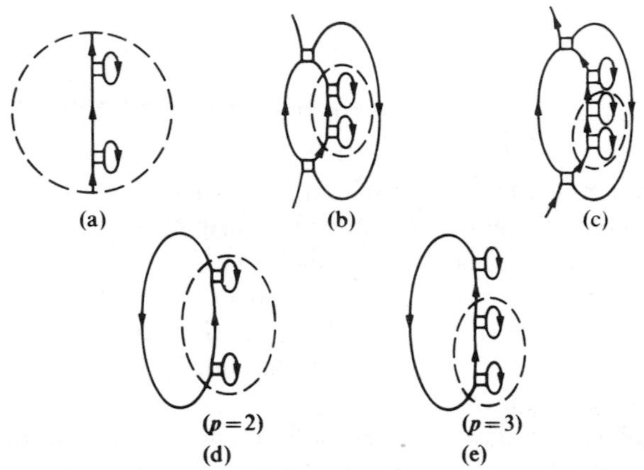

FIG. 3.3.

only a formal advantage. However, G will soon be seen to have a significance in its own right, and it is thus of great convenience to be able to isolate a part of a diagram, and work out its contribution, to be used wherever that part appears.

Exercises

3.1.1. Show that the first-order contributions to $\Sigma(\mathbf{k}, i\omega_l)$ is

$$\sum_{\mathbf{k}_1} [\bar{v}(0)+\varepsilon\bar{v}(|\mathbf{k}_1-\mathbf{k}|)]\langle n_{\mathbf{k}_1}\rangle_0$$

giving $\bar{G}(\mathbf{k}, i\omega_l) = \{i\omega_l - \varepsilon_{\mathbf{k}} + \mu - \sum_{\mathbf{k}_1} [\bar{v}(0)+\varepsilon\bar{v}(|\mathbf{k}_1-\mathbf{k}|)]\langle n_{\mathbf{k}_1}\rangle_0\}^{-1}$.

3.1.2. Show that the second-order contribution to Σ shown in eqn (3.1b) is

$$\Sigma^{(2)}(\mathbf{k}, i\omega_l) = -\frac{\varepsilon}{\beta} \sum_{\substack{\mathbf{k}_1\mathbf{k}_2\mathbf{k}_3 \\ l_1 l_2 l_3}} (-1)^5 \Gamma^0_{\mathbf{k}\mathbf{k}_3\mathbf{k}_1\mathbf{k}_2} \Gamma^0_{\mathbf{k}_1\mathbf{k}_2\mathbf{k}\mathbf{k}_3} \tfrac{1}{2}\beta^2 \delta_{l_1+l_2, l_3+l} \times$$
$$\times (-\beta)^{-3}[i\omega_{l_1} - \varepsilon_{\mathbf{k}_1} + \mu]^{-1}[i\omega_{l_2} - \varepsilon_{\mathbf{k}_2} + \mu]^{-1} \times$$
$$\times [i\omega_{l_3} - \varepsilon_{\mathbf{k}_3} + \mu]^{-1}.$$

Suppose that $\Gamma^0_{\mathbf{kk}_3\mathbf{k}_1\mathbf{k}_2} = \alpha\delta_{\mathbf{k}_1+\mathbf{k}_2,\mathbf{k}+\mathbf{k}_3}$ where α is a constant. That is, the interactions conserve momentum and spin but are otherwise constant. Then

$$\Sigma^{(2)}(\mathbf{k}, \mathrm{i}\omega_l) = \varepsilon\alpha^2/(2\beta^2) \sum_{\substack{\mathbf{k}_3,\mathbf{q},\\ l_2,l_3}} [\mathrm{i}\omega_{l_3} - \varepsilon_{\mathbf{k}_3} + \mu]^{-1}[\mathrm{i}\omega_{l_2} - \varepsilon_{\mathbf{k}+\mathbf{q}} + \mu]^{-1} \times$$
$$\times [\mathrm{i}(\omega_l + \omega_{l_3} - \omega_{l_2}) - \varepsilon_{\mathbf{k}_3-\mathbf{q}} + \mu]^{-1}.$$

Using the trick described in Appendix D, the sum may be performed since, with C as in Appendix D:

$$\mathscr{I} = \frac{\beta}{2\pi\mathrm{i}} \int_C \mathrm{d}\omega[\exp(\beta\omega) - \varepsilon]^{-1}[\omega - \varepsilon_{\mathbf{k}_3} + \mu]^{-1}[\omega + \mathrm{i}\omega_l - \mathrm{i}\omega_{l_2} - \varepsilon_{\mathbf{k}_3-\mathbf{q}} + \mu]^{-1}$$
$$= 0$$
$$= \varepsilon \sum_{l_3} [\mathrm{i}\omega_{l_3} - \varepsilon_{\mathbf{k}_3} + \mu]^{-1}[\mathrm{i}(\omega_{l_3} + \omega_l - \omega_{l_2}) - (\varepsilon_{\mathbf{k}_3-\mathbf{q}} - \mu)]^{-1} +$$
$$+ \beta\{\exp[\beta(\varepsilon_{\mathbf{k}_3} - \mu)] - \varepsilon\}^{-1}\{\mathrm{i}\omega_l - \mathrm{i}\omega_{l_2} - \varepsilon_{\mathbf{k}_3-\mathbf{q}} + \varepsilon_{\mathbf{k}_3}\}^{-1} -$$
$$- \beta\{\exp[\beta(\varepsilon_{\mathbf{k}_3-\mathbf{q}} - \mu)] - \varepsilon\}^{-1}\{\mathrm{i}\omega_l - \mathrm{i}\omega_{l_2} - \varepsilon_{\mathbf{k}_3-\mathbf{q}} + \varepsilon_{\mathbf{k}_3}\}^{-1}.$$

(Remember $\exp[\mathrm{i}(\omega_l - \omega_{l_2})] = 1$.)

Perform the sum over l_2 in a similar way to show that

$$\Sigma^{(2)}(\mathbf{k}, \mathrm{i}\omega_l) = -\varepsilon(\alpha^2/2) \sum_{\mathbf{k}_3,\mathbf{q}} \{\langle n_{\hat{\mathbf{k}}+\mathbf{q}}\rangle_0 \langle n_{\mathbf{k}_3}\rangle_0 + \langle n_{\mathbf{k}_3-\mathbf{q}}\rangle_0 \langle n_{\mathbf{k}_3}\rangle_0 -$$
$$- \langle n_{\mathbf{k}_3-\mathbf{q}}\rangle_0 \langle n_{\mathbf{k}+\mathbf{q}}\rangle_0 + \varepsilon\langle n_{\mathbf{k}_3}\rangle_0\} \times$$
$$\times \{\mathrm{i}\omega_l - \varepsilon_{\mathbf{k}+\mathbf{q}} - \varepsilon_{\mathbf{k}_3-\mathbf{q}} + \varepsilon_{\mathbf{k}_3} + \mu\}^{-1}.$$

3.1.3. Show that, when there is no '**k**-conservation' rule for the interactions, eqn (3.1) becomes

$$\bar{G}(\mathbf{k}_1, \mathbf{k}_2, \mathrm{i}\omega_n) = \bar{g}(\mathbf{k}_1, \mathrm{i}\omega_n)\delta_{\mathbf{k}_1,\mathbf{k}_2} + \sum_{\mathbf{k}_3} \bar{g}(\mathbf{k}_1, \mathrm{i}\omega_n)\Sigma(\mathbf{k}_1, \mathbf{k}_3, \mathrm{i}\omega_n)\bar{G}(\mathbf{k}_3, \mathbf{k}_2, \mathrm{i}\omega_n).$$

This equation is no longer easy to solve. If we regard $\bar{G}(\mathbf{k}_3, \mathbf{k}_2, \mathrm{i}\omega_n)$ as the $\mathbf{k}_1, \mathbf{k}_2$ element of the matrix $\bar{\mathbf{G}}$ and similarly for $\bar{\mathbf{g}}$ and $\mathbf{\Sigma}$ we may write the equation,

$$\bar{\mathbf{G}} = \bar{\mathbf{g}} + \bar{\mathbf{g}}\mathbf{\Sigma}\bar{G}$$
$$\bar{\mathbf{G}} = (\bar{\mathbf{g}}^{-1} - \mathbf{\Sigma})^{-1}$$

Although this equation looks simple, it involves the inversion of the matrix $(\bar{\mathbf{g}}^{-1} - \mathbf{\Sigma})$, and this is not possible in general.

3.1.4. For the electron–phonon Hamiltonian of exercise 1.2.4, show that the lowest order diagram for the electron self-energy is:

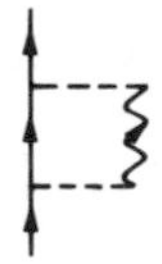

and evaluate its contribution.

3.1.5. In the case that $\varepsilon_{\mathbf{k}}$ = constant, c, (independent of $\mathbf{k}$), it is possible to sum the series:

$G_{\mathbf{k}}(\tau_1 - \tau_2) =$

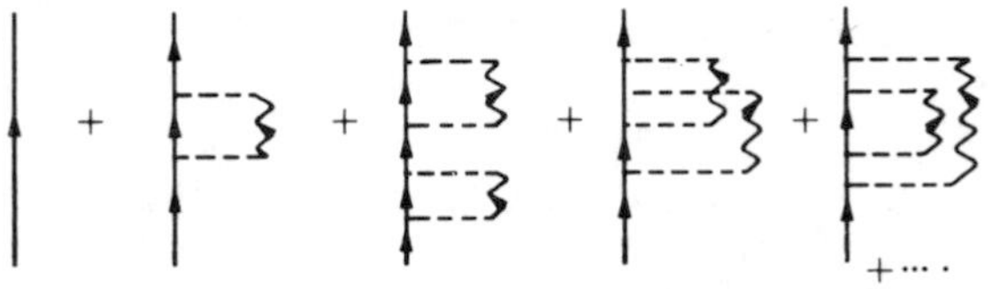

That is, all the terms involving just one electron line. Show that, in this case, for $\tau > 0$

$$G_{\mathbf{k}}(\tau) = [-\langle n_{\mathbf{k}}\rangle_0 + 1]\exp[-(c' - \mu)\tau]\exp\left\{\sum_{\mathbf{q}} \frac{(F(\mathbf{q}))^2}{\omega_{\mathbf{q}}^2}[N_{\mathbf{q}}(e^{\omega_{\mathbf{q}}\tau} - 1) + \right.$$

$$\left. + (N_{\mathbf{q}} + 1)(e^{-\omega_{\mathbf{q}}\tau} - 1)]\right\},$$

where $N_{\mathbf{q}} = [\exp(\beta\omega_{\mathbf{q}}) - 1]^{-1}$ and c' is a constant. This is the static scalar field model of field theory applied to statistical mechanics.

3.2. The thermodynamic potential in terms of G

(The rest of the chapter is largely independent of this section which might therefore be omitted at first reading.)

The most obvious way in which G could be used is to help to reduce the number of diagrams for Ω. Thus the diagram in Fig. 3.4(b) includes (a) and all those in (c) to (e) and many more. The basic diagrams with which we are left after this reduction, such as (a) or (f) in Fig. 3.4, with no self-energy insertions, are known as skeleton diagrams. None of the diagrams in (c) can be a skeleton diagram, since each can be obtained from (a) by the insertion of a contribution to the proper self-energy part. At first sight it appears that to find Ω we need to draw all skeleton diagrams and, for each full line, include a factor for the full Green function, rather than the non-interacting Green function in rule (ii). Unfortunately, as we have seen at the end of the previous

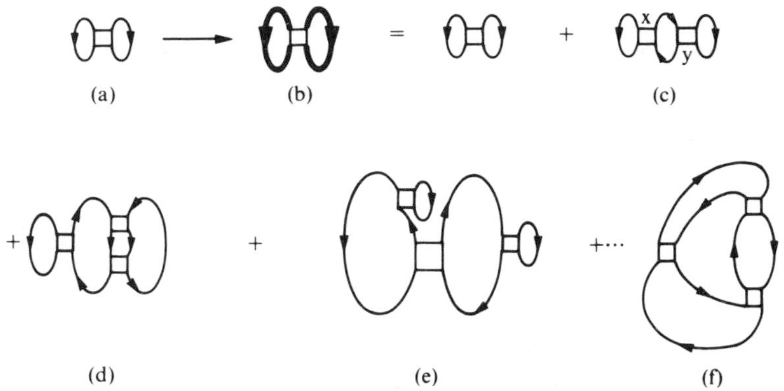

FIG. 3.4. Replacing the gs in (a) by Gs leads to (b), which includes the diagrams in (c), (d), and (e), and many others. (f) is not included.

section, we shall not always obtain the correct numerical factor for the contribution of a diagram by this procedure. Further, the difficult question of overcounting arises. The diagram in Fig. 3.4(c) could be obtained in two ways from (b): with x as the vertex appearing in (b) and all to the right of x as part of a full Green function or with y as the vertex in (b) with all to the left of y as part of a full Green function.

Consider any nth-order diagram for Ω. This will contain $2n$ full lines, and if we break open any one of these, we shall obtain a diagram for G, which may be proper or improper:

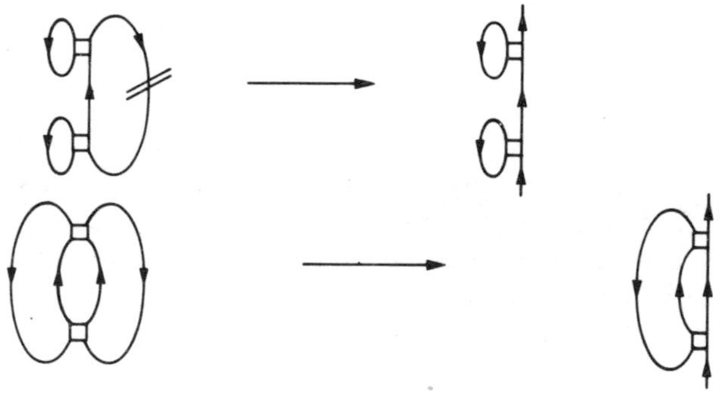

FIG. 3.5.

By this procedure, we obtain each diagram for G at least once. If $p = 1$, and $n_e = 1$ for the diagram for Ω we are considering, we obtain a unique diagram for G. If $p \neq 1$, we shall obtain the same diagram for G p times. (If, in the diagram for Ω, the vertex at τ_a is connected with that at τ_m, and at τ_b with τ_n, and interchanging τ_a with τ_b and τ_m with τ_n produces the same set of contractions, then the diagram for G obtained by breaking the line from τ_a to τ_m

will be identical with that obtained by breaking the line from τ_b to τ_n. (See Fig. 3.6((a) to (d)).) This will cancel the $1/p$ factor in Ω. Similarly, if $n_e \neq 0$, we shall obtain the same diagram for G twice for each pair of equivalent arms; on the other hand, we shall reduce the number of such pairs by one in breaking open one line of an equivalent arm. (See Fig. 3.6((e) to (f)).)

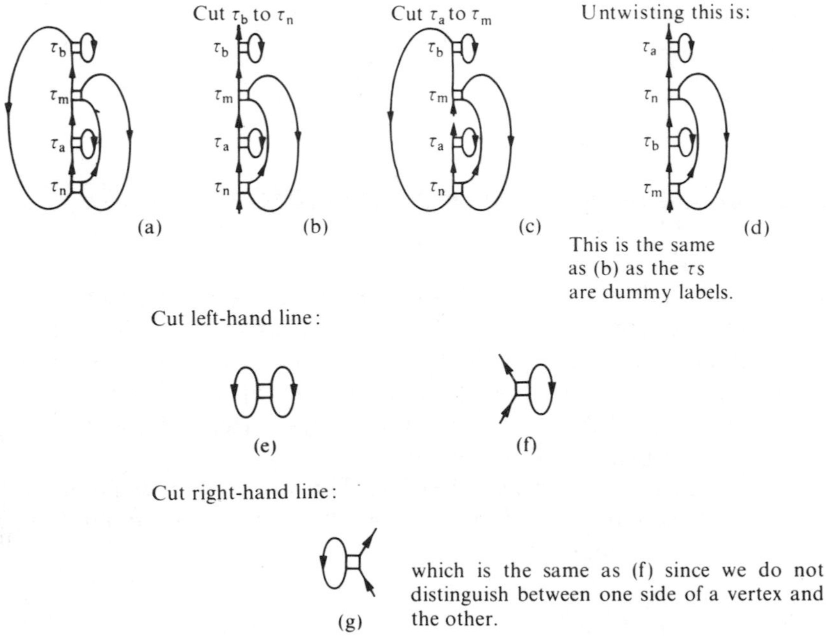

Cut τ_b to τ_n
Cut τ_a to τ_m
Untwisting this is:
τ_b
τ_m
τ_a
τ_n
(a)
(b)
(c)
(d)
This is the same as (b) as the τs are dummy labels.
Cut left-hand line:
(e)
(f)
Cut right-hand line:
(g)
which is the same as (f) since we do not distinguish between one side of a vertex and the other.

FIG. 3.6.

Finally, we reduce the number of closed fermion loops by one each time we break open a fermion line. Thus, if we break open in turn each full line in all diagrams for Ω in nth order, we shall obtain every nth-order diagram for G exactly once, each with the correct weighting. If, on the other hand, we take every nth-order diagram for G and join together the ingoing and outgoing lines, we shall obtain every nth-order diagram for Ω $2n$ times, as there are $2n$ lines to break open in Ω_n. Thus, for the nth-order term in Ω†,

$$\Omega_n = -\varepsilon/(2n) \operatorname*{Lt}_{\substack{\tau\to 0\\ \tau<0}} \sum_{\mathbf{k},l} \exp(-\mathrm{i}\omega_l\tau)\bar{G}_n(\mathbf{k}, \mathrm{i}\omega_l)/[\beta\bar{g}(\mathbf{k}, \mathrm{i}\omega_l)], \tag{3.5}$$

where $\bar{G}_n$ is the nth-order term in $\bar{G}$. (In the diagram for $\bar{G}(\mathbf{k}, \mathrm{i}\omega_l)$ obtained by breaking a line labelled $\mathbf{k}, l$, we need *two* factors of $\bar{g}(\mathbf{k}, \mathrm{i}\omega_l)$: this explains the $1/g$ factor in eqn (3.5). The factor $-\beta^{-1}$ comes from rule (vii) of section 2.8.

† As usual, the $\exp(-\mathrm{i}\omega_l\tau)$ factor, with $\tau \to 0$, $\tau < 0$, is to make sure the sum converges to the correct limit. See, also, the paragraph below eqn. (2.50).

If we multiply each interaction vertex by λ and differentiate with respect to λ, we shall rid ourselves of the awkward $1/n$ factor in this equation, obtaining

$$\lambda\frac{\partial\Omega_n}{\partial\lambda} = -\frac{\varepsilon}{2}\underset{\substack{\tau\to 0\\ \tau<0}}{\mathrm{Lt}}\sum_{\mathbf{k},l}\exp(-\mathrm{i}\omega_l\tau)\bar{G}_n(\mathbf{k},\mathrm{i}\omega_l)/[\beta\bar{g}(\mathbf{k},\mathrm{i}\omega_l)]. \tag{3.6}$$

Hence

$$\lambda\frac{\partial\Omega}{\partial\lambda} = -\frac{\varepsilon}{2\beta}\underset{\substack{\tau\to 0\\ \tau<0}}{\mathrm{Lt}}\sum_{\mathbf{k},l}[\bar{G}(\mathbf{k},\mathrm{i}\omega_l)-\bar{g}(\mathbf{k},\mathrm{i}\omega_l)]\exp(-\mathrm{i}\omega_l\tau)/\bar{g}(\mathbf{k},\mathrm{i}\omega_l). \tag{3.7}$$

We have $\bar{G}-\bar{g}$ in the numerator, rather than just $\bar{G}$, since the zero-order diagram for G can never be obtained by cutting lines in diagrams for Ω. Using eqn (3.2), we may rewrite this:

$$\lambda\frac{\partial\Omega}{\partial\lambda} = -\frac{\varepsilon}{2\beta}\underset{\substack{\tau\to 0\\ \tau<0}}{\mathrm{Lt}}\sum_{\mathbf{k},l}\bar{G}(\mathbf{k},\mathrm{i}\omega_l)\Sigma(\mathbf{k},\mathrm{i}\omega_l)\exp[-\mathrm{i}\omega_l\tau]. \tag{3.8}$$

We now define

$$\bar{\Omega} = \Omega_0-\frac{\varepsilon}{\beta}\underset{\substack{\tau\to 0\\ \tau<0}}{\mathrm{Lt}}\sum_{\mathbf{k},l}\exp(-\mathrm{i}\omega_l\tau)\{\ln[\bar{G}(\mathbf{k},\mathrm{i}\omega_l)\bar{g}^{-1}(\mathbf{k},\mathrm{i}\omega_l)]-$$
$$-\bar{g}^{-1}(\mathbf{k},\mathrm{i}\omega_l)\bar{G}(\mathbf{k},\mathrm{i}\omega_l)+1\}+\Omega' \tag{3.9}$$

where Ω' is the contribution of all closed linked skeleton diagrams computed according to rules (i) to (vii) of section 2.7 or 2.8 but with $\bar{g}$ replaced by $\bar{G}$. We shall show that $\bar{\Omega} = \Omega$.

To do this, call ν the number of interactions which appear explicitly in some skeleton diagram in Ω'_ν. Following the same arguments as above, we may write, omitting the Lt sign, for simplicity

$$\Omega' = -\frac{\varepsilon}{2\beta}\sum_{\nu,\mathbf{k},l}\frac{1}{\nu}\bar{G}_\nu(\mathbf{k},\mathrm{i}\omega_l)/\bar{g}(\mathbf{k},\mathrm{i}\omega_l), \tag{3.10}$$

where $\bar{G}_\nu(\mathbf{k},\mathrm{i}\omega_l)$ is the νth-order contribution to $\bar{G}$, and where only the explicit interactions in the skeleton diagram are counted, in determining the order of the diagram for $\bar{G}$. Similarly, differentiating Ω' with respect to $\bar{G}(\mathbf{k},\mathrm{i}\omega_l)$ has the effect of removing each in turn of the 2ν lines of a νth-order diagram, leaving a unique diagram for $\Sigma(\mathbf{k},\mathrm{i}\omega_l)$, apart from factors. This will be a proper diagram, since the original diagram for Ω' was a skeleton diagram. Thus

$$\frac{\partial\Omega'}{\partial\bar{G}(\mathbf{k},\mathrm{i}\omega_l)} = -\frac{\varepsilon}{\beta}\Sigma(\mathbf{k},\mathrm{i}\omega_l). \tag{3.11}$$

Thus, differentiating eqn (3.9) and using eqn (3.2),

$$\frac{\partial\bar{\Omega}}{\partial\bar{G}(\mathbf{k}, \mathrm{i}\omega_l)} = 0. \tag{3.12}$$

Now consider $\lambda\partial\bar{\Omega}/\partial\lambda$. By eqn (3.12) we may ignore the dependence of $\bar{G}$ on λ, whence the only dependence comes from the explicit dependence of Ω' on λ in the interactions. Thus we have, from eqn (3.10),

$$\begin{aligned}\lambda\frac{\partial\bar{\Omega}}{\partial\lambda} &= -\frac{\varepsilon}{2\beta}\sum_{\nu,\mathbf{k},l}\bar{G}_\nu(\mathbf{k}, \mathrm{i}\omega_l)/\bar{g}(\mathbf{k}, \mathrm{i}\omega_l)\\ &= -\frac{\varepsilon}{2\beta}\sum_{\mathbf{k},l}[\bar{G}(\mathbf{k}, \mathrm{i}\omega_l)-\bar{g}(\mathbf{k}, \mathrm{i}\omega_l)]/\bar{g}(\mathbf{k}, \mathrm{i}\omega_l)\\ &= \lambda\frac{\partial\Omega}{\partial\lambda}.\end{aligned} \tag{3.13}$$

But, from eqn (3.9),

$$\bar{\Omega}(\lambda = 0) = \Omega_0. \tag{3.14}$$

Hence

$$\bar{\Omega} = \Omega. \tag{3.15}$$

We have thus an expression for Ω in terms of the sum of the contributions of skeleton diagrams. For fermions, using the method described in Appendix D:

$$\begin{aligned}\gamma &\equiv \frac{1}{\beta}\underset{\substack{\tau\to 0\\ \tau<0}}{\mathrm{Lt}}\sum_{\mathbf{k},l}\ln\bar{g}(\mathbf{k}, \mathrm{i}\omega_l)\exp[-\mathrm{i}\omega_l\tau]\\ &= -\frac{1}{2\pi\mathrm{i}}\underset{\substack{\tau\to 0\\ \tau<0}}{\mathrm{Lt}}\sum_{\mathbf{k}}\int_\Gamma \mathrm{d}\omega\ln(\omega-\varepsilon_{\mathbf{k}}+\mu)\frac{\exp(-\omega\tau)}{\exp(\beta\omega)+1},\end{aligned}$$

where Γ is the contour in the ω-plane which goes from $-\infty$ to ∞ just below the real axis and returns just above the real axis. Thus

$$\begin{aligned}\gamma &= -\frac{1}{2\pi\mathrm{i}}\underset{\substack{\tau\to 0\\ \tau<0}}{\mathrm{Lt}}\sum_{\mathbf{k}}\frac{1}{\beta}\int_\Gamma\frac{\exp(-\omega\tau)}{\omega-\varepsilon_{\mathbf{k}}+\mu}\ln[1+\exp(-\beta\omega)]\,\mathrm{d}\omega\\ &= -\frac{1}{\beta}\sum_{\mathbf{k}}\ln\{1+\exp[-\beta(\varepsilon_{\mathbf{k}}-\mu)]\}\\ &= \Omega_0.\end{aligned}$$

An alternative form of eqn (3.9) is thus, for fermions,

$$\Omega = \frac{1}{\beta} \underset{\substack{\tau \to 0 \\ \tau < 0}}{\mathrm{Lt}} \sum_{\mathbf{k},l} \exp(-\mathrm{i}\omega_l \tau)\{\ln \bar{G}(\mathbf{k}, \mathrm{i}\omega_l) - \Sigma(\mathbf{k}, \mathrm{i}\omega_l)\bar{G}(\mathbf{k}, \mathrm{i}\omega_l)\} + \Omega'. \quad (3.9a)$$

We shall use this form in an example in section 3.9.

Two other important relations have been obtained in the course of this proof. Eqn (3.8) is a direct connection between the Green function and the thermodynamic properties of our system, and at first sight appears to be more convenient than eqn (3.9), since it does not involve any further summation over diagrams, once we have determined $\bar{G}$ (and hence Σ). Unfortunately, it involves an integration over interaction strengths, which is often inconvenient in realistic calculations in which one is making approximations for the Green functions. These approximations, if reasonable at all, are usually so only for certain ranges of interaction strength. Hence an integration over this strength may involve ranges for which the approximation is poor.

Eqn (3.12) when combined with eqn (3.15) gives us a variational principle. The thermodynamic potential is stationary with respect to changes in $\bar{G}$. We shall use this property in the example at the end of section 3.9.

Exercise

3.2.1. Follow through the arguments of this section with particular first-order and second-order diagrams.

3.3. The algebraic expression for the full Green function

We now wish to return to our statement that G is the fully interacting Green function. So far, G has been defined purely in terms of diagrams. If we are to take seriously the statement that G is the analogue of g in the presence of interactions, we should expect to be able to write it, in the τ-representation:

$$\begin{aligned} G_{\mathbf{k}}(\tau_1, \tau_2) &= \langle a^{\mathrm{H}}_{\mathbf{k}}(\tau_1)\tilde{a}^{\mathrm{H}}_{\mathbf{k}}(\tau_2)\rangle \qquad \tau_1 > \tau_2 \\ &= \varepsilon\langle \tilde{a}^{\mathrm{H}}_{\mathbf{k}}(\tau_2)a^{\mathrm{H}}_{\mathbf{k}}(\tau_1)\rangle \qquad \tau_1 \leqslant \tau_2, \end{aligned} \quad (3.16)$$

where, for any operator A,

$$A^{\mathrm{H}}(\tau) = \exp[(H - \mu\hat{N})\tau]A \exp[-(H - \mu\hat{N})\tau], \quad (3.17)$$

and

$$\tilde{a}^{\mathrm{H}}(\tau) = \exp[(H - \mu\hat{N})\tau]a^{+} \exp[-(H - \mu\hat{N})\tau]. \quad (3.17a)$$

(The superscript H is to remind us of the similarity of this definition to that of operators in the Heisenberg representation). Note that the averages in eqn (3.16) are taken over the interacting ensemble.

Since

$$\mathrm{Tr}\{\exp[-\beta(H-\mu\hat{N})]\exp[(H-\mu\hat{N})\tau_1]a_{\mathbf{k}}\exp[(H-\mu\hat{N})(\tau_2-\tau_1)]a_{\mathbf{k}}^+ \times$$
$$\times\exp[-(H-\mu\hat{N})\tau_2]\} = \mathrm{Tr}\{\exp[-\beta(H-\mu\hat{N})]\exp[(H-\mu\hat{N})\times$$
$$\times(\tau_1-\tau_2)]a_{\mathbf{k}}\exp[(H-\mu\hat{N})(\tau_2-\tau_1)]a_{\mathbf{k}}^+\},$$

it is easy to see that $G_{\mathbf{k}}(\tau_1, \tau_2)$ is a function only of $\tau_1-\tau_2$, and we write it as $G_{\mathbf{k}}(\tau_1-\tau_2)$.

To prove eqn (3.16), we define

$$G_{\mathbf{k}}^{(1)}(\tau_1, \tau_2) = \langle\mathscr{T}\{a_{\mathbf{k}}(\tau_1)\bar{a}_{\mathbf{k}}(\tau_2)S^{(\beta)}\}\rangle_0/\langle S^{(\beta)}\rangle_0 \tag{3.18}$$

with $S^{(\beta)}$ as defined in eqn (2.11). There are two stages in our proof: we first show that $G^{(1)}$ is equal to the Green function defined in section 3.1 in terms of diagrams, and then that it is equal to the G defined in eqn (3.16).

Consider the diagrammatic expansion of the numerator of $G^{(1)}$. To obtain this, we would use our expansion for $S^{(\beta)}$ (eqns (2.22) and (2.27)). A typical term would be

$$(n!)^{-1}\langle\mathscr{T}\left\{a_{\mathbf{k}}(\tau_1)\bar{a}_{\mathbf{k}}(\tau_2)\int_0^\beta\cdots\int_0^\beta \mathrm{d}\tau_1'\,\mathrm{d}\tau_2'\ldots\mathrm{d}\tau_n'\,\mathscr{T}[H_1(\tau_1')H_1(\tau_2')\ldots H_1(\tau_n')]\right\}\rangle_0 \tag{3.19}$$

We now define the interpretation of the $\mathscr{T}$-ordering operator in eqn (3.18) so that it refers to all the τs in the expansion of $S^{(\beta)}$ as well as the two τs explicit in the definition. In this case, the second τ-ordering operator in eqn (3.19) is superfluous, since its job is already done by the first ordering operator. Thus, we have to evaluate expectation values of integrals of products of $\mathscr{T}$-ordered operators, a problem with which we are familiar, from the work described in Chapter 2. Following the methods described in that chapter, we shall use Wick's theorem to develop a diagrammatic expansion of the expectation values. In an nth-order diagram for $G_{\mathbf{k}}^{(1)}(\tau_1, \tau_2)$, there will be $n+2$ τ-points: τ_2, at which a line labelled $\mathbf{k}$ will enter the diagram, τ_1, at which a line labelled $\mathbf{k}$ will leave the diagram and n vertices for the n H_1s (see Fig. 3.7). To take the sum of all possible contractions, we shall join these $n+2$ points in all possible ways. From Fig. 3.7, we see that by this procedure we shall obtain all the diagrams for G, and many more besides. The extra diagrams are unlinked ones, such as those in (c) and (f). In fact, by following an argument very similar to that used in the proof of the linked-cluster theorem, one can show that, with each linked diagram appearing in $G_{\mathbf{k}}^{1}(\tau_1-\tau_2)$, we shall obtain a series consisting of that diagram together with all the diagrams for $\langle S^{(\beta)}\rangle_0$. As in Chapter 2, the unlinked parts of a diagram separate into independent factors. Hence, the

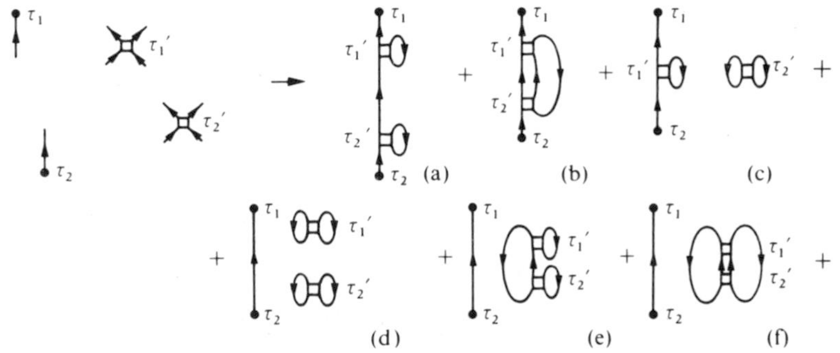

+ diagrams obtained from (a), (b), and (c) by interchanging τ_1' and τ_2'

FIG. 3.7. Second-order diagrams for the numerator of eqn (3.18).

numerator of eqn (3.18) is $G_{\mathbf{k}}(\tau_1-\tau_2)\times\langle S^{(\beta)}\rangle_0$ and $G_{\mathbf{k}}^{(1)}(\tau_1-\tau_2) = G_{\mathbf{k}}(\tau_1-\tau_2)$ as defined diagrammatically.

To complete the proof, we have to sort out the algebraic effect of the τ-ordering operator in eqn (3.18). We define

$$S(\tau_1,\tau_2) = \mathscr{T}\left\{\exp\left[-\int_{\tau_2}^{\tau_1} H_1(\tau)\,\mathrm{d}\tau\right]\right\} \tag{3.20}$$

If $\tau_1 > \tau' > \tau_2$, we have

$$\begin{aligned} S(\tau_1,\tau_2) &= \mathscr{T}\left\{\exp\left[-\int_{\tau_2}^{\tau'} H_1(\tau)\,\mathrm{d}\tau - \int_{\tau'}^{\tau_1} H_1(\tau)\,\mathrm{d}\tau\right]\right\} \\ &= \mathscr{T}\left\{\exp\left[-\int_{\tau_2}^{\tau'} H_1(\tau)\,\mathrm{d}\tau\right]\exp\left[-\int_{\tau'}^{\tau_1} H_1(\tau)\,\mathrm{d}\tau\right]\right\} \\ &= S(\tau_1,\tau')S(\tau',\tau_2), \end{aligned} \tag{3.21}$$

since the labels of all the operators in the second factor are less than those in the first factor. Thus, for $\tau_1 > \tau_2$,

$$S(\tau_1,0)[S(\tau_2,0)]^{-1} = S(\tau_1,\tau_2) \tag{3.22}$$

and

$$S^{(\beta)} = S(\beta,0) = S(\beta,\tau)S(\tau,0), \tag{3.23}$$

and hence

$$S(\beta,\tau) = S^{(\beta)}[S(\tau,0)]^{-1}. \tag{3.24}$$

Further,

$$\exp[-\tau(H-\mu\hat{N})] = \exp[-\tau(H_0-\mu\hat{N})]S(\tau,0) \tag{3.25}$$

from eqn (2.10) and (2.11): hence

$$\exp[-\tau(H-\mu\hat{N})][S(\tau,0)]^{-1} = \exp[-\tau(H_0-\mu\hat{N})]. \tag{3.26}$$

Therefore,

$$\begin{aligned}
&\langle\mathscr{T}\{a_{\mathbf{k}}(\tau_1)\bar{a}_{\mathbf{k}}(\tau_2)S^{(\beta)}\}\rangle_0/\langle S^{(\beta)}\rangle_0 \\
&= \langle S(\beta,\tau_1)a_{\mathbf{k}}(\tau_1)S(\tau_1,\tau_2)\bar{a}_{\mathbf{k}}(\tau_2)S(\tau_2,0)\rangle_0/\langle S^{(\beta)}\rangle_0 \qquad \text{for } \tau_1 > \tau_2 \\
&= \mathrm{Tr}\{\exp[-\beta(H_0-\mu\hat{N})]S^{(\beta)}[S(\tau_1,0)]^{-1}a_{\mathbf{k}}(\tau_1)S(\tau_1,0)\times \\
&\quad \times[S(\tau_2,0)]^{-1}\bar{a}_{\mathbf{k}}(\tau_2)S(\tau_2,0)\}/\mathrm{Tr}\{\exp[-\beta(H_0-\mu\hat{N})S^{(\beta)}\} \\
&= \mathrm{Tr}\{\exp[-\beta(H-\mu\hat{N})]a^{\mathrm{H}}_{\mathbf{k}}(\tau_1)\tilde{a}^{\mathrm{H}}_{\mathbf{k}}(\tau_2)\}/\mathrm{Tr}\{\exp[-\beta(H-\mu\hat{N})]\} \\
&= \langle a^{\mathrm{H}}_{\mathbf{k}}(\tau_1)\tilde{a}^{\mathrm{H}}_{\mathbf{k}}(\tau_2)\rangle.
\end{aligned} \tag{3.27}$$

A very similar proof may be constructed for $\tau_1 < \tau_2$. Thus the G defined in terms of diagrams in section 3.1 is equal to the G defined by eqn (3.16).

A rough-and-ready interpretation of the rules for evaluating the interacting Green function may now be given: we add to our system at τ_2 a particle of momentum $\mathbf{k}$ and remove it at τ_1. Between these 'times' (really, inverse temperatures) we allow the added particle to interact with all the other particles in our system. It can interact not at all, or once, or twice, ... corresponding to the zeroth, first-, second- ... order terms in the perturbation expansion. Thus we can say that the added particle interacts twice with a particle in the system in Fig. 3.7(b), whereas in 3.7(a) we can say that it interacts once with two different particles. We must not rely too heavily on this interpretation, but it sometimes helps to give a physical feeling for what is going on in the rather thick undergrowth of mathematics we are developing.

(The rest of this section should be omitted at first reading).

We may slightly extend the work in this section to give us the rules for evaluating the expectation value in the interacting ensemble of more complicated $\mathscr{T}$-ordered products. Suppose, for example, that for some reason we wished to calculate

$$\begin{aligned}
G(\mathbf{k}_1,\mathbf{k}_2,\ldots\mathbf{k}_m;\mathbf{k}'_1,\mathbf{k}'_2,\ldots\mathbf{k}'_m) &= \langle\mathscr{T}\{a^{\mathrm{H}}_{\mathbf{k}_1}(\tau_1)a^{\mathrm{H}}_{\mathbf{k}_2}(\tau_2)\ldots a^{\mathrm{H}}_{\mathbf{k}_m}(\tau_m)\times \\
&\quad\times\tilde{a}^{\mathrm{H}}_{\mathbf{k}'_1}(\tau'_1)\tilde{a}^{\mathrm{H}}_{\mathbf{k}'_2}(\tau'_2)\ldots\tilde{a}^{\mathrm{H}}_{\mathbf{k}'_m}(\tau'_m)\}\rangle.
\end{aligned} \tag{3.28}$$

We should first repeat a slightly more complicated version of the work starting at eqn (3.27) to show that this may be written:

$$\langle\mathscr{T}\{a_{\mathbf{k}_1}(\tau_1)\ldots a_{\mathbf{k}_m}(\tau_m)\bar{a}_{\mathbf{k}'_1}(\tau'_1)\ldots a_{\mathbf{k}'_m}(\tau'_m)S^{(\beta)}\}\rangle_0/\langle S^{(\beta)}\rangle_0. \tag{3.29}$$

Once $G(\mathbf{k}_1 \ldots \mathbf{k}'_m)$ is in this form, we may expand S and use Wick's theorem to express the expansion of the numerator diagrammatically. Each diagram will now have $2m$ external lines, each with a τ-label, corresponding to the $2m$ operators in G. An nth order diagram will have n vertices each with a τ-label over which we have eventually to integrate. The external lines and the four lines from each vertex will be connected up in all possible ways. Some ways of connecting up the external lines and the vertices will lead to diagrams in which there are vertices unconnected with an external line. (See Fig. 3.8.) These diagrams are called disconnected. The disconnected parts may be

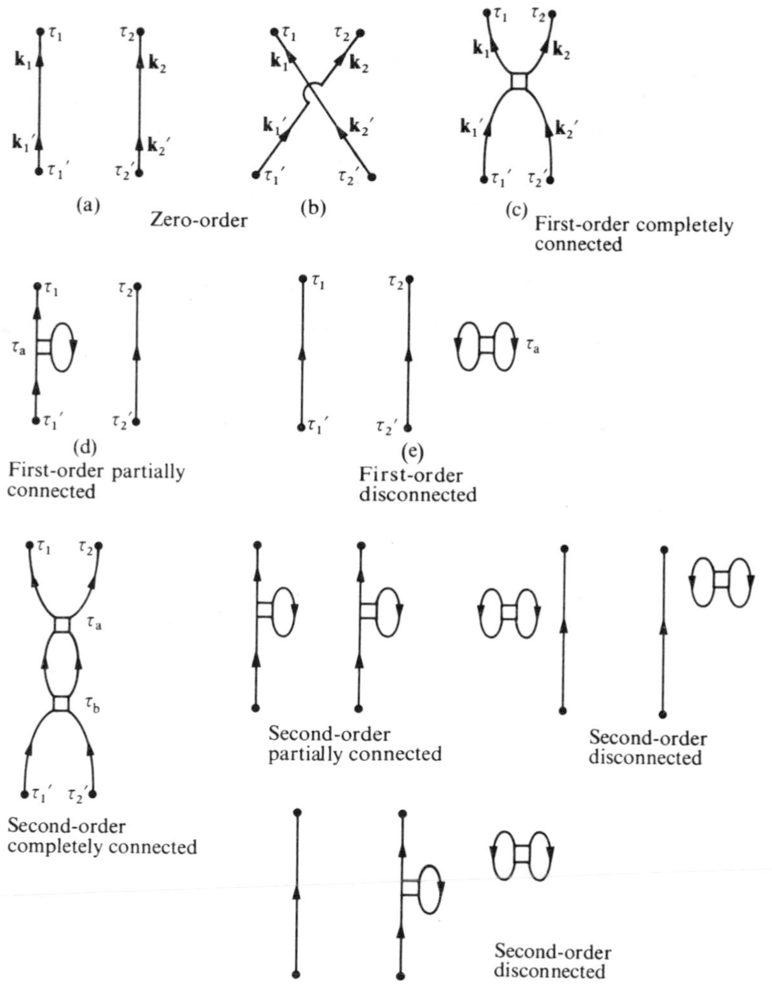

FIG. 3.8. Some diagrams for $G(\mathbf{k}_1, \mathbf{k}_2, \mathbf{k}'_1, \mathbf{k}'_2)$.

factorized as before, and eventually cancel the factor $\langle S^{(\beta)}\rangle_0$ in the denominator of eqn (3.29). The remainder are either completely connected diagrams, in which each vertex is connected to every external line, or partially connected diagrams, those neither completely connected, nor disconnected. (See Fig. 3.8.)

Sometimes two of the operators in eqn (3.28) have the same τ-label. In this case, the $\mathscr{T}$-ordering operator leaves them in the same order as they appeared in the original product. So that we may use the same rules as above for evaluating the Green function, it is usual to consider only Gs in which operators with the same τ-label are arranged with creation operators to the left of the annihilation operators. Thus we would work with $\langle\mathscr{T}\{\tilde{a}_1^{\mathrm{H}}(\tau)\times$ $\times a_2^{\mathrm{H}}(\tau)\tilde{a}_3^{\mathrm{H}}(\tau')a_4^{\mathrm{H}}(\tau')\}\rangle$ rather than $\langle\mathscr{T}\{a_2^{\mathrm{H}}(\tau)\tilde{a}_1^{\mathrm{H}}(\tau)a_4^{\mathrm{H}}(\tau')\tilde{a}_3^{\mathrm{H}}(\tau')\}\rangle$.

The rules for evaluating such a generalized Green function are thus, in the τ-representation:

For the function $G(\mathbf{k}_1 \dots \mathbf{k}_m; \mathbf{k}'_1 \dots \mathbf{k}'_m)$ draw all different diagrams with lines, each with the appropriate $\mathbf{k}$ label, entering at $\tau'_1 \dots \tau'_m$ and leaving at $\tau_1 \dots \tau_m$. The factors making up the contribution of a diagram to G are then identical to those on p. 61. Some care has to be exercised in determining the sign of a diagram such as that in Fig. 3.8(b), which carries an extra factor of $(-\varepsilon)$. This is best determined by a direct use of Wick's theorem.

It is also possible to work in the ω-representation. We shall consider only the case most frequently needed, with two pairs of operators at different times:

$$G(\mathbf{k}_1, \mathbf{k}_2, \tau, \mathbf{k}_3, \mathbf{k}_4, \tau') = \langle\mathscr{T}\tilde{a}_{\mathbf{k}_1}^{\mathrm{H}}(\tau)a_{\mathbf{k}_2}^{\mathrm{H}}(\tau)\tilde{a}_{\mathbf{k}_3}^{\mathrm{H}}(\tau')a_{\mathbf{k}_4}^{\mathrm{H}}(\tau')\rangle \tag{3.30}$$

It is easy to prove that $G(\tau, \tau')$ is a function only of $\tau-\tau'$ by a slight modification of the same proof for the single-particle Green function. Thus, this Green function may also be expressed in a Fourier series:

$$G(\mathbf{k}_1, \mathbf{k}_2; \mathbf{k}_3, \mathbf{k}_4; \tau-\tau') = -\frac{1}{\beta}\sum_n \bar{G}(\mathbf{k}_1, \mathbf{k}_2; \mathbf{k}_3, \mathbf{k}_4, \mathrm{i}\omega_n)\,\mathrm{e}^{-\mathrm{i}\omega_n(\tau-\tau')}, \tag{3.31}$$

where

$$\bar{G}(\mathbf{k}_1, \mathbf{k}_2; \mathbf{k}_3, \mathbf{k}_4; \mathrm{i}\omega_n) = -\frac{1}{2}\int_{-\beta}^{\beta} G(\mathbf{k}_1, \mathbf{k}_2; \mathbf{k}_3, \mathbf{k}_4; \tau-\tau')\,\mathrm{e}^{+\mathrm{i}\omega_n(\tau-\tau')}\mathrm{d}(\tau-\tau'). \tag{3.32}$$

If we express all the zero-order Green functions for the full lines in the τ-representation, the τ-dependence of the whole diagram will be

$$\exp[-\mathrm{i}\tau(\omega_m-\omega_l)]$$

with the notation of Fig. 3.9. Since the total ω is conserved at each vertex, $\omega_m+\omega_{l'} = \omega_{m'}+\omega_l$, and the τ'-dependence will be

$$\exp[\mathrm{i}\tau'(\omega_{m'}-\omega_{l'})] = \exp[\mathrm{i}\tau'(\omega_m-\omega_l)].$$

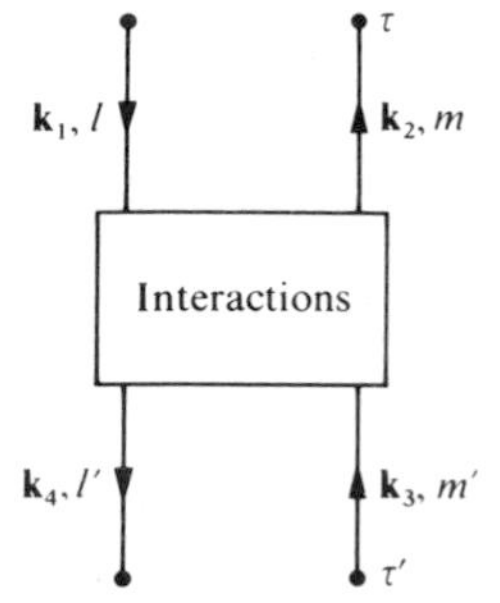

(Mixed representation)

FIG. 3.9.

Thus

$$\omega_n = \omega_m - \omega_l = \omega_{m'} - \omega_{l'}. \tag{3.33}$$

Thus, the rules for calculating $-\beta^{-1}\bar{G}(\mathbf{k}_1, \mathbf{k}_2; \mathbf{k}_3, \mathbf{k}_4; i\omega_n)$ are:

Draw all different diagrams with lines labelled $\mathbf{k}_1, l, \mathbf{k}_3, m'$ entering and $\mathbf{k}_2, m, \mathbf{k}_4, l'$ leaving, where $\omega_n = \omega_m - \omega_l$. The factors making up the contribution of a diagram to $-\beta^{-1}\bar{G}$ are then identical to those on p. 61. We notice that if the creation and annihilation operators in G are all for fermions or all for bosons eqn (3.33) implies that ω_n is π/β times an even number. This may also be proved by showing that $G(\tau - \tau' + \beta) = G(\tau - \tau')$ for $\tau < \tau'$ in a manner similar to that for the single particle Green function. Thus the contribution to $-\beta^{-1}\bar{G}(\mathbf{k}_1, \mathbf{k}_2; \mathbf{k}_3, \mathbf{k}_4; i\omega_n)$ of the simple series of diagrams

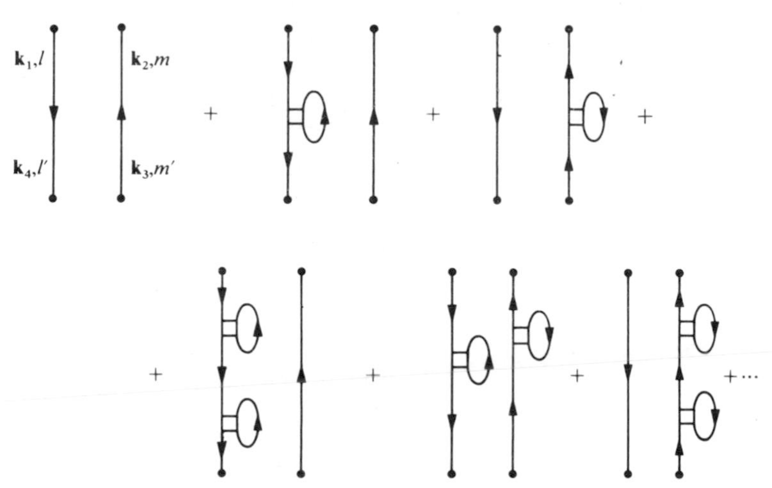

FIG. 3.10.

shown in Fig. 3.10 is

$$\frac{\varepsilon}{\beta^2}\sum_{\substack{m,m'\\l,l'}}\delta_{\mathbf{k}_2,\mathbf{k}_3}\delta_{\mathbf{k}_1,\mathbf{k}_4}\delta_{m,m'}\delta_{l,l'}\delta_{n,m-l}[\mathrm{i}\omega_m-(\varepsilon_{\mathbf{k}_2}-\mu)-\Sigma^{(a)}(\mathbf{k}_2,\mathrm{i}\omega_m)]^{-1}\times$$
$$\times[\mathrm{i}\omega_l-(\varepsilon_{\mathbf{k}_1}-\mu)-\Sigma^{a}(\mathbf{k}_1,\mathrm{i}\omega_l)]^{-1}$$

with the notation of Fig. 3.10. $\Sigma^{(a)}$ is defined in eqn (3.1). We shall show how to calculate the contribution of a more complex series of diagrams in sections 4.2 and 7.4.

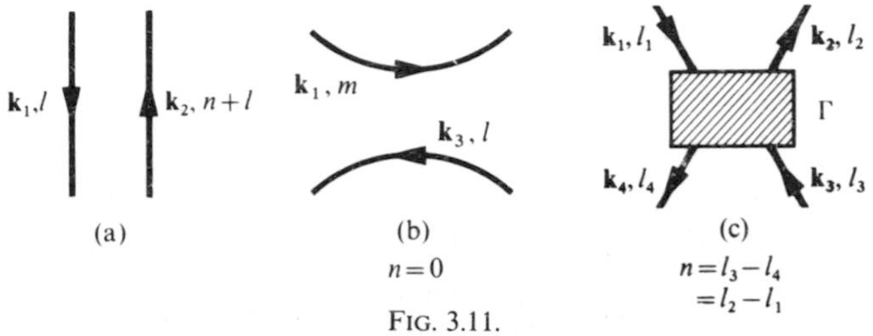

FIG. 3.11.

We notice that all diagrams for $\bar{G}(\mathbf{k}_1, \mathbf{k}_2; \mathbf{k}_3, \mathbf{k}_4; \mathrm{i}\omega_n)$ can be classified as shown in Fig. 3.11. The full lines, as usual, represent fully interacting single-particle Green functions, so that (a) and (b) sum all the partially connected diagrams and (c) all the completely connected diagrams. The shaded square, called the complete vertex part and written

$$-\Gamma(\mathbf{k}_4, \mathrm{i}\omega_{l_4}, \mathbf{k}_2, \mathrm{i}\omega_{l_2}, \mathbf{k}_1, \mathrm{i}\omega_{l_1}, \mathbf{k}_3, \mathrm{i}\omega_{l_3})\beta\delta_{l_1+l_2,l_3+l_4}$$

represents the sum of all connected diagrams with two lines entering and two leaving such that no part can be disconnected by cutting a single line. It is thus an analogue for this more complicated Green function of the proper self-energy part of the single-particle Green function (see, however, exercise 4.3.1). As for the proper self-energy part, we do not include in our definition of Γ the factors for the in- and out-going lines. Thus the simplest diagram for Γ is Γ^0. A few other simple diagrams for Γ are shown in Fig. 3.12. Fig. 3.13 shows some diagrams not to be included in Γ.

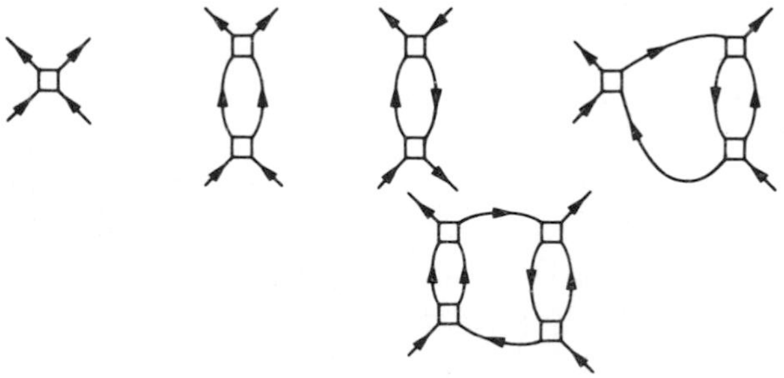

FIG. 3.12. Some simple diagrams for Γ.

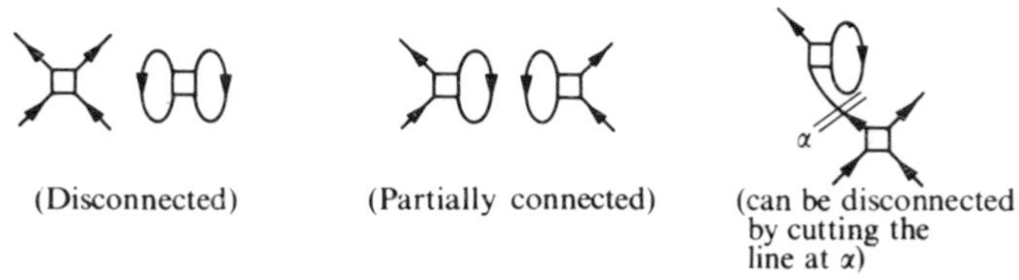

FIG. 3.13. Some diagrams not to be included in Γ.

We may derive a relationship between the one-particle Green function, $\langle \mathscr{T} a_{\mathbf{k}}^{\mathrm{H}}(\tau)\tilde{a}_{\mathbf{k}}^{\mathrm{H}}(\tau')\rangle$, and the two-particle Green function $\langle \mathscr{T} \tilde{a}_{\mathbf{k}_1}^{\mathrm{H}}(\tau)a_{\mathbf{k}_2}^{\mathrm{H}}(\tau) \times \times \tilde{a}_{\mathbf{k}_3}^{\mathrm{H}}(\tau')a_{\mathbf{k}_4}^{\mathrm{H}}(\tau')\rangle$ or, more simply, between their Fourier coefficients $\bar{G}(\mathbf{k}, \mathrm{i}\omega_n)$ and $\bar{G}(\mathbf{k}_1, \mathbf{k}_2, \mathbf{k}_3, \mathbf{k}_4, \mathrm{i}\omega_n)$, by considering the structure of diagrams for $\Sigma(\mathbf{k}, \mathrm{i}\omega_n)$. These may be split into two classes; in the first, (Fig. 3.14(a)), the outgoing line is attached to the same vertex as the ingoing line; in the second, it is attached to a different vertex (Fig. 3.14(b)). For diagrams of the second class, the vertex of the ingoing line may have two of its other three lines attached by a full Green function (Fig. 3.14(c)); this leads to an improper diagram for Σ, and therefore is not allowed. The only other possibility is shown in Fig. 3.14(d); the square with a cross inside it represents the sum of all connected diagrams with two lines entering and two leaving such that no part may be disconnected by cutting a single line. (Such parts are already included by the full Green functions in the lines (i), (ii) and (iii) of (d), or, if on line (iv), would lead to improper diagrams for Σ.) But this is just the definition of the full vertex function, which we have decided to represent by a shaded square. Thus we should redraw (d) as (e).

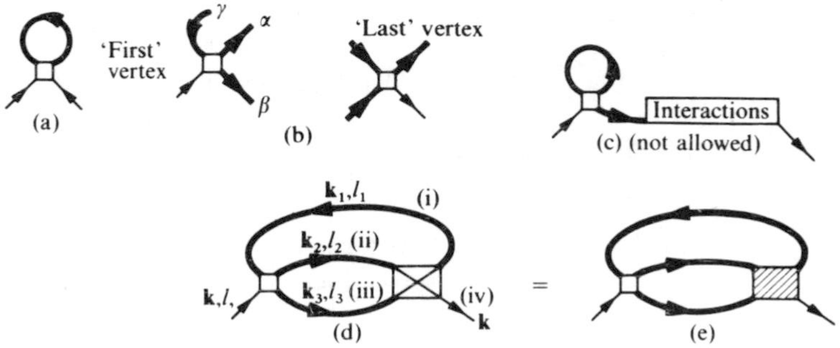

FIG. 3.14. The structure of diagrams for Σ. In (b) either line α or β must be attached either to γ giving a diagram of type (c), or not, giving a diagram of type (d).

The connection between and one- and two-particle Green functions is thus given diagrammatically by Figs. 3.11 and 3.14, and eqn (3.3). Algebraically, we have:†

$$\bar{G}(\mathbf{k}, \mathrm{i}\omega_l) = [\mathrm{i}\omega_l - (\varepsilon_{\mathbf{k}} - \mu) - \Sigma(\mathbf{k}, \mathrm{i}\omega_l)]^{-1} \tag{3.3}$$

$$-\beta\Sigma(\mathbf{k}, \mathrm{i}\omega_l) = \frac{1}{\beta}\sum_m \sum_{\mathbf{k}_1} \varepsilon\beta\Gamma^0(\mathbf{k}, \mathbf{k}_1, \mathbf{k}, \mathbf{k}_1)\bar{G}(\mathbf{k}_1, \mathrm{i}\omega_m) -$$

$$-\frac{\varepsilon}{2}\sum_{\substack{\mathbf{k}_1\mathbf{k}_2\mathbf{k}_3 \\ l_1 l_2 l_3}} \beta^2\Gamma^0(\mathbf{k}, \mathbf{k}_1, \mathbf{k}_3, \mathbf{k}_2)\Gamma(\mathbf{k}_2, \mathrm{i}\omega_{l_2}, \mathbf{k}_3, \mathrm{i}\omega_{l_3}, \mathbf{k}_1, \mathrm{i}\omega_{l_1}, \mathbf{k}, \mathrm{i}\omega_l) \times$$

$$\times\, \delta_{l+l_1, l_2+l_3}\beta^{-3}\bar{G}(\mathbf{k}_1, \mathrm{i}\omega_{l_1})\bar{G}(\mathbf{k}_2, \mathrm{i}\omega_{l_2})\bar{G}(\mathbf{k}_3, \mathrm{i}\omega_{l_3}) \text{ (Fig. 3.14)},$$

$$-\beta^{-1}\bar{G}(\mathbf{k}_1, \mathbf{k}_2, \mathbf{k}_3, \mathbf{k}_4, \mathrm{i}\omega_n)$$

$$= \frac{\varepsilon}{\beta^2}\sum_l \bar{G}(\mathbf{k}_3, \mathrm{i}\omega_n + \mathrm{i}\omega_l)\bar{G}(\mathbf{k}_1, \mathrm{i}\omega_l)\delta_{\mathbf{k}_1,\mathbf{k}_4}\delta_{\mathbf{k}_2,\mathbf{k}_3} +$$

$$+\frac{1}{\beta^2}\delta_{n,0}\sum_{ml} \bar{G}(\mathbf{k}_1, \mathrm{i}\omega_m)\bar{G}(\mathbf{k}_3, \mathrm{i}\omega_l)\delta_{\mathbf{k}_1,\mathbf{k}_2}\delta_{\mathbf{k}_3,\mathbf{k}_4} -$$

$$-\frac{\varepsilon}{\beta^4}\sum_{l_1 l_2 l_3 l_4} \bar{G}(\mathbf{k}_1, \mathrm{i}\omega_{l_1})\bar{G}(\mathbf{k}_2, \mathrm{i}\omega_{l_2})\bar{G}(\mathbf{k}_3, \mathrm{i}\omega_{l_3})\bar{G}(\mathbf{k}_4, \mathrm{i}\omega_{l_4}) \times$$

$$\times\, \beta\Gamma(\mathbf{k}_4, \mathrm{i}\omega_{l_4}, \mathbf{k}_2, \mathrm{i}\omega_{l_2}, \mathbf{k}_1, \mathrm{i}\omega_{l_1}, \mathbf{k}_3, \mathrm{i}\omega_{l_3})\delta_{l_1+l_3, l_2+l_4}\delta_{n, l_2-l_1}. \text{ (Fig. 3.11)}.$$

Exercise

3.3.1. Repeat the work above for the two-particle Green function:

$$\langle \mathscr{T} \tilde{a}^{\mathrm{H}}_{\mathbf{k}_1}(\tau)\tilde{a}^{\mathrm{H}}_{\mathbf{k}_2}(\tau)a^{\mathrm{H}}_{\mathbf{k}_3}(\tau')a^{\mathrm{H}}_{\mathbf{k}_4}(\tau')\rangle.$$

3.4. The analytic properties of *G*

In this section, we wish to show that the many apparently different single-particle Green functions defined in the literature are only different manifestations of the same object. We shall also derive certain exact relations which are useful in checking and interpreting the Green functions we obtain as a result of approximations in real problems. Unfortunately, this means that we have to work through rather a lot of mathematics.

We have already shown that $G(\tau_1, \tau_2)$ is a function only of $\tau_1 - \tau_2$. It may therefore be expanded in a Fourier series, just as g was in section 2.8. $G(\tau_1 - \tau_2)$ is defined in the range $-\beta$ to β, and we have

$$G_{\mathbf{k}}(\tau_1 - \tau_2) \equiv G_{\mathbf{k}}(\tau) = -\frac{1}{\beta}\sum_l \exp[-\mathrm{i}\pi l\tau/\beta]\bar{G}^l_{\mathbf{k}}, \quad l \text{ integral},$$

$$= -\frac{1}{\beta}\sum_l \exp[-\mathrm{i}\omega_l\tau]\bar{G}(\mathbf{k}, \mathrm{i}\omega_l) \tag{3.34}$$

† For ease of writing, we have omitted limit signs for lines leaving and entering the same vertex. The contribution of Fig. 3.11b should carry these factors, since it contains lines starting and finishing at the same τ.

where $\bar{G}(\mathbf{k}, \mathrm{i}\omega_l)$ are Fourier coefficients. For $\tau = \tau_1 - \tau_2 < 0$, $\tau + \beta > 0$, and

$$G_{\mathbf{k}}(\tau+\beta) = G_{\mathbf{k}}(\tau+\beta, 0) = \mathrm{Tr}\{\exp[-\beta(H-\mu\hat{N})]\exp[(\tau+\beta)(H-\mu\hat{N})]a_{\mathbf{k}} \times$$
$$\times \exp[-(\tau+\beta)(H-\mu\hat{N})]a_{\mathbf{k}}^{+}\}/\mathrm{Tr}\{\exp[-\beta(H-\mu\hat{N})]\}$$
$$= \frac{\mathrm{Tr}\{\exp[\tau(H-\mu\hat{N})]a_{\mathbf{k}}\exp[-\beta(H-\mu\hat{N})]\exp[-\tau(H-\mu\hat{N})]a_{\mathbf{k}}^{+}\}}{\mathrm{Tr}\{\exp[-\beta(H-\mu\hat{N})]\}}$$
$$= \frac{\mathrm{Tr}\{\exp[-\beta(H-\mu\hat{N})]\exp[-\tau(H-\mu\hat{N})]a_{\mathbf{k}}^{+}\exp[\tau(H-\mu\hat{N})]a_{\mathbf{k}}\}}{\mathrm{Tr}\{\exp[-\beta(H-\mu\hat{N})]\}}$$

using the invariance of the trace under cyclic permutation. Thus

$$\begin{aligned} G_{\mathbf{k}}(\tau+\beta) &= \varepsilon G_{\mathbf{k}}(0, -\tau) \\ &= \varepsilon G_{\mathbf{k}}(\tau). \end{aligned} \tag{3.35}$$

Thus, we have Fourier coefficients only at

$$\begin{aligned} \omega_l &= (2l+1)\pi/\beta \quad \text{for fermions} \\ \omega_l &= 2l\pi/\beta \quad\quad\;\; \text{for bosons.} \end{aligned} \tag{3.36}$$

By comparing the rules for the determination of G in the τ-representation and in the ω-representation, it can be seen that the Fourier coefficients are in fact the Green function in the ω-representation, multiplied by $-\beta^{-1}$, as usual.

It is now a straightforward, but rather tedious, task to determine these coefficients in terms of matrix elements of $a_{\mathbf{k}}$ and $a_{\mathbf{k}}^{+}$, using the normal formula for Fourier series expansions. Thus

$$\begin{aligned} \bar{G}(\mathbf{k}, \mathrm{i}\omega_l) = &-\tfrac{1}{2}\int_{-\beta}^{0} \mathrm{d}\tau\, \varepsilon\langle \tilde{a}_{\mathbf{k}}^{\mathrm{H}}(-\tau)a_{\mathbf{k}}^{\mathrm{H}}(0)\rangle \exp[\mathrm{i}\omega_l\tau] - \\ &-\tfrac{1}{2}\int_{0}^{\beta} \mathrm{d}\tau\langle a_{\mathbf{k}}^{\mathrm{H}}(\tau)a_{\mathbf{k}}^{+}\rangle \exp[\mathrm{i}\omega_l\tau]. \end{aligned} \tag{3.37}$$

Unfortunately, we know rather little about the τ-dependence of the integrands. However, we can obtain a useful expression for $\bar{G}$ by writing out the traces in the expectation values in eqn (3.37) in terms of eigenvectors and eigenvalues of $H - \mu\hat{N}$:

$$\begin{aligned} (H-\mu\hat{N})|\nu, N\rangle &= (E_{N\nu} - \mu N)|\nu, N\rangle \\ \hat{N}|\nu N\rangle &= N|\nu N\rangle. \end{aligned} \tag{3.38}$$

Then, for example,

$$\begin{aligned}
&\mathrm{Tr}\{\exp[-\beta(H-\mu\hat{N})]a_{\mathbf{k}}^{\mathrm{H}}(\tau)a_{\mathbf{k}}^{+}\} \\
&= \sum_{\nu N} \langle \nu N|\exp[-\beta(H-\mu\hat{N})]\exp[(H-\mu\hat{N})\tau]a_{\mathbf{k}}\exp[-(H-\mu\hat{N})\tau]\times \\
&\qquad \times a_{\mathbf{k}}^{+}|\nu, N\rangle \\
&= \sum_{\substack{\nu N \\ \nu' N'}} \langle \nu N|\exp[-\beta(H-\mu\hat{N})]\exp[(H-\mu\hat{N})\tau]a_{\mathbf{k}}\times \\
&\qquad \times \exp[-(H-\mu\hat{N})\tau]|\nu' N'\rangle\langle \nu' N'|a_{\mathbf{k}}^{+}|\nu N\rangle \\
&= \sum_{\substack{\nu N \\ \nu' N'}} \langle \nu N|a_{\mathbf{k}}|\nu' N'\rangle\langle \nu' N'|a_{\mathbf{k}}^{+}|\nu N\rangle \exp[-\beta(E_{N\nu}-\mu N)]\times \\
&\qquad \times \exp[\tau(E_{N\nu}-E_{N'\nu'}-\mu N+\mu N')],
\end{aligned} \tag{3.39}$$

where we have used the fact that the set of states $|\nu N'\rangle$ is complete, so that

$$1 = \sum_{\nu' N'} |\nu' N'\rangle\langle N'\nu'|, \qquad \text{(cf. eqn (1.3))}$$

and that

$$\langle \nu N|\exp[-\beta(H-\mu\hat{N})]\exp[\tau(H-\mu\hat{N})] = \exp[-\beta(E_{N\nu}-\mu N)]\exp[\tau(E_{N\nu}-\mu N)]\times\langle \nu N|.$$

Further, since $a_{\mathbf{k}}$ destroys one particle, $N' = N+1$.

The τ-dependence of each term in the sum in eqn (3.39) is now simple, so that, assuming suitable convergence properties, the series may be integrated term by term. After some rearranging, we obtain

$$\begin{aligned}
\bar{G}(\mathbf{k}, \mathrm{i}\omega_l) = \frac{1}{Q}\sum_{\nu,\nu',N} \exp[-\beta(E_{N\nu}-\mu N)]\Bigg\{&\frac{|\langle \nu' N+1|a_{\mathbf{k}}^{+}|\nu N\rangle|^2}{\mathrm{i}\omega_l - E_{N+1,\nu'} + E_{N\nu} + \mu} - \\
&-\varepsilon\frac{|\langle \nu N|a_{\mathbf{k}}^{+}|\nu' N-1\rangle|^2}{\mathrm{i}\omega_l - E_{N\nu} + E_{N-1,\nu'} + \mu}\Bigg\}.
\end{aligned} \tag{3.40}$$

This rather clumsy expression is of great importance in discussing the properties of $\bar{G}$. It tells us what $\bar{G}$ is at an infinite set of points, $\{\mathrm{i}\omega_l\}$, along the imaginary ω-axis. Consider the function $\bar{G}(\mathbf{k}, \omega)$ obtained by writing eqn (3.40) for all real and complex values of the argument ω, with $\mathrm{i}\omega_l \to \omega$:

$$\begin{aligned}
\bar{G}(\mathbf{k}, \omega) \equiv \frac{1}{Q}\sum_{\nu\nu' N} \exp[-\beta(E_{N\nu}-\mu N)]\Bigg\{&\frac{|\langle \nu' N+1|a_{\mathbf{k}}^{+}|\nu N\rangle|^2}{\omega - E_{N+1\nu'} + E_{N\nu} + \mu} - \\
&-\varepsilon\frac{|\langle \nu N|a_{\mathbf{k}}^{+}|\nu' N-1\rangle|^2}{\omega - E_{N\nu} + E_{N-1\nu'} + \mu}\Bigg\}.
\end{aligned} \tag{3.41}$$

This new $\bar{G}$ is equal to the old $\bar{G}$ at the infinite set of points along the imaginary ω-axis. It follows from Carleman's theorem (Titchmarsh, 1939, p. 131) that it is the only function which:

(i) is identical to the original $\bar{G}$ at the infinite set of points $\{i\omega_l\}$,
(ii) is analytic except possibly along the real axis, and
(iii) tends to zero along any line in the upper or lower half-plane.

Property (ii) follows from the form of eqn (3.41); since μ and the eigenvalues of H are real, $\bar{G}$ has a series of discontinuities along the real axis, and elsewhere is analytic. (See Baym and Mermin, 1961.)

The fact that there is only one function satisfying conditions (i) to (iii) is important, as it means that if we have evaluated $\bar{G}(\mathbf{k}, i\omega_l)$ by some means, we may obtain $\bar{G}(\mathbf{k}, \omega)$ merely by making the substitution $i\omega_l \to \omega$, provided that we ensure that the function so obtained satisfies (ii) and (iii). This is not necessarily a trivial proviso, since $\exp(i\omega_l) = \pm 1$ at all the points for which we know $\bar{G}(\mathbf{k}, i\omega_l)$. We may therefore have to add factors of $\pm\exp(i\omega_l)$ within our function in order to obtain that particular function which satisfies (ii) and (iii).

3.5. Time-temperature-dependent Green functions

From $\bar{G}(\mathbf{k}, \omega)$ we may obtain three new Green functions which are much used in the literature of the many-body problem. Consider first:

$$\mathscr{G}_r(\mathbf{k}, t-t') = \operatorname*{Lt}_{\delta\to 0} \frac{1}{2\pi} \int_{-\infty+i\delta}^{\infty+i\delta} \bar{G}(\mathbf{k}, \omega) \exp[-i\omega(t-t')]. \tag{3.42}$$

The integral in this equation is to be taken along a line just above the real axis, as shown in Fig. 3.15(a). The subscript r stands for 'retarded', and $\mathscr{G}_r$ is

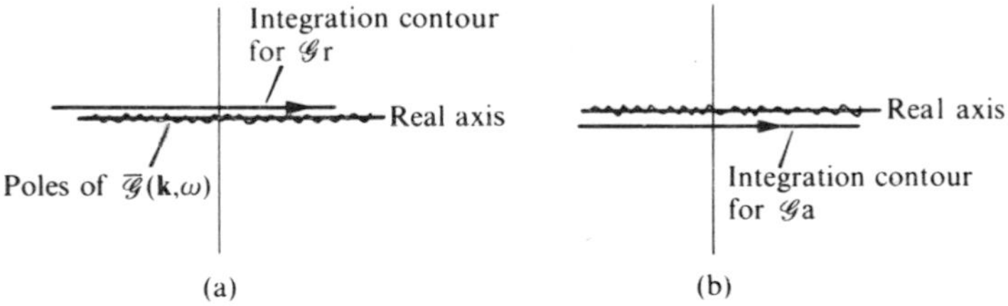

FIG. 3.15.

called the retarded Green function. The reason for this is clear once the integral is performed. For, if $t-t' < 0$, the contour can be closed in the upper half by the semicircle at infinity, where the integrand is zero. But the integral round the whole contour is zero, by Cauchy's theorem, since the integrand is analytic within the contour. Thus

$$\mathscr{G}_r(\mathbf{k}, t-t') = 0, \qquad t < t'. \tag{3.43}$$

For $t > t'$, we may close the contour in the lower half plane by the semicircle at infinity, where again the integrand is zero. However, our contour now contains all the poles in $\bar{G}$ along the real axis. Thus, using Cauchy's theorem,

$$\begin{aligned}
\mathscr{G}_r(\mathbf{k}, t-t') &= -\frac{i}{Q}\sum_{\nu,\nu',N} \exp[-\beta(E_{N\nu}-\mu N)] \times \{|\langle \nu', N+1|a_{\mathbf{k}}^{+}|\nu N\rangle|^2 \times \\
&\quad \times \exp[-i(E_{N+1,\nu'}-E_{N\nu}-\mu)(t-t')] - \varepsilon|\langle \nu N|a_{\mathbf{k}}^{+}|\nu' N-1\rangle|^2 \times \\
&\quad \times \exp[-i(E_{N\nu}-E_{N-1,\nu'}-\mu)(t-t')]\} \\
&= -\frac{i}{Q}\sum_{\nu\nu' N} \exp[-\beta(E_{N\nu}-\mu N)]\{\langle \nu N|\exp[i(E_{N\nu}-\mu N)t] \times \\
&\quad \times a_{\mathbf{k}} \exp[-i(E_{N+1,\nu'}-\mu(N+1)t]|\nu' N+1\rangle \times \\
&\quad \times \langle \nu' N+1|\exp[i(E_{N+1,\nu'}-\mu(N+1))t']a_{\mathbf{k}}^{+}\exp[-i(E_{N\nu}- \\
&\quad -\mu N)t']|\nu N\rangle - \varepsilon\langle \nu N|\exp[i(E_{N\nu}-\mu N)t']a_{\mathbf{k}}^{+}\exp[-i(E_{N-1,\nu'}- \\
&\quad -\mu(N-1))t'|\nu', N-1\rangle\langle \nu', N-1|\exp[i(E_{N-1,\nu'}- \\
&\quad -\mu(N-1))t]a_{\mathbf{k}}\exp[-i(E_{N\nu}-\mu N)t]|\nu N\rangle\} \\
&= -i\langle [a_{\mathbf{k}}(t), a_{\mathbf{k}}^{+}(t')]_{-\varepsilon}\rangle, \qquad (3.43a)
\end{aligned}$$

where

$$A(t) = \exp[i(H-\mu\hat{N})t]A\exp[-i(H-\mu\hat{N})t], \qquad (3.44)$$

so that the operators are now in the normal Heisenberg representation, except that H is replaced by $H-\mu\hat{N}$. Combining eqns (3.43) and (3.43a) we may write:

$$\mathscr{G}_r(\mathbf{k}, t-t') = -i\theta(t-t')\langle [a_{\mathbf{k}}(t), a_{\mathbf{k}}^{+}(t')]_{-\varepsilon}\rangle, \qquad (3.45)$$

where $\theta(t)$ is the usual step function:

$$\begin{aligned}
\theta(t) &= 0 \qquad t < 0 \\
&= 1 \qquad t > 0.
\end{aligned}$$

This Green function is not defined for $t = t'$. It is sometimes called the retarded commutator.

We may obtain an 'advanced' Green function, or advanced commutator, by integrating along a line just below the real axis (Fig. 3.15(b)).
Thus,

$$\begin{aligned}
\mathscr{G}_a(\mathbf{k}, t-t') &= \operatorname*{Lt}_{\delta\to 0}\frac{1}{2\pi}\int_{-\infty-i\delta}^{+\infty-i\delta} d\omega\bar{G}(\mathbf{k},\omega)\exp[-i\omega(t-t')] \\
&= i\theta(t'-t)\langle [a_{\mathbf{k}}(t), a_{\mathbf{k}}^{+}(t')]_{-\varepsilon}\rangle, \qquad (3.46)
\end{aligned}$$

If we define the Fourier transforms of $\mathscr{G}_r$ and $\mathscr{G}_a$ as

$$\mathscr{G}_{\substack{r\\a}}(\mathbf{k}, t-t') = \frac{1}{2\pi}\int_{-\infty}^{\infty} \mathrm{d}E\,\overline{\mathscr{G}}_{\substack{r\\a}}(\mathbf{k}, E)\exp[-\mathrm{i}E(t-t')], \tag{3.47}$$

we have, comparing eqn (3.47) with eqns (3.42) and (3.46),

$$\overline{\mathscr{G}}_{\substack{r\\a}}(\mathbf{k}, E) = \operatorname*{Lt}_{\delta\to 0} \bar{G}(\mathbf{k}, E\pm\mathrm{i}\delta). \tag{3.48}$$

Lastly, for completeness, we define a 'causal' or time-ordered Green function. We shall use this very little in what follows. To obtain it, we need to do something a little more complicated. We take the first term in $\bar{G}$ of eqn (3.41) multiplied by $\exp[-\mathrm{i}\omega(t-t')]$ and integrate the product along a line just above the real axis, and the second term, again multiplied by $\exp[-\mathrm{i}\omega(t-t')]$, but integrate this product along a line just below the real axis. A neater way of achieving the same result is to move the poles in the first term to just below the real axis and those in the second term to just above the real axis, and then to integrate along the real axis:

$$\begin{aligned}
\mathscr{G}_c(\mathbf{k}, t-t') = &\operatorname*{Lt}_{\delta\to 0}\frac{1}{2\pi}\int_{-\infty}^{\infty}\mathrm{d}\omega\frac{1}{Q}\sum_{\nu,\nu',N}\exp[-\beta(E_{N\nu}-\mu N)]\times\\
&\times\left\{\frac{|\langle\nu', N+1|a_{\mathbf{k}}^{+}|\nu N\rangle|^2}{\omega+\mathrm{i}\delta-E_{N+1,\nu'}+E_{N\nu}+\mu}-\right.\\
&\left.-\varepsilon\frac{|\langle\nu N|a_{\mathbf{k}}^{+}|\nu', N-1\rangle|^2}{\omega-\mathrm{i}\delta-E_{N\nu}+E_{N-1,\nu'}+\mu}\right\}\exp[-\mathrm{i}\omega(t-t')]\\
= &-\mathrm{i}\langle a_{\mathbf{k}}(t)a_{\mathbf{k}}^{+}(t')\rangle \qquad t>t'\\
= &-\mathrm{i}\varepsilon\langle a_{\mathbf{k}}^{+}(t')a_{\mathbf{k}}(t)\rangle \qquad t<t'.
\end{aligned} \tag{3.49}$$

This is very similar to the definition of $G_{\mathbf{k}}(\tau)$, but the ordering of the operators is now a t-ordering. We write it

$$\mathscr{G}_c(\mathbf{k}, t-t') = -\mathrm{i}\langle T\{a_{\mathbf{k}}(t)a_{\mathbf{k}}^{+}(t')\}\rangle$$

where T is a time-ordering operator. Notice that if we have obtained an expression for $\bar{G}(\mathbf{k}, \omega)$ by some method, we can use eqns (3.36 and 3.41) to obtain $\mathscr{G}_{\substack{r\\a}}(\mathbf{k}, t-t')$ directly. We cannot yet find $\mathscr{G}_c(\mathbf{k}, t-t')$, however, since we shall not know which part of our function corresponds to the first term in eqn (3.41), and which to the second; thus we shall be unable to perform the separation leading to eqn (3.49). A method for obtaining $\mathscr{G}_c(\mathbf{k}, t-t')$ from $\bar{G}(\mathbf{k}, \omega)$ is given later in the chapter.

3.6. Correlation functions

So far, there seems to be little reason for the introduction of time–temperature-dependent Green functions. However, in the next two sections, we shall show that they offer new ways of obtaining thermodynamic functions from many-body calculations, and that they allow us to calculate quantities of experimental interest other than purely thermodynamic ones. Further, in Chapter 5, we shall show that they offer a new way of calculating $\bar{G}(\mathbf{k}, \omega)$.

We start this programme by generalising our definitions of time–temperature Green functions:

$$\begin{aligned}\mathscr{G}_{\mathrm{r}}^{AB}(t-t') &\equiv \langle\langle A(t); B(t')\rangle\rangle_{\mathrm{r}} \\ &\equiv -\mathrm{i}\theta(t-t')\langle[A(t), B(t')]_{\pm}\rangle,\end{aligned} \tag{3.50}$$

$$\begin{aligned}\mathscr{G}_{\mathrm{a}}^{AB}(t-t') &\equiv \langle\langle A(t); B(t')\rangle\rangle_{\mathrm{a}} \\ &\equiv \mathrm{i}\theta(t'-t)\langle[A(t), B(t')]_{\pm}\rangle,\end{aligned} \tag{3.51}$$

$$\begin{aligned}\mathscr{G}_{\mathrm{c}}^{AB}(t-t') &\equiv \langle\langle A(t); B(t')\rangle\rangle_{\mathrm{c}} \\ &= -\mathrm{i}\langle A(t)B(t')\rangle \qquad t > t' \\ &= -\mathrm{i}\varepsilon\langle B(t')A(t)\rangle \qquad t < t',\end{aligned} \tag{3.52}$$

where A and B are arbitrary operators in the Heisenberg representation, and

$$[A, B]_{\pm} = AB - \varepsilon BA \tag{3.53}$$

$\varepsilon = \pm 1$, according to our *choice*. If, for example, $A = a_{\mathbf{k}}$ and $B = a_{\mathbf{k}}^{+}$, where a^{+} and a are creation and annihilation operators for fermions, the choice $\varepsilon = -1$ would give us the Green functions of eqns (3.45, 46, and 49). Following an analysis very similar to the one in the previous section, we see that these functions may be obtained by integrating along suitable contours the product $\bar{G}^{AB}(\omega)\exp[-\mathrm{i}\omega(t-t')]/2\pi$, where

$$\begin{aligned}\bar{G}^{AB}(\omega) = \frac{1}{Q}\sum_{\substack{\nu\nu'\\NN'}} \exp[-\beta(E_{N\nu}-\mu N)] \times \\ \times \left\{\frac{\langle \nu N|A|\nu' N'\rangle\langle \nu' N'|B|\nu N\rangle}{\omega - E_{N'\nu'} + E_{N\nu} + \mu(N'-N)} - \right. \\ \left. -\varepsilon\frac{\langle \nu N|B|\nu' N'\rangle\langle \nu' N'|A|\nu N\rangle}{\omega - E_{N\nu} + E_{N'\nu'} - \mu(N'-N)}\right\}.\end{aligned} \tag{3.54}$$

The generalized Green functions may be calculated at the usual infinite set of points along the imaginary axis according to the rules in section 3.3. We may then extend the function to the whole of the complex plane, to obtain $\bar{G}^{AB}(\omega)$.

We are now going to trace the connections between such Green functions and correlation functions. We should emphasize that the mathematical details of these connections are unimportant. The reader should be aware that such results exist, and should know how to use them, but, especially on first reading, is probably best advised to give the algebra leading to them rather scant attention.

It is convenient to define functions $\rho^{AB}(E)$ and $J^{BA}(E)$ through the equations

$$\mathrm{i}\rho^{AB}(E) = \mathrm{i}[\exp(\beta E)-\varepsilon]J^{BA}(E) = \operatorname*{Lt}_{\gamma\to 0}\,[\bar{G}^{AB}(E-\mathrm{i}\gamma)-\bar{G}^{AB}(E+\mathrm{i}\gamma)]. \qquad (3.55)$$

(This equation does not define $J^{AB}(0)$ in the case $\varepsilon = 1$; this point is discussed below.) In this equation, E is real, so that J is proportional to the change in $\bar{G}$ across the real axis: this is different from zero because of the non-analyticity of $\bar{G}$ on the real axis. By using eqn (3.54) and

$$\delta(x) = \frac{1}{2\pi \mathrm{i}}\operatorname*{Lt}_{\gamma\to 0}\,[(x-\mathrm{i}\gamma)^{-1}-(x+\mathrm{i}\gamma)^{-1}] \quad \text{(see Appendix E)}$$

we obtain

$$\begin{aligned}&[\exp(\beta E)-\varepsilon]J^{BA}(E)\\ &\quad=\frac{2\pi}{Q}\sum_{\substack{\nu\nu'\\ NN'}}\exp[-\beta(E_{N\nu}-\mu N)]\times\\ &\quad\times\{\langle\nu N|A|\nu'N'\rangle\langle\nu'N'|B|\nu N\rangle\delta(E-E_{N'\nu'}+E_{N\nu}+\mu(N'-N))-\\ &\quad-\varepsilon\langle\nu N|B|\nu'N'\rangle\langle\nu'N'|A|\nu N\rangle\delta(E-E_{N\nu}+E_{N'\nu'}-\mu(N'-N)). \quad (3.56)\end{aligned}$$

Interchanging ν and ν', and N and N', in the first term on the right-hand side, we obtain

$$\begin{aligned}&[\exp(\beta E)-\varepsilon]J^{BA}(E)\\ &\quad=\frac{2\pi}{Q}\sum_{\substack{\nu\nu'\\ NN'}}\langle\nu'N'|A|\nu N\rangle\langle\nu N|B|\nu'N'\rangle\delta(E-E_{N\nu}+E_{N'\nu'}-\mu(N'-N))\times\\ &\quad\times\{\exp[-\beta(E_{N'\nu'}-\mu N')]-\varepsilon\exp[-\beta(E_{N\nu}-\mu N)]\}. \qquad (3.57)\end{aligned}$$

But because of the δ-function, we may write the last factor of this equation as

$$\exp[-\beta(E_{N\nu}-\mu N)]\{\exp(-\beta E)-\varepsilon\},$$

and hence

$$J^{BA}(E) = \frac{2\pi}{Q} \sum_{\substack{\nu\nu' \\ NN'}} \exp[-\beta(E_{N\nu} - \mu N)] \langle \nu N|B|\nu' N'\rangle \langle \nu' N'|A|\nu N\rangle \times$$
$$\times\, \delta(E - E_{N\nu} + E_{N'\nu'} - \mu(N'-N)) \tag{3.58}$$

for all E in the case $\varepsilon = -1$, and for all $E \neq 0$ when $\varepsilon = 1$. Since eqn (3.55) does not define $J^{AB}(0)$ in the latter case, we are free to choose eqn (3.58) to be true for all values of E. By following a very similar procedure to that leading from eqn (3.41) to eqn (3.43), it is now easy to show that

$$\frac{1}{2\pi} \int_{-\infty}^{\infty} \mathrm{d}E\, J^{BA}(E) \exp[-\mathrm{i}E(t-t')] = \langle B(t')A(t)\rangle. \tag{3.59}$$

Thus $J^{BA}(-E)$ is the Fourier transform of the correlation function $\langle B(t')A(t)\rangle$.

Further, we may rewrite eqn (3.58), using the δ-function, in the form

$$J^{BA}(E) = \frac{2\pi}{Q} \exp(-\beta E) \sum_{\substack{\nu\nu' \\ NN'}} \exp[-\beta(E_{N'\nu'} - \mu N')] \times$$
$$\times \langle \nu' N'|A|\nu N\rangle \langle \nu N|B|\nu' N'\rangle \delta(E - E_{N\nu} + E_{N'\nu'} + \mu(N-N')). \tag{3.60}$$

Thus

$$\frac{1}{2\pi} \int_{-\infty}^{\infty} \mathrm{d}E \exp(\beta E) J^{BA}(E) \exp[-\mathrm{i}E(t-t')] = \langle A(t)B(t')\rangle \tag{3.61}$$

and we see that $\exp(\beta E)J^{BA}(E)$ is the Fourier transform of the correlation function $\langle A(t)B(t')\rangle$.

Such correlation functions are often required in many-body calculations. Later in this chapter, we shall use them to establish a direct connection between our Green functions and thermodynamics. In Chapter 7, we shall need them in the theory of linear response and of relaxation processes; the correlation function $\langle \rho(x', t')\rho(x, t)\rangle$ where $\rho(x, t)$ is the particle-density operator, is needed in the theory of neutron, and of X-ray scattering from a many-body system (see, for example, Kittel, 1963, Chapter 19), and many other examples could be given. We now have a way of calculating such correlation functions from the corresponding Green functions. To make full use of this, we shall need to use the methods of calculating general Green functions described at the end of section 3.3, or methods to be described in Chapter 5.

There is one difficulty associated with this procedure, and this is connected with the point, mentioned above, that eqn (3.55) does not determine

$J^{BA}(0)$ when $\varepsilon = 1$. If $J^{BA}(0)$ is finite, it does not contribute to the integral in eqn (3.59) or (3.61). There is the possibility, however, that $J^{BA}(E)$ contains a term proportional to $\delta(E)$, $2\pi\alpha\,\delta(E)$ say. In this case, $\langle B(t')A(t)\rangle$ will contain a constant term, α. As $t'-t \to \infty$ for most realistic systems, the correlations between $B(t')$ and $A(t)$ will have disappeared (see, for example, the discussion on p. 104) and

$$\langle B(t')A(t)\rangle \to \langle B(t')\rangle\langle A(t)\rangle.$$

In many cases one can see that $\langle A\rangle$ or $\langle B\rangle$ are zero since

$$\langle A\rangle = \frac{1}{Q}\sum_{\nu N}\exp[-\beta(E_{N\nu}-\mu N)]\langle \nu N|A|\nu N\rangle$$

so that A must conserve particle number if $\langle A\rangle$ is to differ from zero. In other cases, one may know $\langle A\rangle$ and $\langle B\rangle$, for example when A is the number density operator. Methods for dealing with cases in which the long-time behaviour of $\langle B(t')A(t)\rangle$ are neither negligible nor known are given by Stevens and Toombs (1965) and Bloomfield and Nafari (1972) and references therein.

Since we know $\langle A(t)B(t')\rangle$ and $\langle B(t')A(t)\rangle$ for all t and t', it is possible to construct all the Green functions (3.50–52) from J^{BA}, and hence from G^{AB}, using eqn (3.55). This is particularly convenient for the causal Green function, for which

$$\mathscr{G}_c^{AB}(t-t') = \frac{1}{2\pi}\int_{-\infty}^{\infty} dE\,\overline{\mathscr{G}}_c^{AB}(E)\exp[-i(E(t-t')], \tag{3.62}$$

where

$$\overline{\mathscr{G}}_c^{AB}(E) = \operatorname*{Lt}_{\gamma\to 0}\int_{-\infty}^{\infty} dE'\left[\frac{\varepsilon J^{BA}(E')}{E'-E+i\gamma} - \frac{J^{BA}(E')\exp(\beta E')}{E'-E-i\gamma}\right]. \tag{3.63}$$

For $t > t'$, we may close the contour in eqn (3.62) in the lower half-plane. Thus, inserting the expression (3.63) for $\mathscr{G}^{AB}(E)$, we obtain a contribution only from the second term, and

$$\mathscr{G}_c^{AB}(t-t') = -i\langle A(t)B(t')\rangle, \qquad t > t'$$

as we expect. Similarly, for $t < t'$, only the first term of eqn (3.63) gives a contribution to the integral in eqn (3.62) and

$$\mathscr{G}_c^{AB}(t-t') = -i\varepsilon\langle B(t')A(t)\rangle.$$

We may similarly show that

$$\overline{\mathscr{G}}_r^{AB}(E) = \frac{1}{2\pi}\operatorname*{Lt}_{\gamma\to 0}\int_{-\infty}^{\infty} dE'\frac{[\exp(\beta E')-\varepsilon]J^{BA}(E')}{E-E'+i\gamma}, \tag{3.64}$$

and

$$\bar{\mathscr{G}}_{\mathrm{a}}^{AB}(E) = \frac{1}{2\pi} \underset{\gamma\to 0}{\mathrm{Lt}} \int_{-\infty}^{\infty} \mathrm{d}E' \frac{[\exp(\beta E') - \varepsilon] J^{BA}(E')}{E - E' - \mathrm{i}\gamma}. \tag{3.65}$$

These equations may also be obtained by noticing that

$$\bar{G}^{AB}(\omega) = \frac{1}{2\pi} \int_{-\infty}^{\infty} \mathrm{d}E' \frac{[\exp(\beta E') - \varepsilon] J^{BA}(E')}{\omega - E'} \tag{3.66}$$

$$= \frac{1}{2\pi} \int_{-\infty}^{\infty} \frac{\rho^{AB}(E')}{\omega - E'} \mathrm{d}E', \tag{3.66a}$$

which may be proved by inserting eqn (3.56) into the integral; using the δ-function to do the integral, one obtains eqn (3.54). Eqns (3.64) and (3.65) then follow from the generalization of eqn (3.48):

$$\bar{\mathscr{G}}_{\substack{\mathrm{r}\\ \mathrm{a}}}^{AB}(E) = \underset{\gamma\to 0}{\mathrm{Lt}}\, \bar{G}^{AB}(E \pm \mathrm{i}\gamma). \tag{3.67}$$

The expression for $\bar{G}$ in eqn (3.66a) is called the spectral representation of $\bar{G}$.

Lastly, consider the integral,

$$\frac{1}{2\pi \mathrm{i}} \int_{\mathrm{C}} \frac{\bar{G}^{AB}(\omega')}{\omega' - \omega} \mathrm{d}\omega', \qquad \mathrm{Im}\, \omega > \gamma,$$

taken round the contour consisting of the infinite semi-circle in the upper half-plane, and a line just above the real axis. Since $\bar{G}^{AB}(\omega') \to 0$ at least as $1/\omega'$ as $\omega' \to \infty$ (eqn (3.54)), the integral round the semi-circle vanishes. Further, $\bar{G}^{AB}(\omega')$ is analytic within the contour. Thus, using Cauchy's theorem and eqn (3.67):

$$\bar{G}^{AB}(\omega) = \underset{\gamma\to 0}{\mathrm{Lt}} \frac{1}{2\pi \mathrm{i}} \int_{-\infty+\mathrm{i}\gamma}^{\infty+\mathrm{i}\gamma} \frac{\bar{G}^{AB}(\omega')}{\omega' - \omega} \mathrm{d}\omega'$$

$$= \underset{\gamma\to 0}{\mathrm{Lt}} \frac{1}{2\pi \mathrm{i}} \int_{-\infty}^{\infty} \frac{\bar{\mathscr{G}}_{\mathrm{r}}^{AB}(E')\, \mathrm{d}E'}{E' - \omega + \mathrm{i}\gamma}, \tag{3.68}$$

since

$$\underset{\gamma\to 0}{\mathrm{Lt}}\, \bar{G}^{AB}(E' + \mathrm{i}\gamma) = \bar{\mathscr{G}}_{\mathrm{r}}^{AB}(E').$$

Putting $\omega = E+\mathrm{i}\gamma'$, with $\gamma' > \gamma$, and using

$$\mathop{\mathrm{Lt}}_{\gamma\to 0}\int_{-\infty}^{\infty} \mathrm{d}x\frac{f(x)}{x\pm\mathrm{i}\gamma} = P\int_{-\infty}^{\infty} \mathrm{d}x\frac{f(x)}{x}\mp\mathrm{i}\pi f(0),$$

where P denotes the principal part of the integral (see Appendix E), we have

$$\overline{\mathscr{G}}_{\mathrm{r}}^{AB}(E) = -\frac{\mathrm{i}}{\pi}P\int_{-\infty}^{\infty}\frac{\overline{\mathscr{G}}_{\mathrm{r}}^{AB}(E')\,\mathrm{d}E'}{E'-E}. \tag{3.69}$$

Taking real and imaginary parts of this equation:

$$\mathrm{Re}\{\overline{\mathscr{G}}_{\mathrm{r}}^{AB}(E)\} = \frac{P}{\pi}\int_{-\infty}^{\infty}\mathrm{d}E'\frac{\mathrm{Im}\{\overline{\mathscr{G}}_{\mathrm{r}}^{AB}(E')\}}{E'-E} \tag{3.70}$$

$$\mathrm{Im}\{\overline{\mathscr{G}}_{\mathrm{r}}^{AB}(E)\} = -\frac{P}{\pi}\int_{-\infty}^{\infty}\mathrm{d}E'\frac{\mathrm{Re}\{\overline{\mathscr{G}}_{\mathrm{r}}^{AB}(E')\}}{E'-E}. \tag{3.71}$$

Similarly:

$$\overline{G}^{AB}(\omega) = -\frac{1}{2\pi\mathrm{i}}\mathop{\mathrm{Lt}}_{\gamma\to 0}\int_{-\infty}^{\infty}\frac{\overline{\mathscr{G}}_{\mathrm{a}}^{AB}(E')\,\mathrm{d}E'}{E'-\omega-\mathrm{i}\gamma}\qquad \mathrm{Im}\,\omega < -\gamma \tag{3.72}$$

$$\overline{\mathscr{G}}_{\mathrm{a}}^{AB}(E) = \frac{\mathrm{i}}{\pi}P\int_{-\infty}^{\infty}\frac{\overline{\mathscr{G}}_{\mathrm{a}}^{AB}(E')\,\mathrm{d}E'}{E'-E}, \tag{3.73}$$

$$\mathrm{Re}\{\overline{\mathscr{G}}_{\mathrm{a}}(E)\} = -\frac{1}{\pi}P\int_{-\infty}^{\infty}\frac{\mathrm{Im}\{\overline{\mathscr{G}}_{\mathrm{a}}^{AB}(E')\}\,\mathrm{d}E'}{E'-E}, \tag{3.74}$$

$$\mathrm{Im}\{\overline{\mathscr{G}}_{\mathrm{a}}(E)\} = \frac{1}{\pi}P\int_{-\infty}^{\infty}\frac{\mathrm{Re}\{\overline{\mathscr{G}}_{\mathrm{a}}^{AB}(E')\}\,\mathrm{d}E'}{E'-E}. \tag{3.75}$$

Exercises

3.6.1. Find $G(k, \omega)$ with Σ correct to first order by using the results of exercise 3.1.1. Show that the spectral density ρ for this Green function (eqn 3.55) is a δ-function displaced from its original position for the non-interacting case.

3.6.2. Repeat this calculation using Σ correct to second order (exercise 3.1.2).

Notice that near the real axis

$$\Sigma^{(2)}(\mathbf{k}, E \pm \mathrm{i}\gamma) = -\frac{\varepsilon}{2}\alpha^2 \sum_{\mathbf{k}_3\mathbf{q}} \{\langle n\rangle_0 \text{ factors}\} \frac{1}{E \pm \mathrm{i}\gamma - (\varepsilon_{\mathbf{k}+\mathbf{q}} + \varepsilon_{\mathbf{k}_3-\mathbf{q}} - \varepsilon_{\mathbf{k}_3} - \mu)}$$

$$= -\frac{\varepsilon\alpha^2}{2} \sum_{\mathbf{k}_3\mathbf{q}} \{\langle n\rangle_0 \text{ factors}\} \left[\frac{P}{E - (\varepsilon_{\mathbf{k}+\mathbf{q}} + \varepsilon_{\mathbf{k}_3-\mathbf{q}} - \varepsilon_{\mathbf{k}_3} - \mu)} \mp \right.$$

$$\left. \mp \mathrm{i}\pi\, \delta(E - \varepsilon_{\mathbf{k}+\mathbf{q}} - \varepsilon_{\mathbf{k}_3-\mathbf{q}} + \varepsilon_{\mathbf{k}_3} + \mu) \right]$$

(see Appendix E)

$$\equiv K_{\mathbf{k}}(E) \mp \mathrm{i} L_{\mathbf{k}}(E).$$

Thus $\rho_{\mathbf{k}}(E)$ will have a peak when

$$E - \varepsilon_{\mathbf{k}} + \mu - K_{\mathbf{k}}(E) = 0.$$

Let this be at $E = E_{\mathbf{k}}$. Then, provided $L_{\mathbf{k}}(E)$ is not quickly varying, the width of the peak will be $\approx L_{\mathbf{k}}(E_{\mathbf{k}})$. Since we are still dealing with a weakly interacting gas $K_{\mathbf{k}}(E)$ will be small, so that $E_{\mathbf{k}} \approx \varepsilon_{\mathbf{k}}$ and the width of the peak will be approximately equal to $L_{\mathbf{k}}(\varepsilon_{\mathbf{k}})$.

3.6.3. Discuss the forms of the time-dependence of the correlation function $\langle a_{\mathbf{k}}(t)a_{\mathbf{k}}^{+}(t')\rangle$ obtained from the approximate Green functions of 3.6.1 and 3.6.2. (See section 3.10.)

3.6.4. Derive an expression for the variation with energy and wave-number change of the scattering cross-section for neutrons or X-rays scattered from a many-body-system of non-interacting particles. (Refer to Kittel (1963), Chapter 19.) How could interactions between particles be taken into account?

3.6.5. A condensing system is very sensitive to the application of external fields: the application of the earth's gravitational field to a mixture of liquid and saturated vapour causes the liquid to fall to the bottom of the containing vessel. Thus $\rho(\mathbf{r}) = \sum_{\mathbf{k},\mathbf{q}} \langle a_{\mathbf{k}}^{+} a_{\mathbf{k}+\mathbf{q}}\rangle \exp(\mathrm{i}\mathbf{q}\,.\,\mathbf{r})$ would change dramatically if we applied a small external field. Suppose the external field may be written $\lambda \sum_{\mathbf{r}_j} \exp[\mathrm{i}\mathbf{q}_0\,.\,\mathbf{r}_j]$ + complex conjugate, which in second-quantized notation becomes

$$\lambda \sum_{\mathbf{k}} [a_{\mathbf{k}}^{+} a_{\mathbf{k}+\mathbf{q}_0} + a_{\mathbf{k}}^{+} a_{\mathbf{k}-\mathbf{q}_0}]$$

The single-particle Green function will no longer conserve momentum.

If we consider first-order effects in λ, the Green function

$$\bar{G}_\lambda(\mathbf{k}, \mathbf{q}_0, i\omega_n) = \langle\langle a_\mathbf{k}^+ ; a_{\mathbf{k}+\mathbf{q}_0} : i\omega_n \rangle\rangle$$

will be non-zero.

In the limit of very small q_0, corresponding to applied fields of very long-wavelength, show that we may write, to first order:

$$\begin{aligned}\bar{G}_\lambda(\mathbf{k}, \mathbf{q}_0, i\omega_n) &= \lambda \sum_{\mathbf{k}', i\omega_{n'}} \frac{\partial \bar{G}(\mathbf{k}, i\omega_n)}{\partial \bar{g}(\mathbf{k}', i\omega_{n'})} \bar{g}^2(\mathbf{k}', i\omega_{n'}) \\ &= -\lambda \frac{\partial \bar{G}(\mathbf{k}, i\omega_n)}{\partial \mu}.\end{aligned}$$

Thus show that, if the application of a very small external field is to have a finite effect on the density, $\partial \bar{N}/\partial \mu$ must be very large. Note that for a condensing system $(\partial \bar{N}/\partial \mu)/V$ is infinite in the infinite volume limit.

3.7. Further connections between Green functions and thermodynamic quantities

We are now in a position to derive another important direct relation between Green functions and thermodynamics. So far, we have eqn (3.9), the use of which involves the summation of the contribution of all skeleton diagrams even after we have determined the fully interacting Green function, $\bar{G}(\mathbf{k}, i\omega_n)$, and eqn (3.8), which suffers from the disadvantage mentioned in section 3.2.

Another example of a formula of this type is

$$\frac{\partial \Omega}{\partial \mu} = -\bar{N} = \varepsilon \beta^{-1} \underset{\substack{\tau \to 0 \\ \tau < 0}}{\mathrm{Lt}} \sum_l \sum_\mathbf{k} \exp(-i\omega_l \tau) \bar{G}(\mathbf{k}, i\omega_l), \tag{3.76}$$

which theoretically could be used to determine Ω by integrating over μ. Further, from the definition of Ω, it is easy to prove that

$$\left(\frac{\partial \Omega}{\partial m}\right)_{T,V,\mu} = \left\langle \frac{\partial H}{\partial m} \right\rangle \tag{3.77}$$

$$= -\frac{\varepsilon}{\beta m} \underset{\substack{\tau \to 0 \\ \tau < 0}}{\mathrm{Lt}} \sum_l \sum_\mathbf{k} \exp(-i\omega_l \tau) \bar{G}(\mathbf{k}, i\omega_l) \times \frac{k^2}{2m}. \tag{3.78}$$

Again, this could be used to find Ω, by integrating over m from a value for which $\Omega(m)$ is known. None of these formulae, however, has much practical use.

A more useful relation may be derived by considering the differential equation governing the time-development of the correlation function

$\langle a_{\mathbf{k}}^{+}(0)a_{\mathbf{k}}(t)\rangle$. Since, in the case of a Hamiltonian of the form in eqn (2.9),

$$\mathrm{i}\,\frac{\mathrm{d}a_{\mathbf{k}}(t)}{\mathrm{d}t} = [a_{\mathbf{k}}, H-\mu\hat{N}]$$

$$= \left(\frac{k^2}{2m}-\mu\right)a_{\mathbf{k}} + \sum_{\mathbf{k}_1\mathbf{q}} \bar{v}(q)a_{\mathbf{k}_1}^{+}a_{\mathbf{k}_1+\mathbf{q}}a_{\mathbf{k}-\mathbf{q}}, \tag{3.79}$$

we have

$$\operatorname*{Lt}_{t\to 0}\,\mathrm{i}\frac{\mathrm{d}}{\mathrm{d}t}\langle a_{\mathbf{k}}^{+}(0)a_{\mathbf{k}}(t)\rangle = \left(\frac{k^2}{2m}-\mu\right)\langle a_{\mathbf{k}}^{+}a_{\mathbf{k}}\rangle + \sum_{\mathbf{k}_1\mathbf{q}} \bar{v}(q)\langle a_{\mathbf{k}}^{+}a_{\mathbf{k}_1}^{+}a_{\mathbf{k}_1+\mathbf{q}}a_{\mathbf{k}-\mathbf{q}}\rangle,$$

and hence

$$2\langle H\rangle = 2\sum_{\mathbf{k}}\frac{k^2}{2m}\langle a_{\mathbf{k}}^{+}a_{\mathbf{k}}\rangle + \sum_{\mathbf{k}\mathbf{k}_1\mathbf{q}} \bar{v}(q)\langle a_{\mathbf{k}}^{+}a_{\mathbf{k}_1}^{+}a_{\mathbf{k}_1+\mathbf{q}}a_{\mathbf{k}-\mathbf{q}}\rangle$$

$$= \operatorname*{Lt}_{t\to 0}\sum_{\mathbf{k}}\left[\mathrm{i}\frac{\mathrm{d}}{\mathrm{d}t}\langle a_{\mathbf{k}}^{+}(0)a_{\mathbf{k}}(t)\rangle + \left(\frac{k^2}{2m}+\mu\right)\langle a_{\mathbf{k}}^{+}(0)a_{\mathbf{k}}(t)\rangle\right] \tag{3.80}$$

Thus, if $J_{\mathbf{k}}(E)$ is the Fourier transform of the correlation function $\langle a_{\mathbf{k}}^{+}(0)a_{\mathbf{k}}(t)\rangle$,

$$\langle H\rangle = \tfrac{1}{2}\operatorname*{Lt}_{t\to 0}\sum_{\mathbf{k}}\int_{-\infty}^{\infty}\mathrm{d}E\,\mathrm{e}^{-\mathrm{i}Et}\left(E+\frac{k^2}{2m}+\mu\right)J_{\mathbf{k}}(E) \tag{3.81}$$

$$= \frac{1}{2\mathrm{i}}\operatorname*{Lt}_{t\to 0}\sum_{\mathbf{k}}\int_{-\infty}^{\infty}\mathrm{d}E\,\mathrm{e}^{-\mathrm{i}Et}\left(E+\frac{k^2}{2m}+\mu\right)[\exp(\beta E)-\varepsilon]^{-1}\times$$

$$\times \operatorname*{Lt}_{\gamma\to 0}[\bar{G}(\mathbf{k}, E-\mathrm{i}\gamma)-\bar{G}(\mathbf{k}, E+\mathrm{i}\gamma)]. \tag{3.82}$$

This gives us a direct connection between the expectation value of the energy and the full Green functions. From the energy, many thermodynamic quantities may easily be calculated.

Exercises

3.7.1. Find the energy of a gas interacting through pairwise central forces to the first order by using the results of exercise 3.1.1 and eqn (3.82).

3.7.2. Repeat the arguments leading to eqn (3.82) for the electron–phonon Hamiltonian of exercise 1.2.4.

3.8. Exact results satisfied by Green functions and correlation functions

In a field in which approximations have almost always to be made in any realistic calculation, an exact relation can be extremely useful, for example in checking how far astray our approximations may have led us. In this section we shall very briefly discuss several types of exact relation. The first type follows from the form of the single-particle Green function, eqn (3.41). We see at once, for example, that

$$\begin{aligned}\operatorname*{Lt}_{\omega\to\infty} \bar{G}(\mathbf{k},\omega) \\ &= \frac{1}{Q}\sum_{\nu\nu'N}\exp[-\beta(E_{N\nu}-\mu N)]\left\{\frac{\langle \nu N|a_{\mathbf{k}}|\nu' N+1\rangle\langle \nu' N+1|a^{+}_{\mathbf{k}}|\nu N\rangle}{\omega}-\right.\\ &\left.\quad -\varepsilon\frac{\langle \nu N|a^{+}_{\mathbf{k}}|\nu' N+1\rangle\langle \nu' N+1|a_{\mathbf{k}}|\nu N\rangle}{\omega}\right\} \text{ from eqn (3.41)}\\ &= \omega^{-1}\langle a_{\mathbf{k}}a^{+}_{\mathbf{k}}-\varepsilon a^{+}_{\mathbf{k}}a_{\mathbf{k}}\rangle\\ &= \omega^{-1}.\end{aligned} \tag{3.83}$$

This implies that:

$$\operatorname*{Lt}_{\omega\to\infty} \Sigma(\mathbf{k},\omega) = \text{const.} \tag{3.84}$$

A more general form of eqn (3.83) is

$$\operatorname*{Lt}_{\omega\to\infty} \bar{G}^{AB}(\omega) = \omega^{-1}\langle AB-\varepsilon BA\rangle. \tag{3.85}$$

Further examples of this type of relation are provided by extending to all parts of the complex plane the definition of the proper self-energy part:

$$\bar{G}(\mathbf{k},\omega) = [\omega-(\varepsilon_{\mathbf{k}}-\mu)-\Sigma(\mathbf{k},\omega)]^{-1} \tag{3.86}$$

so that

$$\Sigma(\mathbf{k},\omega) = -[\bar{G}(\mathbf{k},\omega)]^{-1}+\omega-(\varepsilon_{\mathbf{k}}-\mu). \tag{3.87}$$

Since $\bar{G}(\mathbf{k},\omega)$ is a real function of its argument,† so is $\Sigma(\mathbf{k},\omega)$. Thus if

$$\operatorname*{Lt}_{\gamma\to 0} \Sigma(\mathbf{k},E+\mathrm{i}\gamma) = K_{\mathbf{k}}(E)-\mathrm{i}L_{\mathbf{k}}(E), \quad E \text{ real}, \tag{3.88}$$

then:

$$\operatorname*{Lt}_{\gamma\to 0} \Sigma(\mathbf{k},E-\mathrm{i}\gamma) = K_{\mathbf{k}}(E)+\mathrm{i}L_{\mathbf{k}}(E), \quad E \text{ real}. \tag{3.89}$$

Further, for fermions, from eqn (3.66),

$$\bar{G}(\mathbf{k},\omega) = \frac{1}{2\pi}\int_{-\infty}^{\infty} \mathrm{d}E\frac{[\exp(\beta E)+1]}{\omega-E}\, J^{a^{+}_{\mathbf{k}}a_{\mathbf{k}}}(E)$$

† i.e. $[\bar{G}(\mathbf{k},\omega)]^{*} = \bar{G}(\mathbf{k},\omega^{*})$.

Since, from eqn (3.60), $J^{a_k^+ a_k}(E)[\exp(\beta E)+1] \geqslant 0$, and

$$\operatorname{Im}\{\bar{G}(k, x+iy)\} = -\frac{y}{2\pi}\int_{-\infty}^{\infty} J^{a_k^+ a_k}\frac{\exp(\beta E)+1}{(x-E)^2+y^2}\,dE, \tag{3.90}$$

we have:

$$\operatorname{Im}\{\bar{G}(\mathbf{k}, x+iy)\} \begin{matrix} >0 & \text{if } y<0 \\ <0 & \text{if } y>0. \end{matrix} \tag{3.91}$$

Thus

$$L_{\mathbf{k}}(x) \geqslant 0. \tag{3.92}$$

We shall show in an example below that

$$L_{\mathbf{k}}(x) = C_{\mathbf{k}}x^2 \quad \text{for } x \to 0,\ T \to 0. \tag{3.93}$$

This result has interesting consequences for the shape of the Fermi distribution function and for the low-temperature specific heat (see next section). Results similar to eqns (3.90) and (3.91) are given by Luttinger (1961) and by Parry and Turner (1963).

As an example of the second type of exact relationship, a sum rule, we shall consider the correlation function $S(\mathbf{k}, \omega)$:

$$S(\mathbf{k}, \omega) = \frac{1}{Q}\sum_{\nu,\nu',N} \exp[-\beta(E_{N\nu}-\mu N)]|\langle \nu N|\rho_{\mathbf{k}}|\nu' N\rangle|^2\delta(\omega - E_{N\nu'}+E_{N\nu}) \tag{3.94}$$

which is the Fourier transform with respect to t of the correlation function mentioned in previous sections,

$$\langle \rho_{\mathbf{k}}(t)\rho_{-\mathbf{k}}(0)\rangle,$$

where, as usual,

$$\rho_{\mathbf{k}} = \sum_{\mathbf{q}} a_{\mathbf{q}}^+ a_{\mathbf{k}+\mathbf{q}} \tag{3.95}$$

so that

$$\rho_{\mathbf{k}}^+ = \rho_{-\mathbf{k}}.$$

From eqn (3.94) we have

$$\int_{-\infty}^{\infty} \omega S(\mathbf{k}, \omega)\,d\omega = \frac{1}{Q}\sum_{\nu\nu' N} \exp[-\beta(E_{N\nu}-\mu N)]|\langle \nu N|\rho_{\mathbf{k}}|\nu' N\rangle|^2(E_{N\nu'}-E_{N\nu}). \tag{3.96}$$

The right-hand side of this equation may be rewritten as

$$\tfrac{1}{2}\langle[\rho_{\mathbf{k}}, [\rho_{-\mathbf{k}}, H]]\rangle,$$

which, by direct evaluation of the commutator, is seen to give

$$\int_{-\infty}^{\infty} \omega S(\mathbf{k}, \omega)\, \mathrm{d}\omega = \bar{N}k^2/2m. \tag{3.97}$$

(See exercises (7.2.1) and (7.3.1) for a further application of this sum rule.)

This procedure may be generalized by a trick introduced by Puff (1965). From eqn (3.64)

$$\begin{aligned}\bar{G}^{AB}(\omega) &= \frac{1}{2\pi}\int_{-\infty}^{\infty} \mathrm{d}\omega' \frac{\rho^{AB}(\omega')}{\omega-\omega'} \\ &= \frac{1}{2\pi}\int_{-\infty}^{\infty} \rho^{AB}(\omega')\frac{1}{\omega}\left[1+\frac{\omega'}{\omega}+\frac{\omega'^2}{\omega^2}+\cdots\right]\end{aligned} \tag{3.98}$$

assuming suitable convergence properties for the series. But from the definition of $\bar{G}^{AB}(\omega)$, eqn (3.54),

$$\begin{aligned}\bar{G}^{AB}(\omega) = \frac{1}{Q}\sum_{\substack{\nu\nu' \\ NN'}} \exp[-\beta(E_{N\nu}-\mu N)]\Big\{&\langle \nu N|A|\nu' N'\rangle\langle \nu' N'|B|\nu N\rangle\frac{1}{\omega}\times \\ &\times\left[1+\frac{E_{N'\nu'}-E_{N\nu}-\mu(N'-N)}{\omega}+(\cdots)^2+\cdots\right]- \\ &-\varepsilon\langle \nu N|B|\nu' N'\rangle\langle \nu' N'|A|\nu N\rangle\frac{1}{\omega}\times \\ &\times\left[1+\frac{E_{N\nu}-E_{N'\nu'}+\mu(N'-N)}{\omega}+(\cdots)^2+\cdots\right]\Big\} \\ = \frac{1}{\omega}\langle AB-\varepsilon BA\rangle+\frac{1}{\omega^2}\langle [A, (H-\mu N)]B&-\varepsilon B[A, H-\mu N]\rangle+\cdots\end{aligned} \tag{3.99}$$

again, provided the series exists. Comparing the two series:

$$\frac{1}{2\pi}\int_{-\infty}^{\infty} \rho^{AB}(\omega')\,\mathrm{d}\omega' = \langle AB-\varepsilon BA\rangle \tag{3.100}$$

$$\frac{1}{2\pi}\int_{-\infty}^{\infty} \omega'\rho^{AB}(\omega')\,\mathrm{d}\omega' = \langle [[A, (H-\mu N)], B]_{-\varepsilon}\rangle \tag{3.101}$$

etc. By no means all such sum rules are useful. Puff has discussed the application of such generalized sum rules to a Bose system. Similar rules have been used, for example, by Nozières and Pines (1958) in discussing the electron gas, and by Miller, Pines and Nozières (1962), Brenig and Parry (1963) in discussing liquid helium.

The third type of exact result follows from an examination of each term in a perturbation expansion. We refer in an example below to Luttinger's result (1961) on the imaginary part of the proper self-energy part for fermions (eqn (3.93). The physical consequences of this result concern (a) the distribution function for the interacting system, mentioned at the end of this section, and (b) the low-temperature specific heat which we work out in an example at the end of section 3.8. A further example is provided in exercise 4.5.1, this time for the Bose system. Eqns (3.11) and (3.8) may also be considered to belong to this category; several similar results may be obtained by using the tricks described when obtaining these equations; differentiation with respect to the strength of the interaction essentially counts the number of vertices in a diagram and differentiation with respect to G counts the number of lines. Similar results relating differentials of the proper self-energy part to the full vertex function, defined in section 3.3, have been obtained by Ward (1950*a*, *b*) in electrodynamics. Their use in the many-body problem has been discussed by Pitaevski (1960), Nozières and Luttinger (1962) and Abrikosov *et al.* (1965).

Finally, inequalities are exact relations which are often useful. Some general inequalities of use in Statistical Mechanics are given by Okubo (1971). A review of the use of inequalities in the many-body problem is given by Huber (1969).

3.9. Example. The discontinuity in the Fermi distribution and the specific heat of a Fermi gas at low temperatures

At zero temperature, the distribution function $\langle n_{\mathbf{k}} \rangle_0$ for an ideal gas of fermions shows a discontinuity at the Fermi surface $\varepsilon_{\mathbf{k}} = \mu_0$. It is of interest to discuss whether or not this discontinuity remains for the interacting gas, or whether the distribution function falls smoothly to zero. The problem was first discussed in detail by Luttinger (1960).

The expectation value of the number of particles in the state $\mathbf{k}$ is, for fermions,

$$\langle n_{\mathbf{k}} \rangle = -\underset{\substack{\tau\to 0\\ \tau<0}}{\mathrm{Lt}}\, G_{\mathbf{k}}(\tau) = \frac{1}{\beta}\underset{\substack{\tau\to 0\\ \tau<0}}{\mathrm{Lt}} \sum_l \frac{\exp(-\mathrm{i}\omega_l \tau)}{\mathrm{i}\omega_l - (\varepsilon_{\mathbf{k}} - \mu) - \Sigma(\mathbf{k}, \mathrm{i}\omega_l)} \tag{3.102}$$

from eqn (3.34). In the limit of zero temperature, the sum in eqn (3.102) goes over to an integral:

$$\langle n_{\mathbf{k}} \rangle = \underset{\substack{\tau\to 0\\ \tau<0}}{\mathrm{Lt}} \frac{1}{2\pi\mathrm{i}} \int_{-\mathrm{i}\infty}^{\mathrm{i}\infty} \mathrm{d}\zeta \frac{\exp(-\zeta\tau)}{\zeta - (\varepsilon_{\mathbf{k}} - \mu) - \Sigma(\mathbf{k}, \zeta)}. \tag{3.103}$$

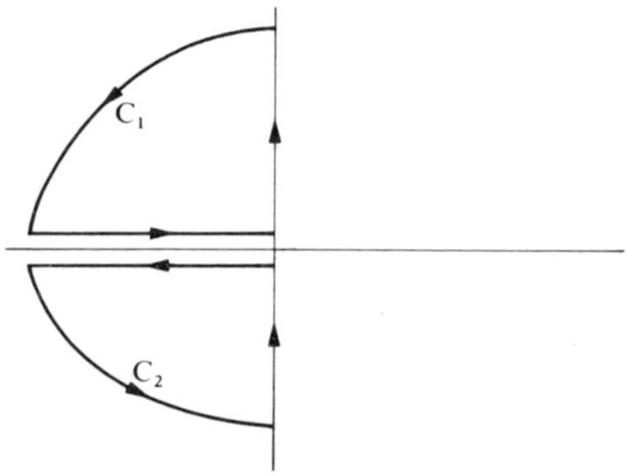

FIG. 3.16.

The integrand tends to zero more quickly than $1/\zeta$ on a semicircle of radius ζ in the left half-plane. Moreover, the integrand is analytic except on the real axis. Thus, with the notation of Fig. (3.16):

$$0 = \mathop{\mathrm{Lt}}_{\substack{\tau\to 0\\ \tau<0}} \frac{1}{2\pi \mathrm{i}} \int_{C_1+C_2} \mathrm{d}\zeta \frac{\exp(-\zeta\tau)}{\zeta-(\varepsilon_{\mathbf{k}}-\mu)-\Sigma(\mathbf{k},\zeta)}$$

$$= \langle n_{\mathbf{k}}\rangle - \frac{1}{2\pi \mathrm{i}}\left[\mathop{\mathrm{Lt}}_{\substack{\tau\to 0\\ \tau<0}} \int_{-\infty-\mathrm{i}\delta}^{0-\mathrm{i}\delta} \mathrm{d}\zeta \frac{\exp(-\zeta\tau)}{\zeta-(\varepsilon_{\mathbf{k}}-\mu)-\Sigma(\mathbf{k},\zeta)} - \text{complex conjugate}\right]$$

i.e.

$$\langle n_{\mathbf{k}}\rangle = \frac{1}{2\pi \mathrm{i}}\left[\int_{-\infty}^{0} \mathrm{d}x[x-(\varepsilon_{\mathbf{k}}-\mu)-K_{\mathbf{k}}(x)-\mathrm{i}L_{\mathbf{k}}(x)]^{-1} - \text{complex conjugate}\right]. \tag{3.104}$$

We shall show that if we define the Fermi surface $|\mathbf{k}| = k_{\mathrm{F}}$ of the interacting system to be that surface in momentum space which satisfies

$$\mu-\varepsilon_{\mathbf{k}}-K_{\mathbf{k}}(0) = 0, \tag{3.105}$$

then there is a discontinuity in the distribution function at the Fermi surface. We assume that, for values of $\mathbf{k}$ near the Fermi surface, there exists a solution $x = E_{\mathbf{k}}-\mu$ of the equation

$$x-(\varepsilon_{\mathbf{k}}-\mu)-K_{\mathbf{k}}(x) = 0, \tag{3.106}$$

and that there is no other solution of this equation. We use the expansion, for x near $E_{\mathbf{k}}-\mu$:

$$x-(\varepsilon_{\mathbf{k}}-\mu)-K_{\mathbf{k}}(x) = Z_{\mathbf{k}}^{-1}(x-E_{\mathbf{k}}+\mu)+\cdots, \tag{3.107}$$

where

$$Z_{\mathbf{k}}^{-1} = 1 - \left[\frac{\partial K_{\mathbf{k}}}{\partial x}\right]_{x = E_{\mathbf{k}} - \mu} \tag{3.108}$$

We split up the integral in eqn (3.104) into two parts:

$$\int_{-\infty}^{0} = \int_{-\infty}^{-\eta} + \int_{-\eta}^{0}, \qquad \eta > 0. \tag{3.109}$$

The first term cannot give rise to a discontinuity, since the integrand is always regular, as we have assumed that there is only one solution to eqn (3.106) and there is no reason to suppose that $K_{\mathbf{k}}(x)$ and $L_{\mathbf{k}}(x)$ will not be smooth functions of **k**. The contribution from the second term, $\langle n_{\mathbf{k}}\rangle'$ can be written, using eqn (3.107) and (3.93) (to be proved below):

$$\langle n_{\mathbf{k}}\rangle' = \frac{1}{2\pi \mathrm{i}}\left[\int_{-\eta}^{0} \mathrm{d}x\{Z_{\mathbf{k}}^{-1}(x - E_{\mathbf{k}} + \mu) - \mathrm{i}C_{\mathbf{k}}x^2\}^{-1} - \text{complex conjugate}\right], \tag{3.110}$$

where we have chosen η to be small.

Making the substitution

$$x = -|E_{\mathbf{k}} - \mu|/s,$$

we have

$$\langle n_{\mathbf{k}}\rangle' = \int_{|E_{\mathbf{k}} - \mu|/\eta}^{\infty} \mathrm{d}s \frac{|E_{\mathbf{k}} - \mu| C_{\mathbf{k}}/\pi}{Z_{\mathbf{k}}^{-2} s^2 (1 \pm s)^2 + C_{\mathbf{k}}^2 (E_{\mathbf{k}} - \mu)^2} \tag{3.111}$$

where the minus sign is to be taken for $E_{\mathbf{k}} < \mu$, i.e. $k < k_{\mathrm{F}}$, and the plus sign for $k > k_{\mathrm{F}}$. As we approach the Fermi surface, $E_{\mathbf{k}} \to \mu$ and the integrand approximates more and more closely to a δ-function. Thus, as $k \to k_{\mathrm{F}}$

$$\langle n_{\mathbf{k}}\rangle' \to Z_{\mathbf{k}_{\mathrm{F}}} \int_{0}^{\infty} \delta(s \pm 1)\,\mathrm{d}s$$

Thus if k is just less than k_{F}, $\langle n_{\mathbf{k}}\rangle' = Z_{\mathbf{k}_{\mathrm{F}}}$; but if k is just greater than k_{F}, $\langle n_{\mathbf{k}}\rangle' = 0$. There is thus a discontinuity in the Fermi distribution function of magnitude $Z_{\mathbf{k}_{\mathrm{F}}}$.

We have still to prove eqn (3.93). The first-order diagrams for $\Sigma(\mathbf{k}, \omega)$ give contributions which are independent of ω, and are real, and which hence do not contribute to $L_{\mathbf{k}}(x)$. The simplest diagram contributing to L is thus the second-order diagram shown in eqn 3.1b. Using the rules given in section 3.1, we see that its contribution to $\Sigma(\mathbf{k}, \mathrm{i}\omega_l)$ is proportional to,

$$\sum_{\mathbf{k}_1\mathbf{k}_2\mathbf{k}_3} |\Gamma^0_{\mathbf{k}\mathbf{k}_3\mathbf{k}_1\mathbf{k}_2}|^2 \frac{[\langle n_{\mathbf{k}_1}\rangle_0\langle n_{\mathbf{k}_3}\rangle_0 + \langle n_{\mathbf{k}_2}\rangle_0\langle n_{\mathbf{k}_3}\rangle_0 - \langle n_{\mathbf{k}_1}\rangle_0\langle n_{\mathbf{k}_2}\rangle_0 - \langle n_{\mathbf{k}_3}\rangle_0]}{\mathrm{i}\omega_l - (\varepsilon_{\mathbf{k}_1} + \varepsilon_{\mathbf{k}_2} - \varepsilon_{\mathbf{k}_3} - \mu)} \tag{3.112}$$

(see exercise 3.1.2). Thus its contribution, $L^{(\mathrm{b})}_{\mathbf{k}}(x)$, to the imaginary part of Σ when this function is analytically continued on to the real axis is

$$L^{(\mathrm{b})}_{\mathbf{k}}(x) \propto \sum_{\mathbf{k}_1\mathbf{k}_2\mathbf{k}_3} |\Gamma^0_{\mathbf{k}\mathbf{k}_3\mathbf{k}_1\mathbf{k}_2}|^2 \{\text{factors of } \langle n\rangle_0 \text{ as above}\}\delta(x - \varepsilon_{\mathbf{k}_1} - \varepsilon_{\mathbf{k}_2} + \varepsilon_{\mathbf{k}_3} + \mu). \tag{3.113}$$

It is easy to check that this gives the correct analytic continuation, from the properties of G. $L^{(\mathrm{b})}_{\mathbf{k}}$ may be rewritten, for $T = 0$:

$$\begin{aligned} L^{(\mathrm{b})}_{\mathbf{k}}(x) \propto \sum_{\mathbf{k}_1\mathbf{k}_2\mathbf{k}_3} & |\Gamma^0_{\mathbf{k}\mathbf{k}_3\mathbf{k}_1\mathbf{k}_2}|^2\, \delta(x - \varepsilon_{\mathbf{k}_1} - \varepsilon_{\mathbf{k}_2} + \varepsilon_{\mathbf{k}_3} + \mu) \times \\ & \times [\theta^-(\varepsilon_{\mathbf{k}_1} - \mu)\theta^-(\varepsilon_{\mathbf{k}_2} - \mu)\theta^+(\varepsilon_{\mathbf{k}_3} - \mu) + \\ & + \theta^+(\varepsilon_{\mathbf{k}_1} - \mu)\theta^+(\varepsilon_{\mathbf{k}_2} - \mu)\theta^-(\varepsilon_{\mathbf{k}_3} - \mu)] \end{aligned} \tag{3.114}$$

where

$$\theta^+(s) \begin{array}{ll} = 0 & s < 0 \\ = 1 & s > 0 \end{array} \quad \text{and} \quad \theta^-(s) = 1 - \theta^+(s).$$

For the first term in eqn (3.114), the θs impose the condition:

$$\varepsilon_{\mathbf{k}_1} + \varepsilon_{\mathbf{k}_2} - \varepsilon_{\mathbf{k}_3} \leqslant \mu.$$

Thus, because of the δ-function, this term gives nothing if $x > 0$. For $x < 0$, we introduce new variables $s_1 = (\mu - \varepsilon_{\mathbf{k}_1})/x$, $s_2 = (\mu - \varepsilon_{\mathbf{k}_2})/x$, and $s_3 = (\varepsilon_{\mathbf{k}_3} - \mu)/x$. Then, for small x, the contribution of this term is proportional to

$$x^2 \int_0^{-1}\int_0^{-1}\int_0^{-1} \mathrm{d}s_1\, \mathrm{d}s_2\, \mathrm{d}s_3\, \delta(1 + s_1 + s_2 + s_3) \propto x^2, \tag{3.115}$$

where we have replaced the sums over $\mathbf{k}$ by integrals in the usual way. The second term in eqn (3.114) may similarly be shown to vanish for $x > 0$ and to behave as x^2 for $x < 0$ and small.

To complete the proof, this work has now to be repeated for diagrams for Σ of arbitrary complexity. These are of two types: the first can be obtained from the diagram considered above by the insertion of self-energy parts to the internal lines; the second type cannot so be obtained (they have different skeleton diagrams). The proof for the first type is accomplished by inserting full Green functions for the internal lines, and using the spectral representation of eqn (3.66a) for these Green functions. The mathematics is then very similar to that given above. The proof for the second type is rather more involved, and we refer the interested reader to the original paper, which we suggest should be read after section 8.2.

We may also derive the form of the specific heat at low temperatures. To do this, we use eqns (3.9a) and (3.15):

$$\Omega = \frac{1}{\beta} \operatorname*{Lt}_{\substack{\tau \to 0 \\ \tau < 0}} \sum_{\mathbf{k},l} \exp(-\mathrm{i}\omega_l \tau)\{\ln \bar{G}(\mathbf{k}, \mathrm{i}\omega_l) - \Sigma(\mathbf{k}, \mathrm{i}\omega_l)\bar{G}(\mathbf{k}, \mathrm{i}\omega_l)\} + \Omega' \quad (3.116)$$

where Ω' is the contribution of all closed linked skeleton diagrams with the free-particle Green function g replaced by the full Green function G. We also know, from eqn (3.12), that $\partial\Omega/\partial\bar{G} = 0$. We shall derive the form of the lowest-order change in Ω, in powers of the temperature, as we go from zero to finite temperature. Ω will change for three reasons; in eqn (3.9) Ω_0 will change; all sums of the form $(1/\beta)\Sigma_l$ will change†, $(\omega_l = (2l+1)\pi/\beta)$; and $\bar{G}$ will change. The latter change may be ignored, however, since $\partial\Omega/\partial\bar{G} = 0$ so that in Ω we may use, instead of $\bar{G}$ at temperature T, its value at $T = 0$, $\bar{G}^{(0)}$. The first change is easy to evaluate, and so we must direct our attention to the second. Consider first the effect on Ω'. A general term will have many sums over l in it, but since we are looking for the first-order change, we need to change the contribution of one sum, associated with one line, using the zero-temperature form for the other sums, and then do the same thing for each line in turn. But, as we saw in deriving eqn (3.11), opening up each in turn of the 2ν lines in a ν-order diagram gives us a diagram for Σ of order ν. Thus, to this order

$$\Omega' = \operatorname*{Lt}_{\substack{\tau \to 0 \\ \tau < 0}} \frac{1}{\beta} \sum_{\mathbf{k},l} \{\mathrm{i}\omega_l - \varepsilon_{\mathbf{k}} + \mu - \Sigma^{(0)}(\mathbf{k}, \mathrm{i}\omega_l)\}^{-1}\, \Sigma^{(0)}(\mathbf{k}, \mathrm{i}\omega_l) \times \exp[-\mathrm{i}\omega_l \tau] \quad (3.117)$$

where $\Sigma^{(0)}(\mathbf{k}, \mathrm{i}\omega_l)$ is the proper self-energy evaluated at zero temperature for $\omega = \mathrm{i}\omega_l$. Combining (3.116 and 117) we have

$$\Omega = \beta^{-1} \operatorname*{Lt}_{\substack{\tau \to 0 \\ \tau < 0}} \sum_{\mathbf{k},l} \ln \bar{G}^{(0)}(\mathbf{k}, \mathrm{i}\omega_l) \exp(-\mathrm{i}\omega_l \tau). \quad (3.118)$$

The summation may be converted to an integral in the usual way:

$$\Omega = -\operatorname*{Lt}_{\substack{\tau \to 0 \\ \tau < 0}} \sum_{\mathbf{k}} \frac{1}{2\pi \mathrm{i}} \int \ln[\omega - \varepsilon_{\mathbf{k}} + \mu - \Sigma^{(0)}(\mathbf{k}, \omega)] \frac{\exp[-\omega\tau]}{\exp(\beta\omega) + 1} \quad (3.119)$$

† The reader should check that this gives the correct number of factors of β^{-1}.

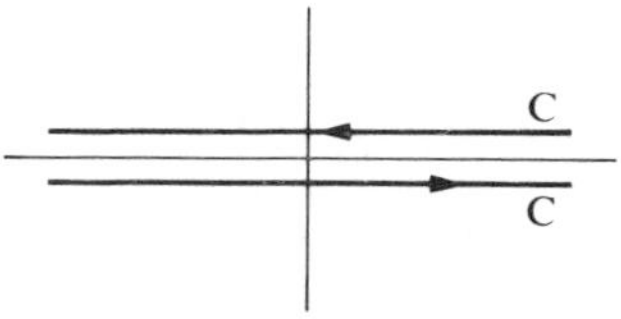

FIG. 3.17.

where the contour C is shown in Fig. 3.17.

But to obtain an expansion in powers of the temperature of integrals of the form of those in eqn (3.119) we use

$$\int f(x)[\exp(\beta x)+1]^{-1}\,\mathrm{d}x = \int f(x)\theta(-x)\,\mathrm{d}x+\frac{\pi^2}{6}(\kappa T)^2 f'(0)+\cdots$$

(see, for example, Sommerfeld (1928)). Thus

$$\Omega = \Omega(\mu, T=0)+\frac{1}{2\pi \mathrm{i}}(\kappa T)^2\frac{\pi^2}{6}\Bigg[\sum_{\mathbf{k}}\frac{\partial}{\partial x}\{\ln[x-\varepsilon_{\mathbf{k}}+\mu-K_{\mathbf{k}}(x)+\mathrm{i}L_{k}(x)]-$$

$$-\ln[x-\varepsilon_{\mathbf{k}}+\mu-K_{\mathbf{k}}(x)-\mathrm{i}L_{\mathbf{k}}(x)]\}\Bigg]_{x=0}$$

$$= \Omega(\mu, T=0)-\tfrac{1}{2}\gamma(\mu)T^2, \quad \text{say.} \tag{3.120}$$

Near $x=0$, $L_{\mathbf{k}}(x)$ is very small (eqn 3.93). Using (3.107) we may thus write,

$$\gamma(\mu) = -\frac{(\pi\kappa)^2}{3}\,\frac{1}{2\pi \mathrm{i}}\sum_{\mathbf{k}}\operatorname*{Lt}_{\delta\to 0}\,[(x-E_{\mathbf{k}}+\mu+\mathrm{i}\delta)^{-1}(x-E_{\mathbf{k}}+\mu-\mathrm{i}\delta)^{-1}]_{x=0}$$

$$= \frac{(\pi\kappa)^2}{3}\sum_{\mathbf{k}}\delta(\mu-E_{\mathbf{k}}). \tag{3.121}$$

Thus:

$$S = -\left(\frac{\partial\Omega}{\partial T}\right)_{\mu V} = \gamma(\mu)T \tag{3.122}$$

and

$$\bar{N} = \frac{\partial\Omega}{\partial\mu}(\mu, T=0)+\frac{1}{2}\frac{\partial\gamma}{\partial\mu}T^2. \tag{3.123}$$

We must now solve eqn (3.123) for μ in terms of $\bar{N}$ and T and substitute it into (3.122). But if we are looking for the lowest-order terms in T, we may replace eqn (3.122) by

$$S = \gamma(\mu^{(0)})T$$

where $\mu^{(0)}$ is the solution of eqn (3.123) at $T=0$. We then have, for the specific heat,

$$C_V = \gamma(\mu^{(0)})T. \tag{3.124}$$

This should be compared with the result for the non-interacting gas:

$$C_V = \gamma_0 T,$$

where

$$\gamma_0 = \frac{(\pi\kappa)^2}{3}\sum_{\mathbf{k}}\delta(\varepsilon_{\mathbf{k}}-\mu). \tag{3.125}$$

We shall discuss these formulae in the next section.

3.10. Quasi-particles and elementary excitations

From eqn (2.48), we know that $\bar{g}(\mathbf{k}, \omega)$, the Green function for the non-interacting system, is $(\omega-\varepsilon_{\mathbf{k}}+\mu)^{-1}$. It is easy to see how eqn (3.41) reduces to this form in this very simple case. For all states $|\nu N\rangle$ and $|\nu', N+1\rangle$ for which $\langle\nu', N+1|a_{\mathbf{k}}^{+}|\nu N\rangle$ is non-zero, $E_{N+1,\nu'}-E_{N\nu} = \varepsilon_{\mathbf{k}}$. Similarly, in the second term, $E_{N\nu}-E_{N-1,\nu} = \varepsilon_{\mathbf{k}}$, so that

$$\begin{aligned}\bar{g}(\mathbf{k}, \omega) &= [\omega-\varepsilon_{\mathbf{k}}+\mu]^{-1}Q^{-1}\sum_{\nu N}\exp[-\beta(E_{N\nu}-\mu N)] \times \\ &\quad\times\langle\nu N|a_{\mathbf{k}}a_{\mathbf{k}}^{+}-\varepsilon a_{\mathbf{k}}^{+}a_{\mathbf{k}}|\nu N\rangle \\ &= [\omega-\varepsilon_{\mathbf{k}}+\mu]^{-1}\end{aligned}$$

where we have used:

$$1 = \sum_{\nu'N'}|\nu'N'\rangle\langle\nu'N'|.$$

We may express this in a slightly different way in terms of the spectral function of G (eqns (3.55) and (3.66a)). For the non-interacting system

$$\rho_0(\mathbf{k}, E') = 2\pi\delta(E'-\varepsilon_{\mathbf{k}}+\mu). \tag{3.126}$$

For a weakly interacting system, the δ-function for ρ is spread into a peak of finite width (see exercise 3.6.2). We interpret this by saying that for most states for which $\langle\nu', N+1|a_{\mathbf{k}}^{+}|\nu N\rangle$ is large, $E_{N+1,\nu'}-E_{N\nu}-\mu$ is approximately the same, $E_{\mathbf{k}}-\mu$, say. The peak is then centred at $E_{\mathbf{k}}-\mu$. Both the width of the peak and $E_{\mathbf{k}}$ will in general be T-dependent. For a weakly-interacting gas, for example, the states $|\nu, N\rangle$ and $|\nu', N+1\rangle$ would differ by the addition of a particle with momentum $\mathbf{k}$. The energy difference would depend slightly on the state of the other particles in the system, with which the additional particle is interacting, that is, it will depend on $|\nu, N\rangle$. Hence the peak will be spread. The quasi-particle with momentum $\mathbf{k}$ would in this case be a "dressed" particle, that is, a particle interacting with its neighbours. Since it will carry along some of the other particles with it as it moves, we would expect its effective mass to be different from that of the bare particle.

Even in strongly interacting systems, ρ may be strongly peaked. Rather generally, changes from one *low-lying* energy level of a many-body system to another may be written in the form

$$\delta E \simeq \sum_s E_s \, \delta v_s. \tag{3.127}$$

We may interpret this in terms of a dilute gas of quasi-particles, with δv_s as the change in the number of quasi-particles having an energy E_s. The phonon description of the energy levels of a crystal provides a familiar example of eqn (3.127).

Before we discuss why this equation should hold, let us consider its effect on the form of $\bar{G}$. At low temperatures, the only states $|vN\rangle$ contributing to the sum in eqn (3.41) will be lowly excited ones, because of the factor $\exp[-\beta(E_{Nv}-\mu N)]$. For each such state, there will be another, $|v, \mathbf{k}, N+1\rangle$ say, which differs from $|vN\rangle$ by the addition of a quasi-particle of momentum $\mathbf{k}$. From the form of eqn (3.127), the energy difference between these two states is approximately $E_\mathbf{k}$ and is almost independent of the state $|vN\rangle$. Thus, when we take the sum in eqn (3.41), we shall pick up rather a lot of contributions near to $E_\mathbf{k}-\mu$ provided that $\langle v, \mathbf{k}, N+1|a_\mathbf{k}^+|vN\rangle$ does not vanish. Thus $\rho(\mathbf{k}, \omega)$ will quite probably be peaked at $E_\mathbf{k}-\mu$, where $E_\mathbf{k}$ is the energy of the quasi-particle of momentum $\mathbf{k}$. The width of the peak will be a measure of the variation of $E_\mathbf{k}$ between the various states connected by the matrix elements of $a_\mathbf{k}$. The proviso that $\langle v, \mathbf{k}, N+1|a_\mathbf{k}^+|vN\rangle$ should not vanish is not trivial; there might well be excitations in our system which are not picked up by the spectral function of the single-particle Green function we are discussing at the moment. We shall have an example of this situation in section 4.2.

A simple non-rigorous argument, which must not be taken too seriously, suggests why we should expect eqn (3.127) to hold generally for lowly excited many-body systems. Consider a lowly excited state with energy $\mathscr{E}_\mathbf{k}$ above the ground state, for given momentum, $\mathbf{k}$. If Ψ_0 is the ground state wavefunction with energy E_0, the wave-function $\Psi_\mathbf{k}$ for this excited state will differ slightly from Ψ_0. By superimposing a number of $\Psi_\mathbf{k}$s of neighbouring $\mathbf{k}$ values, we can obtain a wave-function which is almost an eigenfunction of our Hamiltonian with eigenvalue $E_0+\mathscr{E}_\mathbf{k}$, and yet which differs from Ψ_0 only within a small part of our system (just as we superpose plane-wave-functions to achieve an approximately localized particle with an approximately specified momentum or energy). Elsewhere, loosely speaking, the system is in its ground state. If we have two such excitations, with momentum $\mathbf{k}_1$ and $\mathbf{k}_2$, localized in different parts of our system, the energy of this approximate eigenstate will be very nearly equal to $E_0+\mathscr{E}_{\mathbf{k}_1}+$

$+\mathscr{E}_{\mathbf{k}_2}$, since the two excited regions hardly overlap. We can go on adding excitations to our system until there are so many of them that the excited regions must overlap considerably. Until this point is reached, an equation such as eqn (3.127) will hold, and our system may be described in terms of a dilute gas of elementary excitations, of which some may be particle-like, as discussed above, and others may involve large numbers of particles, such as phonons in a liquid or solid, or plasmons, which will be discussed in the next chapter. The E's may sometimes have negative as well as positive values, as is illustrated by the example of the non-interacting gas of fermions; the positive energies correspond to particles above the Fermi surface, and the negative energies to holes below the surface. The excitation of the above discussion then corresponds to a particle–hole pair, which is most conveniently described in terms of its two components. Further, the excitations are sometimes fermion-like, and sometimes boson-like. It is possible to have boson-like excitations in a fermion system; for example, deuterium gas behaves as if it were made up of bosons in most circumstances, since the constituent atoms, which are fermions, combine to form molecules, which act like bosons. Since fermions carry half-integral spin, however, we cannot expect to observe fermion-like excitations in a system of bosons.

Line-widths of peaks, for example in atomic physics, are usually related to decay times via the uncertainty principle:

$$\Delta E\, \Delta t \sim \hbar \quad (= 1 \text{ in our units}). \tag{3.128}$$

In our present case also, the width of the peak in $\rho(\mathbf{k}, \omega)$ is a measure of the lifetime of the excitations. To see this, we note that in general, a collection of excitations does not correspond to an exact eigenstate of the Hamiltonian. By definition, $\Psi_{\mathbf{k}}$ is an exact eigenfunction of H but if we write:

$$\Psi_{\mathbf{k}} = \phi_{\mathbf{k}} \Psi_0$$

then the operator, $\phi_{\mathbf{k}}$, acting on any other eigenfunction of H will produce not an eigenstate but a superposition of eigenstates with energies close to $E_0 + \mathscr{E}_{\mathbf{k}}$. The spread of these energies is, in the usual way, a measure of the decay rate of the state, via eqn (3.128), and also determines the width of the peak in ρ, as discussed above.

The connection between the width of the peak and a decay rate may also be seen in another way. From eqn (3.55), we see that the width of the peak in ρ will be approximately the same as the width in J. J will thus probably take a form similar to that shown in Fig. 3.18. To take the Fourier transform of J in order to find the correlation function we split J into two parts, the part containing the peak, which we approximate by $A \exp[-(\omega - E_{\mathbf{k}} + \mu)^2/2\sigma_{\mathbf{k}}^2]$,

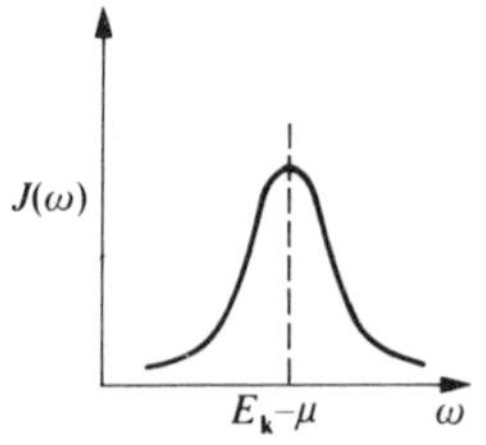

FIG. 3.18.

$\sigma_{\mathbf{k}}$ being the width of the peak, and a second, rather smoothly varying part. The Fourier transform of the latter will die away fairly quickly, and we shall be left with

$$\langle a_{\mathbf{k}}^{+}(t')a_{\mathbf{k}}(t)\rangle \simeq A \int_{-\infty}^{\infty} \exp[-\mathrm{i}\omega(t-t') \exp[-(\omega-E_{\mathbf{k}}+\mu)^2/(2\sigma_{\mathbf{k}}^2)]\, \mathrm{d}\omega$$

$$= A\sqrt{\frac{\pi}{2}}\sigma_{\mathbf{k}} \exp[-\mathrm{i}(E_{\mathbf{k}}-\mu)(t-t')] \exp[-(t-t')^2\sigma_{\mathbf{k}}^2/2]. \qquad (3.129)$$

The correlation function thus dies away with a half-line of about $1/\sigma_{\mathbf{k}}$. (The rather unphysical form of the decay results from our assumption about the form of the peak, chosen for ease of integration, and must not be taken seriously.)

In the weakly interacting case discussed in exercise 3.6.2, we may thus find the energies of the quasi-particle by looking at the position of the peak in ρ:

$$E_{\mathbf{k}}-\mu-\varepsilon_{\mathbf{k}}+\mu-K_{\mathbf{k}}(E_{\mathbf{k}}-\mu) = 0. \qquad (3.130)$$

Assuming $K_{\mathbf{k}}(E)$ is small and slowly varying, we may write this:

$$E_{\mathbf{k}} \simeq k^2/2m+K_{\mathbf{k}}(\varepsilon_{\mathbf{k}}-\mu)$$

$$\simeq k^2/2m^*-k_{\mathrm{F}}^2/2m^* \qquad (3.131)$$

near the Fermi surface, since $K_{\mathbf{k}}(\varepsilon_{\mathbf{k}}-\mu)$ will be a function of $|\mathbf{k}|^2$ for an isotropic system. Here

$$\frac{1}{m^*} = \frac{1}{m} + \left[\frac{\partial}{\partial(k^2)}K_{\mathbf{k}}(\varepsilon_{\mathbf{k}}-\mu)\right]_{\varepsilon_{\mathbf{k}}=\mu}.$$

Similarly the relaxation time of the quasi-particles is $\sigma_{\mathbf{k}}^{-1}$ where

$$\sigma_{\mathbf{k}} \simeq -\frac{\varepsilon\alpha^2}{2}\sum_{\mathbf{k}_3\mathbf{q}} \{\langle n_{\mathbf{k}+\mathbf{q}}\rangle_0\langle n_{\mathbf{k}_3}\rangle_0 + \langle n_{\mathbf{k}_3-\mathbf{q}}\rangle_0\langle n_{\mathbf{k}_3}\rangle_0 - \langle n_{\mathbf{k}_3-\mathbf{q}}\rangle_0\langle n_{\mathbf{k}+\mathbf{q}}\rangle_0 + \\ + \varepsilon\langle n_{\mathbf{k}_3}\rangle_0\}\,\delta(\varepsilon_{\mathbf{k}}+\varepsilon_{\mathbf{k}_3-\mathbf{q}}-\varepsilon_{\mathbf{k}_3}-\varepsilon_{\mathbf{k}+\mathbf{q}}).$$

This should be compared with the expression for the relaxation time derived in kinetic theory, with the use of the Fermi Golden Rule to calculate the probability of a collision.

The single-particle Green function thus gives us information about the energies and lifetimes of some of the excitations. To find out about other excitations, we may have to examine other Green functions with other operators for which the matrix elements between states differing by the addition of the excitation in which we are interested do not vanish. We have to use physical intuition to tell us which Green functions to consider. We shall give an example of such a case in section 4.2.

Having obtained this information about the elementary excitations in our system, we can often use it to give semi-quantitative thermodynamic, and even transport, properties of the system. For, knowing the $E_{\mathbf{k}}$s in eqn (3.127), we can find the value of energy changes in terms of changes in occupation numbers of the quasi-particle states, for which we can use the usual fermion or boson distribution functions for non-interacting particles. From this, we can find the specific heat, and other thermodynamic properties, at low temperatures. Similarly, we can use the half-lives in an elementary kinetic theory of the dilute gas of quasi-particles to give us transport coefficients. Although one can learn a good deal about the system in such ways, their results need to be treated with a good deal of reserve.

A more sophisticated approach to the quasi-particles in a Fermi system was provided by Landau (1956, 1957, 1958b). Landau's initial, and all-important assumption, is that changes in the total energy of a Fermi system may be expressed in terms of changes in the distribution function for quasi-particles $\nu_{\mathbf{k}}$:

$$\delta E = \sum_{\mathbf{k}} E_{\mathbf{k}}\,\delta\nu_{\mathbf{k}}, \tag{3.132}$$

where $E_{\mathbf{k}}$, *defined* as the functional derivative, $\partial E/\partial\nu_{\mathbf{k}}$, of E with respect to $\nu_{\mathbf{k}}$, is the quasiparticle energy. Later, eqn (3.132) is treated as the first term in a Taylor expansion of δE:

$$\delta E = \sum_{\mathbf{k}} E_{\mathbf{k}}\,\delta\nu_{\mathbf{k}} + \sum_{\mathbf{k}\mathbf{k}'} \delta\nu_{\mathbf{k}}\,\delta\nu_{\mathbf{k}'} f(\mathbf{k},\mathbf{k}') + \cdots$$

$f(\mathbf{k},\mathbf{k}')$ is then interpreted in terms of interactions between the quasi-particles in the states $\mathbf{k}$ and $\mathbf{k}'$. Further discussion of the powerful methods introduced by Landau would take us too far from our central task, and we refer the reader to the original papers. For a recent review, showing the connections between the Landau theory and many-body theory, see Brown (1971). Landau and Khalatnikov (Landau (1941, 1944, 1947), Landau and Khalatnikov (1949)) have also given quasi-particle theories of liquid helium four.

The connections between such theories and many-body calculations have been described by many authors. (See, for example, Luttinger and Nozières (1962), Luttinger (1968), Balian and De Dominicis (1971)). Eqn (3.124) is a simple example of exact results otained from such techniques. The quasi-particle approach to the low temperature specific heat of the Fermi gas would lead us to an equation of this form, with

$$\gamma = \frac{(\pi\kappa)^2}{3} \sum_{\mathbf{k}} \delta\left(\frac{k^2}{2m^*} - \mu\right)$$

by analogy with the non-interacting Fermi gas, where m^* is the effective mass of the quasi-particles. But this is what we have obtained in eqn (3.121), by an exact summation of the perturbation series, if we assume that we may expand $E_{\mathbf{k}}$, the solution of eqn (3.106), in the form of eqn (3.131).

Bibliography

The literature on Green functions is vast and still growing. The first papers to use them in statistical mechanics include Landau (1958a), Martin and Schwinger (1959), Fradkin (1959), Abrikosov, Gorkov and Dzyaloshinskii (1959).

Our approach follows Luttinger and Ward (1960) and Luttinger (1960) fairly closely. Details of electron–phonon interactions, mentioned only in exercises here, may be found, for example, in the book of Abrikosov, Gorkov and Dzyaloshinskii (1965).

The problem of calculating Green functions for canonical ensembles is discussed by Kwok and Woo (1971).

4

EXAMPLES IN THE USE OF DIAGRAMMATIC PERTURBATION THEORY

IN this chapter we discuss several approximations which are frequently used in the many-body problem.

4.1. The Hartree–Fock approximation

The simplest approximation we can make for the proper self-energy part is

$$-\beta\,\Sigma^{\mathrm{H}}(\mathbf{k}, \mathrm{i}\omega_n) = \varepsilon \operatorname*{Lt}_{\substack{\tau\to 0\\ \tau<0}} \sum_{\mathbf{k}_1 l} \frac{v_{\mathbf{k}\mathbf{k}_1\mathbf{k}\mathbf{k}_1}}{[\mathrm{i}\omega_l - (\varepsilon_{\mathbf{k}_1} - \mu)]} \exp(-\mathrm{i}\omega_l \tau), \tag{4.1}$$

corresponding to the diagram in Fig. 4.1(a), and giving

$$\Sigma^{\mathrm{H}}(\mathbf{k}, \mathrm{i}\omega_n) = \bar{v}(0) \sum_{\mathbf{k}_1} \langle n_{\mathbf{k}_1} \rangle_0 \tag{4.2}$$

in the case that $v_{\mathbf{k}_1,\mathbf{k}_2,\mathbf{k}_1+\mathbf{q},\mathbf{k}_2-\mathbf{q}} = \bar{v}(|\mathbf{q}|)$. This is known as the Hartree approximation. A similar approximation, the Hartree–Fock approximation, uses for Σ, corresponding to the diagram in Fig. 4.1(b),

$$\begin{aligned}\Sigma^{\mathrm{HF}}(\mathbf{k}, \mathrm{i}\omega_n) &= -\frac{\varepsilon}{\beta} \operatorname*{Lt}_{\substack{\tau\to 0\\ \tau<0}} \sum_{\mathbf{k}_1 l} \frac{\Gamma^0_{\mathbf{k}\mathbf{k}_1\mathbf{k}\mathbf{k}_1}}{[\mathrm{i}\omega_l - (\varepsilon_{\mathbf{k}_1} - \mu)]} \exp(-\mathrm{i}\omega_l \tau) \\ &= \sum_{\mathbf{k}_1} \langle n_{\mathbf{k}_1} \rangle_0 [\bar{v}(0) + \varepsilon \bar{v}(|\mathbf{k}_1 - \mathbf{k}|)]. \end{aligned} \tag{4.3}$$

Alternatively, we may solve the simultaneous equations:

$$\begin{aligned}\Sigma^{\mathrm{SCHF}}(\mathbf{k}, \mathrm{i}\omega_n) &= -\frac{\varepsilon}{\beta} \operatorname*{Lt}_{\substack{\tau\to 0\\ \tau<0}} \sum_{\mathbf{k}_1 l} \Gamma^0_{\mathbf{k}\mathbf{k}_1\mathbf{k}\mathbf{k}_1} \exp(-\mathrm{i}\omega_l \tau) \bar{G}^{\mathrm{SCHF}}(\mathbf{k}_1, \mathrm{i}\omega_l) \\ &= \sum_{\mathbf{k}_1} \langle n_{\mathbf{k}_1} \rangle [\bar{v}(0) + \varepsilon \bar{v}(|\mathbf{k}_1 - \mathbf{k}|)], \end{aligned} \tag{4.4}$$

$$\bar{G}^{\mathrm{SCHF}}(\mathbf{k}, \mathrm{i}\omega_n) = [\mathrm{i}\omega_n - (\varepsilon_{\mathbf{k}} - \mu) - \Sigma^{\mathrm{SCHF}}(\mathbf{k}, \mathrm{i}\omega_n)]^{-1}. \tag{4.5}$$

This corresponds to the summation of the diagrams shown in Fig. 4.1(e), and is known as the self-consistent Hartree–Fock approximation. The solution of these simultaneous equations is by no means easy, but in cases in which Σ is small, an iterative approach should converge quickly. (See, however, exercise 4.1.2.)

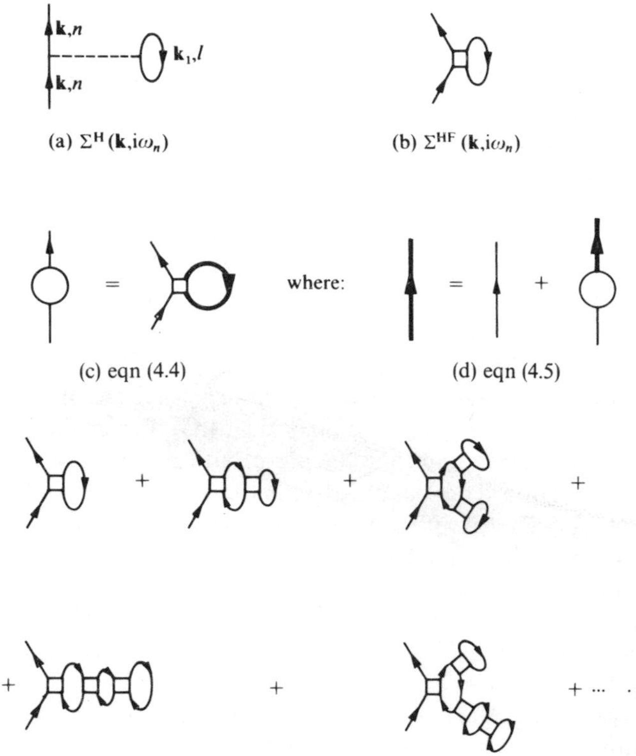

(e) Diagrams included in Σ^{SCHF}

FIG. 4.1.

In all three cases, eqns (4.1), (4.3), and (4.4), Σ is independent of ω_n and may thus be written as $M(\mathbf{k}, T)$ where the temperature dependence arises as a result of the dependence on β of the perturbation-theory rules. The analytic continuation of G into the whole of the complex plane is straightforward:

$$\bar{G}^{SCHF}(\mathbf{k}, \omega) = [\omega - \varepsilon_{\mathbf{k}} + \mu - M(\mathbf{k}, T)]^{-1}. \tag{4.6}$$

since this agrees with eqn (4.5) at the set of points $\{i\omega_n\}$ on the imaginary axis, and tends to zero at infinity. The spectral weight function (eqn 3.55) is thus

$$\rho(\mathbf{k}, E) = 2\pi\delta(E - \varepsilon_{\mathbf{k}} + \mu - M(\mathbf{k}, T)), \tag{4.7}$$

which leads to

$$\langle n_{\mathbf{k}} \rangle = [\exp\{\beta[\varepsilon_{\mathbf{k}_1} - \mu + M(\mathbf{k}, T)]\} - \varepsilon]^{-1}. \tag{4.7a}$$

This describes a quasi-particle with energy $\varepsilon_{\mathbf{k}} + M(\mathbf{k}, T)$ and infinite lifetime. The energy may often be expressed in terms of an effective mass in the manner of eqn (3.131) together with a constant term. The energy includes a term from the mean field set up by all the other particles in the system. It is temperature-dependent, because the mean field depends on the momentum distribution of the particles, which is itself temperature-dependent.

The Hartree and Hartree–Fock approximations are useful in giving one a simple model of how interactions affect a many-body system. The Hartree–Fock approximation will be valid in the case that the interaction is very weak, so that higher orders in the perturbation series need not be taken into account. It is clearly a meaningless approximation as it stands if the interaction potential is such that its matrix elements are not finite, as would be the case, for example, for a potential with a hard-core.

In any problem in which approximations have to be used, it is most important to be able to state the conditions under which the approximations are reasonable, if indeed they ever are. It is clear that if we make the strength of the interaction small enough, and the perturbation expansion converges, a low-order perturbation series has a reasonable chance of success. By comparing the Hartree–Fock term with the zero-order term in the Green function, we might hope that the approximation was a reasonable one if $N\bar{v} \ll \varepsilon$ where $\bar{v}$ is a typical matrix element of the potential, and ε a typical particle energy. This conclusion is supported by an examination of higher-order terms, if we interpret ε as a typical energy difference. (See, for example, exercise 3.1.2.)

The self-consistent Hartree–Fock approximation is not a low-order approximation, however, as may be seen from Fig. 4.1(e). Some sets of diagrams are summed to infinitely high order. Let us first examine the conditions under which the Hartree term will be large compared with the Fock term: $N\bar{v}(0) \gg \sum_{\mathbf{q}}\bar{v}(q)\langle n_{\mathbf{k}_1+\mathbf{q}}\rangle$. This will occur if the matrix element $\bar{v}(q)$ cuts off so quickly as a function of q that the sum over the number of particles picks up only a small fraction of the total number of particles; that is, if the range of $\bar{v}(q)$ is small compared with the range of $\langle n_{\mathbf{q}}\rangle$. This will be the case if the interparticle spacing is small compared with the range of the interaction potential in coordinate space, that is, if we are considering a dense system.

If we now consider higher-order diagrams, we see that those not included by the Hartree self-consistent approximation also have factors of the type $\sum_{\mathbf{q}}\bar{v}(q)\langle n_{\mathbf{k}+\mathbf{q}}\rangle$ or of a more complicated structure, involving products of $\langle n\rangle$s and $(\bar{v}/\varepsilon)$s with mutually-related momentum labels. A similar argument to that above shows that these contributions are small compared with the self-consistent Hartree terms of the same order. Thus, the Hartree self-consistent approximation will be reasonable when the system is dense. Another way of stating this condition is to say that each particle must interact with a large number of other particles, or, loosely, must have a large number of neighbours. We re-emphasize the statement above that the interaction potential matrix elements must be finite for the Hartree approximation to have any meaning.

Notice that we have well-defined quasi-particles in this approximation, which may be valid for strong interactions.

Exercises

4.1.1. Using eqns (3.82), (4.4) and (4.5), find an expression for the total energy $\langle E\rangle$ of the system. Notice that

$$\langle E\rangle \neq \sum_{\mathbf{k}} \langle n_{\mathbf{k}}\rangle [\varepsilon_{\mathbf{k}} + M(\mathbf{k}, T)],$$

whereas, naively, one might expect the equality to hold since $\varepsilon_{\mathbf{k}} + M(\mathbf{k}, T)$ is the quasi-particle energy. However,

$$\delta\langle E\rangle = \sum_{\mathbf{k}} [\varepsilon_{\mathbf{k}} + M(\mathbf{k}, T)]\delta\langle n_{\mathbf{k}}\rangle$$

which should be compared with eqn (3.127).

4.1.2. At $T = 0$, for a gas of fermions eqn (4.7a) tells us that

$$\begin{aligned} \langle n_{\mathbf{k}}\rangle &= 1, \qquad \varepsilon_{\mathbf{k}} + M(\mathbf{k}, 0) < \mu \\ &= 0, \qquad \varepsilon_{\mathbf{k}} + M(\mathbf{k}, 0) > \mu. \end{aligned}$$

Thus if we define k_{F}:

$$\varepsilon_{\mathbf{k}_{\mathrm{F}}} + M(\mathbf{k}_{\mathrm{F}}, 0) = \mu,$$

we have

$$\sum_{\mathbf{k}, |\mathbf{k}| < \mathbf{k}_{\mathrm{F}}} 1 = \bar{N},$$

so that the Fermi momentum is unchanged from the non-interacting value. Eqns (4.4) and (4.5) then become much easier to solve since

$$\Sigma^{\mathrm{SCHF}}(\mathbf{k}, \mathrm{i}\omega_n) = \sum_{\mathbf{k}_1, |\mathbf{k}_1| < \mathbf{k}_{\mathrm{F}}} \Gamma_{\mathbf{k}\mathbf{k}_1\mathbf{k}\mathbf{k}_1} = M(\mathbf{k}, 0)$$

and the self-consistency problems are avoided. This equation has been used, for example, in a very simple approach to nuclear matter calculations. Since, however, the nuclear force contains a hard-core, one cannot hope for quantitatively accurate results.

4.2. The random-phase approximation

We shall next consider an electron gas. This is a system of negatively charged particles interacting through the Coulomb potential, with a uniform background of positive charge to keep the system neutral. It thus provides, for example, a simple model of a metal, in which the ionic charge has been smoothed out. The Hamiltonian is

$$H-\mu\hat{N} = \sum_{\mathbf{k}}\left(\frac{k^2}{2m}-\mu\right)a_{\mathbf{k}}^{+}a_{\mathbf{k}}+\tfrac{1}{2}\sum_{\mathbf{k}_1,\mathbf{k}_2,\mathbf{q}\neq 0}\bar{v}(q)a_{\mathbf{k}_1}^{+}a_{\mathbf{k}_2}^{+}a_{\mathbf{k}_2-\mathbf{q}}a_{\mathbf{k}_1+\mathbf{q}}.$$

$$\bar{v}(q) = V^{-1}e^2\int e^{i\mathbf{q}.\mathbf{r}}r^{-1}\,d^3r = \frac{4\pi e^2}{Vq^2}, \tag{4.8}$$

where the $\mathbf{q} = 0$ term in the interaction potential has been cancelled by the uniform positive background.

We notice at once that, because of the behaviour of $\bar{v}$ for small values of q, we are likely to get divergent contributions in our perturbation expansion. Diagrams with several factors of $\bar{v}(q)$ for the same q will be more highly divergent than those in which each q appears only once. Thus a straightforward application of the Hartree–Fock approximation is unlikely to be at all successful, as many of the terms omitted will be non-finite. The diagrams in Fig. 4.2 will give the most divergent contributions to the proper self-energy in any order. If we take the sum of all such diagrams, however, we

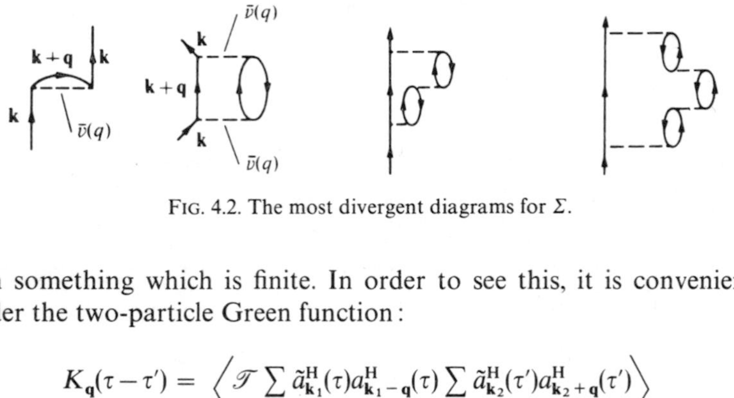

FIG. 4.2. The most divergent diagrams for Σ.

obtain something which is finite. In order to see this, it is convenient to consider the two-particle Green function:

$$K_{\mathbf{q}}(\tau-\tau') = \left\langle \mathscr{T}\sum_{\mathbf{k}_1}\tilde{a}_{\mathbf{k}_1}^{\mathrm{H}}(\tau)a_{\mathbf{k}_1-\mathbf{q}}^{\mathrm{H}}(\tau)\sum_{\mathbf{k}_2}\tilde{a}_{\mathbf{k}_2}^{\mathrm{H}}(\tau')a_{\mathbf{k}_2+\mathbf{q}}^{\mathrm{H}}(\tau')\right\rangle$$

$$= -\frac{1}{\beta}\sum \bar{K}(\mathbf{q}, i\omega_l)\exp[-i\omega_l(\tau-\tau')], \tag{4.9}$$

with Fourier coefficients $\bar{K}(\mathbf{q}, i\omega_l)$, $\omega_l = 2\pi l/\beta$ (cf. eqn (3.33) and the subsequent discussion). Because of the divergent nature of $\bar{v}(q)$, it is convenient to classify the diagrams for K rather differently from the classification in Fig. 3.11. We subtract from K any diagrams in which τ and τ' may be separated by cutting only an interaction (dotted) line (e.g. Fig. 4.3(f), (g)), and call the remainder the irreducible part of $\bar{K}(\mathbf{q}, i\omega_l)$, labelled $\Pi(\mathbf{q}, i\omega_l)$. The

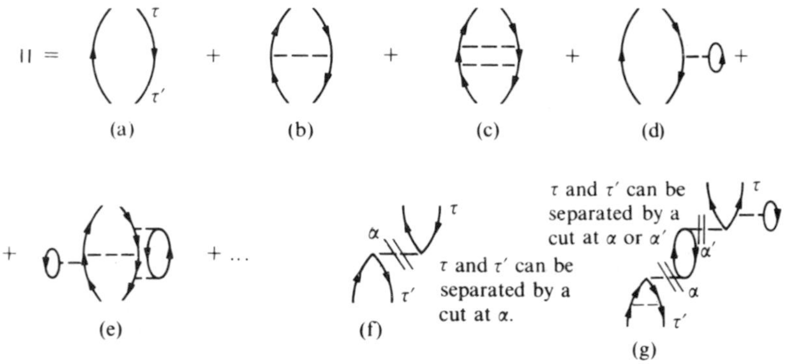

FIG. 4.3.

diagrams for Π are thus of the type (a) to (e). K is then given by the sum

$$-\beta^{-1}\bar{K}(\mathbf{q}, \mathrm{i}\omega_l) = -\beta^{-1}[\Pi(\mathbf{q}, \mathrm{i}\omega_l) + \Pi(\mathbf{q}, \mathrm{i}\omega_l)\bar{v}(q)\Pi(\mathbf{q}, \mathrm{i}\omega_l) + \Pi\bar{v}\Pi\bar{v}\Pi + \cdots], \quad (4.10)$$

or

$$\bar{K}(\mathbf{q}, \mathrm{i}\omega_l) = \frac{\Pi(\mathbf{q}, \mathrm{i}\omega_l)}{1 - \bar{v}(q)\Pi(\mathbf{q}, \mathrm{i}\omega_l)} \quad (4.11)$$

as shown in Fig. 4.4. The unpleasant $\bar{v}(q)$ term is now in the denominator, and $\bar{K}$ will be finite so long as $\bar{v}(q)\Pi(\mathbf{q}, \mathrm{i}\omega_l)$ is different from unity. We shall thus turn our attention to Π.

The simplest contribution, that of diagram (a) of Fig. 4.3, is

$$-\beta^{-1}\Pi^{(a)}(\mathbf{q}, \mathrm{i}\omega_l) = -\beta^{-2}\sum_{\mathbf{k},n}[\mathrm{i}\omega_n - (\varepsilon_{\mathbf{k}} - \mu)]^{-1}[\mathrm{i}(\omega_n - \omega_l) - (\varepsilon_{\mathbf{k}-\mathbf{q}} - \mu)]^{-1}, \quad (4.12)$$

where $\omega_n = (2n+1)\pi/\beta$, since we are dealing with fermions. ω_l, as mentioned above, is $2l\pi/\beta$. Notice that $-\beta^{-1}\Pi^{(a)}$ carries a minus sign since it is effectively a closed loop.

Performing the sum over n in the usual way:

$$\Pi^{(a)}(\mathbf{q}, \mathrm{i}\omega_l) = \sum_{\mathbf{k}}[\langle n_{\mathbf{k}}\rangle_0 - \langle n_{\mathbf{k}-\mathbf{q}}\rangle_0][\varepsilon_{\mathbf{k}} - \varepsilon_{\mathbf{k}-\mathbf{q}} - \mathrm{i}\omega_l]^{-1}. \quad (4.13)$$

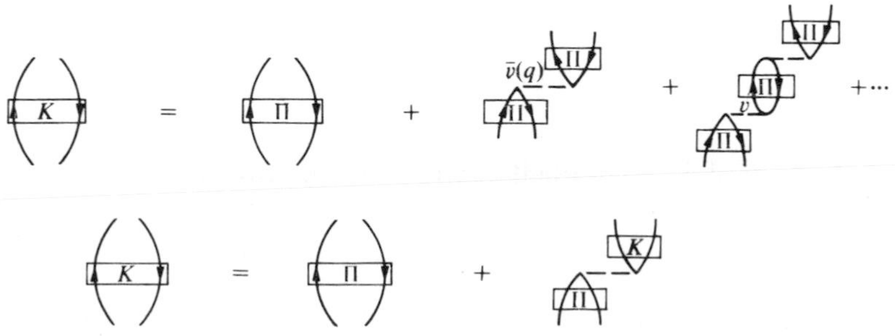

FIG. 4.4.

For small q, this may be written:

$$\Pi^{(a)}(\mathbf{q}, i\omega_l) = \sum_{\mathbf{k}} \frac{\partial \langle n(\varepsilon_{\mathbf{k}})\rangle_0}{\partial \varepsilon_{\mathbf{k}}} \mathbf{u}_{\mathbf{k}} \cdot \mathbf{q}[\mathbf{u}_{\mathbf{k}} \cdot \mathbf{q} - i\omega_l]^{-1}, \tag{4.14}$$

where

$$\varepsilon_{\mathbf{k}} - \varepsilon_{\mathbf{k}-\mathbf{q}} \sim \mathbf{q} \cdot \nabla_{\mathbf{k}} \varepsilon_{\mathbf{k}} \equiv \mathbf{u}_{\mathbf{k}} \cdot \mathbf{q} \tag{4.15}$$

and $\mathbf{u}_{\mathbf{k}} = \mathbf{k}/m$. Similarly, the contribution represented by the diagram 4.3(b) is

$$-\Pi^{(b)}(\mathbf{q}, i\omega_l) = \sum_{\mathbf{k}_1\mathbf{k}_2} \frac{\partial \langle n(\varepsilon_{\mathbf{k}_1})\rangle_0}{\partial \varepsilon_{\mathbf{k}_1}} \frac{\mathbf{u}_{\mathbf{k}_1} \cdot \mathbf{q}}{\mathbf{u}_{\mathbf{k}_1} \cdot \mathbf{q} - i\omega_l} \frac{\partial \langle n(\varepsilon_{\mathbf{k}_2})\rangle_0}{\partial \varepsilon_{\mathbf{k}_2}} \times$$

$$\times \frac{\mathbf{u}_{\mathbf{k}_2} \cdot \mathbf{q}}{\mathbf{u}_{\mathbf{k}_2} \cdot \mathbf{q} - i\omega_l} \bar{v}(|\mathbf{k}_1 - \mathbf{k}_2|) \quad \text{for small } q. \tag{4.16}$$

We shall consider the ratio of $\Pi^{(a)}$ to $\Pi^{(b)}$ for small q in the case that the temperature is low and the density high. In this case the Fermi gas is degenerate and $\partial \langle n_{\mathbf{k}}\rangle_0/\partial \varepsilon_{\mathbf{k}}$ is approximately $-\delta(\varepsilon_{\mathbf{k}} - \mu)$. In this case, as far as orders of magnitude are concerned, $\Pi^{(b)}/\Pi^{(a)} \sim me^2/k_{\mathrm{F}} = (a_0 k_{\mathrm{F}})^{-1}$ where a_0 is the Bohr radius and k_{F} the Fermi momentum. Thus, $\Pi^{(b)}/\Pi^{(a)} \ll 1$ provided $a_0 k_{\mathrm{F}} \gg 1$, i.e. provided the system is dense. In this limit, it may similarly be shown that the contributions of all other diagrams to Π are small compared with that of (a), so that $\Pi^{(a)}$ is good approximation to Π.

Similarly, it may be shown that $\Pi^{(a)}$ is a good approximation in the limit of high temperature and low densities. From now on, we shall consider just these two limits, and take $\Pi^{(a)}$ for Π.

The sum of the series of diagrams shown in Fig. 4.2 is then†

$$-\beta\Sigma(\mathbf{k}, i\omega_l) = \sum_{\mathbf{q},n} \{\bar{v}(q) + [\bar{v}(q)]^2 \bar{K}(\mathbf{q}, i\omega_l - i\omega_n)\} \bar{G}(\mathbf{k}+\mathbf{q}, i\omega_n)$$

$$= \sum_{\mathbf{q},n} v_{\mathrm{eff}}(\mathbf{q}, i\omega_l - i\omega_n) \bar{G}(\mathbf{k}+\mathbf{q}, i\omega_n), \tag{4.17}$$

where

$$v_{\mathrm{eff}}(\mathbf{q}, i\omega_m) = \frac{\bar{v}(q)}{1 - \bar{v}(q)\Pi^{(a)}(\mathbf{q}, i\omega_m)} \quad \text{using eqn (4.11).}$$

We have thus an effective potential v_{eff} which is both q- and ω-dependent. v_{eff} has the form of a Coulomb interaction between two charges in a region with a dielectric constant:

$$\varepsilon(\mathbf{q}, i\omega_m) = 1 - \bar{v}(q)\Pi^{(a)}(\mathbf{q}, i\omega_m). \tag{4.18}$$

† We have used $\bar{G}$ rather than $\bar{g}$ to introduce some self-consistency. This sums a larger class of diagrams than those in Fig. 4.2.

For small q, and $i\omega_m = 0$, we may use eqn (4.14) to show that

$$v_{\text{eff}} = \frac{1}{V}\frac{4\pi e^2}{q^2+\lambda^2}, \tag{4.19}$$

where

$$\lambda^2 = 6\pi n e^2/\mu \tag{4.20}$$

for the low-temperature, high-density limit, where $n = \bar{N}/V$. Similarly

$$\lambda^2 = 4\pi n e^2 \beta \tag{4.21}$$

for the low-density, high-temperature limit. In either case, v_{eff} is the Fourier transform of

$$\frac{e^2}{r}\mathrm{e}^{-\lambda r}, \tag{4.22}$$

a shielded Coulomb interaction.

If we insert the $\omega = 0$ limit into eqn (4.17) we have

$$\Sigma(\mathbf{k}, i\omega_l) = \beta^{-1}\sum_{\mathbf{q},n}\bar{G}(\mathbf{k}+\mathbf{q}, i\omega_n)\frac{4\pi e^2}{q^2+\lambda^2} \tag{4.23}$$

$$= \sum_{\mathbf{q}}\langle n(\mathbf{k}+\mathbf{q})\rangle\frac{4\pi e^2}{q^2+\lambda^2}, \tag{4.24}$$

which is independent of ω. Hence, in this very crude approximation, we have undamped quasi-particles, with energies at

$$\varepsilon_{\mathbf{k}} + \sum_{\mathbf{q}}\langle n_{\mathbf{k}+\mathbf{q}}\rangle \cdot \frac{4\pi e^2}{q^2+\lambda^2}.$$

A proper discussion of the quasi-particle spectrum given by the Σ of eqn (4.17) would take us too far away from our central course, and we refer the interested reader to the literature. (See, for example, Pines and Nozières (1966).)

We may still learn more about our system, however, from our expression for $\bar{K}$. We remarked on p. 103 that Green functions give us information about the elementary excitations of our system, and that if the matrix elements of an operator A were non-zero between states differing by one excitation of a certain type, then a Green function involving that operator would be a good tool for investigating that particular excitation. But $\bar{K}(\mathbf{q}, i\omega_l)$ of eqn (4.9) is a Green function, and since $\sum_{\mathbf{k}_1} a^+_{\mathbf{k}_1}a_{\mathbf{k}_1-\mathbf{q}}$ is the Fourier transform of the density operator, $a^+(\mathbf{r})a(\mathbf{r})$, we would expect $\bar{K}$ to be a good tool with which to examine the energies of excitations involving fluctuations in the density.

We first extend $\bar{K}$ on to the whole of the complex plane. The correct extension is:

$$\bar{K}(\mathbf{q}, \omega) = \frac{\Pi(\mathbf{q}, \omega)}{1 - \bar{v}(q)\Pi(\mathbf{q}, \omega)}, \tag{4.25}$$

$$\Pi(\mathbf{q}, \omega) = \sum_{\mathbf{k}} [\langle n_{\mathbf{k}} \rangle_0 - \langle n_{\mathbf{k}-\mathbf{q}} \rangle_0][\varepsilon_{\mathbf{k}} - \varepsilon_{\mathbf{k}-\mathbf{q}} - \omega]^{-1}. \tag{4.26}$$

We shall have a singularity in our Green function if

$$1 - \bar{v}(q)\Pi(\mathbf{q}, \omega) = 0,$$

i.e. if

$$1 = \bar{v}(q) \sum_{\mathbf{k}} \frac{\partial \langle n_{\mathbf{k}} \rangle_0}{\partial \varepsilon_{\mathbf{k}}} \frac{\mathbf{u}_{\mathbf{k}} \cdot \mathbf{q}}{\mathbf{u}_{\mathbf{k}} \cdot \mathbf{q} - \omega} \tag{4.27}$$

for small q.

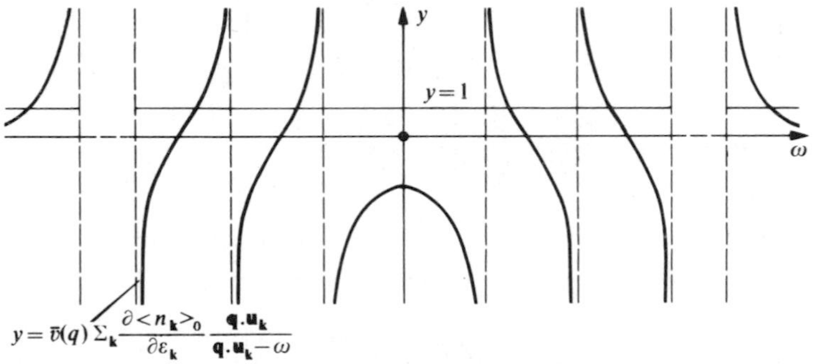

FIG. 4.5.

A plot of both sides of this equation is shown in Fig. 4.5. The right-hand side is infinite at each value of $\mathbf{q} \cdot \mathbf{u}_{\mathbf{k}}$ in the sum, and there is a solution to eqn (4.27) between each such value. Since these values are very close together (as the dotted lines in Fig. 4.5 should be), these solutions form a continuum. These lead in turn to a rather smoothly varying part of the spectral function, which we cannot interpret in terms of an elementary excitation. However, the two outer solutions of eqn (4.27) are isolated. Suppose that the positive solution occurs at $\omega = \omega_q$. Then, from the figure, we see that $\omega_q > \mathbf{q} \cdot \mathbf{u}_{\mathbf{k}}$ so that we may expand the denominator in powers of $\mathbf{q}$:

$$1 = -\bar{v}(q) \sum_{\mathbf{k}_1} \frac{\partial \langle n \rangle_0}{\partial \varepsilon_{\mathbf{k}}} \left[\frac{1}{\omega_q} + \frac{\mathbf{q} \cdot \mathbf{u}_{\mathbf{k}}}{(\omega_q)^2} + \cdots \right] \mathbf{q} \cdot \mathbf{u}_{\mathbf{k}}. \tag{4.28}$$

The first term drops out when the integration over angles is performed, and we have:

$$\omega_q^2 \simeq -\bar{v}(q) \sum_{\mathbf{k}_1} \frac{\partial \langle n \rangle_0}{\partial \varepsilon_{\mathbf{k}}} (\mathbf{q} \cdot \mathbf{u}_{\mathbf{k}})^2. \tag{4.29}$$

After some rearranging, this becomes

$$\omega_q^2 \simeq (4\pi n e^2/m). \tag{4.30}$$

Since this solution is isolated, it will lead to a δ-function in the spectral function, and, according to the discussion on p. 103 this corresponds to the presence of an elementary excitation of energy ω_q. This is called a plasmon. It is a very energetic excitation for the electronic densities typical of metals; for example, for copper ω_q is about 15 ev, whereas the Fermi energy is about 7 ev. For semiconductors ω_q may be smaller by a factor of 10^4. The q-dependence of ω_q follows from taking further terms in eqn (4.28).

So far, we have ignored the spread of the peak, which tells us about the damping of the excitations, and comes from the imaginary part of the denominator of $\bar{K}$ (see exercise 3.6.2) close to the real axis. We have

$$\operatorname{Im} \Pi_{\mathbf{q}}^{(a)}(\omega_q - i\delta) = -\pi \sum_{\mathbf{k}} \mathbf{q} \cdot \mathbf{u}_{\mathbf{k}}\, \delta(\omega_q - \mathbf{q} \cdot \mathbf{u}_{\mathbf{k}}) \frac{\partial \langle n_{\mathbf{k}} \rangle_0}{\partial \varepsilon_{\mathbf{k}}}; \tag{4.31}$$

this gives a damping, called 'Landau damping', proportional to $\exp[-m\beta\omega_q^2/(2q^2)]\omega_q^3/q^3$ for the low-density, high-temperature limit.

At low temperatures, the damping is not given correctly by this formula, since higher-order terms in an expansion in powers of e^2 become important.

Further reading

The Theory of Quantum Liquids, by Pines and Nozières (1966), contains a review of calculations on the electron gas and a list of references.

4.3. The t-matrix

Consider a system of fermions at low temperatures, and suppose that the system is interacting through short-range, but possibly strong, forces, and that it is dilute. The range of the interaction potential in momentum space, q_0, will be large, and we shall consider the situation in which $q_0 \gg k_F$, the Fermi momentum. In this case, a restriction on a momentum label that it should be less than the Fermi momentum is much more important than that it should be less than q_0. Thus, from eqn (2.36), we shall get large contributions from diagrams with a large number of lines with arrows pointing upwards. This, of course, is a very loose statement, since our diagrams include all relative τ-placings of their vertices. However, consider the simple diagram for Σ shown in Fig. 4.6(a), and suppose $\tau_2 > \tau_1$. If we add a further interaction as in (b), the contribution will be large so long as $\tau_2 > \tau_3 > \tau_1$. Adding a further interaction as in (c) will produce a contribution which is large in none of the cases $\tau_2 > \tau_3 > \tau_1$, $\tau_3 < \tau_1$, $\tau_3 > \tau_2$. This leads us to the conclusion that the dominant diagrams for Σ in any order will be the so-called 'ladder diagrams' of Fig. 4.6(d), where the square represents the t-matrix shown in (e). (Compare Fig. 2.16.)

There is a further reason for the importance of the t-matrix. In the introduction to the first chapter, we mentioned the difficulty which arises in the many-body problem from the fact that interatomic forces become very large at small separations. The Fourier transform of the interatomic potential does not exist, so that even at low densities, a low-order perturbation expansion is meaningless. It is essential to take into account at least all those terms which, loosely speaking, describe multiple interactions between two given particles. But these are just the terms summed by the diagrams in Fig. 4.6.

The diagrammatic equation of Fig. 4.6(e) may be written algebraically:

$$\begin{aligned}
-\beta\delta_{\omega_1+\omega_2,\omega_3+\omega_4} & t(\mathbf{k}_1 i\omega_1, \mathbf{k}_2 i\omega_2, \mathbf{k}_3 i\omega_3, \mathbf{k}_4 i\omega_4) \\
&= -\beta\delta_{\omega_1+\omega_2,\omega_3+\omega_4}\Gamma^0_{\mathbf{k}_1\mathbf{k}_2\mathbf{k}_3\mathbf{k}_4} + \\
&\quad + \tfrac{1}{2}\sum_{\substack{\mathbf{k}_5\omega_5 \\ \mathbf{k}_6\omega_6}} \Gamma^0_{\mathbf{k}_1\mathbf{k}_2\mathbf{k}_5\mathbf{k}_6}\bar{g}(\mathbf{k}_5, i\omega_5)\bar{g}(\mathbf{k}_6, i\omega_6) t(\mathbf{k}_5 i\omega_5, \mathbf{k}_6 i\omega_6, \mathbf{k}_3 i\omega_3, \mathbf{k}_4 i\omega_4) \times \\
&\quad \times \delta_{\omega_1+\omega_2,\omega_5+\omega_6}\,\delta_{\omega_5+\omega_6,\omega_3+\omega_4}.
\end{aligned} \tag{4.32}$$

If we had only two particles in our system, and we were working with well-known states, instead of a statistical ensemble, this equation would be the equation for the scattering matrix. In that case, its exact solution is equivalent to the exact solution of the Schrödinger equation for the problem, and t will be finite even if $\bar{v}(q)$ is not (although not in all cases; see, for example, de Dominicis (1957)). Similarly, in the present case, we have good reason to hope that the use of t will lead to finite results for Σ, even in cases in which $\bar{v}(q)$ is not finite. We note that, although only two particles are involved in the scattering, the other particles of the many-body system have some effect in eqn (4.32), on account of the statistical weighting factors hidden in the Green functions.

To attempt to find the exact solution of eqn (4.32) would be a most formidable undertaking. Brueckner and his co-workers (see, for example, Brueckner and Wada (1956), Bethe and Goldstone (1957)) have found approximate numerical solutions for certain cases at zero temperature. Indeed, they have done even more than that. It is possible to make the equation self-consistent by using, in eqn (4.29), instead of the unperturbed propagator, $\bar{g}$, an approximation for the full Green function,

$$[i\omega_l - \varepsilon_{\mathbf{k}} + \mu - \Sigma^{(t)}(\mathbf{k}, i\omega_l)]^{-1},$$

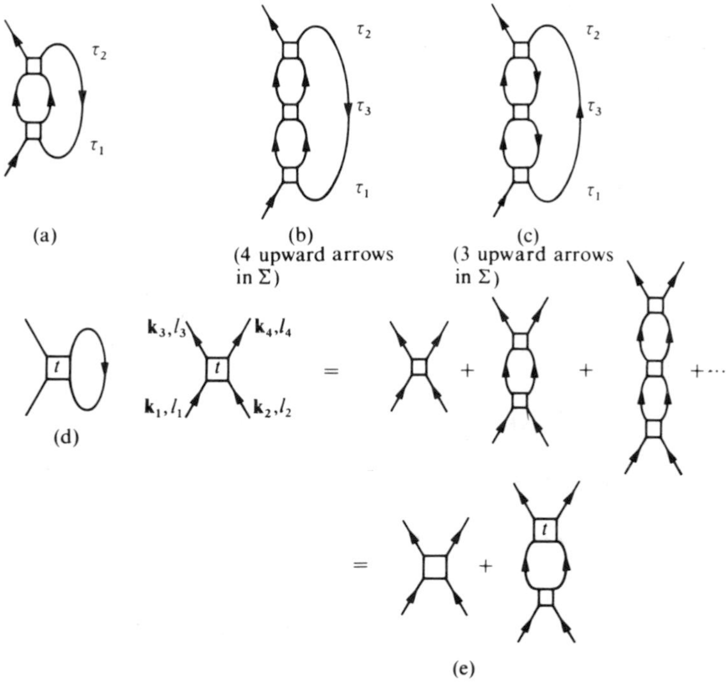

FIG. 4.6.

where $\Sigma^{(t)}(\mathbf{k}, i\omega_2)$ is the proper self-energy calculated by using the t-matrix, as in Fig. 4.7. The equation is then non-linear and even more difficult to solve. It sums diagrams such as those in Fig. 4.7(b).

Further reading

Calculations using the t-matrix are reviewed by Brueckner (1959).

Exercise

4.3.1. Show that t as defined in eqn (4.32) is a function of the ωs only in the form $i(\omega_1+\omega_2) = i(\omega_3+\omega_4) \equiv i\omega$; carry out the sum over the internal ωs, and hence rewrite the equation in the form, for fermions:

$$t(\mathbf{k}_1, \mathbf{k}_2, \mathbf{k}_3, \mathbf{k}_4, i\omega) = \Gamma^0_{\mathbf{k}_1\mathbf{k}_2\mathbf{k}_3\mathbf{k}_4} - \\ -\tfrac{1}{2}\sum_{\mathbf{k}_5\mathbf{k}_6} \Gamma^0_{\mathbf{k}_1\mathbf{k}_2\mathbf{k}_5\mathbf{k}_6} \frac{\langle n_{\mathbf{k}_5}\rangle_0 + \langle n_{\mathbf{k}_6}\rangle_0 - 1}{i\omega - \varepsilon_{\mathbf{k}_5} - \varepsilon_{\mathbf{k}_6} + 2\mu} \times \\ \times t(\mathbf{k}_5, \mathbf{k}_6, \mathbf{k}_3, \mathbf{k}_4, i\omega).$$

(Remember $i\omega = 2\pi n i/\beta$.) If we ignore the ω-dependence of t, the use of

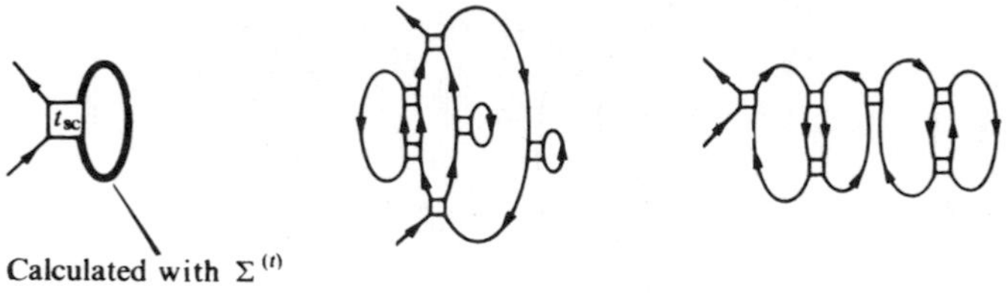

(a) (b)

FIG. 4.7.

the approximation of Fig. 4.7 gives us quasiparticles with energies

$$\varepsilon'_{\mathbf{k}} = \varepsilon_{\mathbf{k}} + \Sigma^{(t)}(\mathbf{k}, \varepsilon'_{\mathbf{k}}),$$

where
$$\Sigma^{(t)}(\mathbf{k}, \varepsilon'_{\mathbf{k}}) = \sum_{\mathbf{k}_1} t(\mathbf{k}, \mathbf{k}_1, \mathbf{k}, \mathbf{k}_1, \varepsilon_{\mathbf{k}} + \varepsilon'_{\mathbf{k}}) \langle n_{\mathbf{k}_1} \rangle.$$

At zero temperature, the factor

$$\begin{aligned} \langle n_{\mathbf{k}_5} \rangle + \langle n_{\mathbf{k}_6} \rangle - 1 &= 1, |\mathbf{k}_5|, |\mathbf{k}_6| < k_{\mathrm{F}} \\ &= -1, |\mathbf{k}_5|, |\mathbf{k}_6| > k_{\mathrm{F}} \\ &= 0, \text{otherwise.} \end{aligned}$$

It would be consistent with the approximations we have already used to neglect the first range compared with the second. Thus at zero temperature, we may write the self consistent equations in the form

$$t(\mathbf{k}_1\mathbf{k}_2\mathbf{k}_3\mathbf{k}_4) = \Gamma^0_{\mathbf{k}_1\mathbf{k}_2\mathbf{k}_3\mathbf{k}_4} + \frac{1}{2} \sum_{\substack{\mathbf{k}_5\mathbf{k}_6 \\ k_5, k_6 > k_{\mathrm{F}}}} \frac{\Gamma^0_{\mathbf{k}_1\mathbf{k}_2\mathbf{k}_5\mathbf{k}_6} t(\mathbf{k}_5\mathbf{k}_6\mathbf{k}_3\mathbf{k}_4)}{\varepsilon'_{\mathbf{k}_1} + \varepsilon'_{\mathbf{k}_2} - \varepsilon'_{\mathbf{k}_5} - \varepsilon'_{\mathbf{k}_6}},$$

where

$$\varepsilon'_{\mathbf{k}} = \varepsilon_{\mathbf{k}} + \sum_{\substack{\mathbf{k}_1 \\ k_1 < k_{\mathrm{F}}}} t(\mathbf{k}\mathbf{k}_1\mathbf{k}\mathbf{k}_1).$$

This equation (at zero temperature) is known as the Bethe–Goldstone equation.

A more general form of this equation, the Bethe–Salpeter equation, may be written down for the complete vertex part (section 3.3):

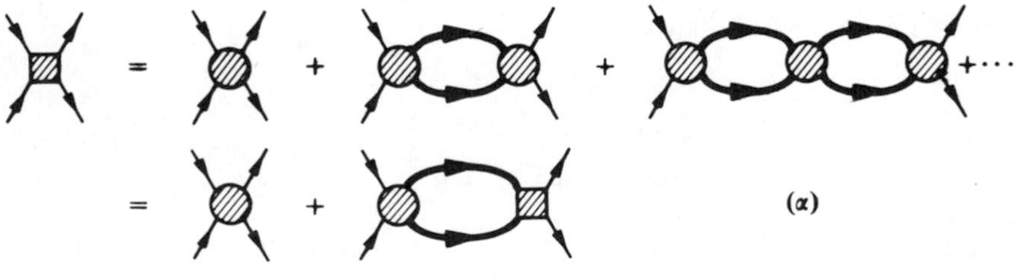

(α)

where the shaded circle, called the irreducible vertex part, denotes the sum of all diagrams which cannot be split into two by cutting two (full) lines with arrows pointing in the same direction. Note that:

is part of the irreducible vertex part, since it can be cut into two only by cutting two lines with arrows pointing in the opposite direction. Show that if one uses the simplest approximation for the irreducible vertex part:

and replaces full Green functions by non-interacting Green functions, one obtains the t-matrix for the approximate complete vertex part.

Find an algebraic form of Fig. 3.11 which uses the irreducible vertex part instead of the full vertex part. Draw some further contributions to the irreducible vertex parts.

The decomposition of diagrams made possible by the introduction of the irreducible vertex part is useful when one wishes to take into account two-particle effects such as bound states of two quasi-particles. Between the two shaded circles in eqn (α), we have two fully dressed particles, interacting with the rest of the system in all possible ways. The shaded circle then describes possible interactions between the two particles, these interactions being modified by the rest of the system.

For systems for which hole–particle interactions are important, it is useful to define the irreducible hole–particle vertex part, and write Γ as:

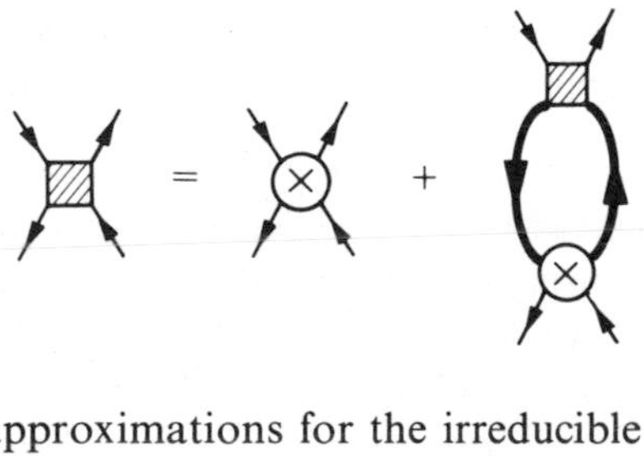

Draw some simple approximations for the irreducible hole–particle vertex part.

4.4. Scattering by static impurities

We now wish to turn our attention to a slightly different application of diagrammatic methods. We shall consider the scattering of particles by a set of static impurities, a problem which has received much attention recently, and which we shall consider in Chapter 7 when discussing transport properties. It has an obvious application to the scattering of electrons by lattice impurities in a metal, and to random alloys, and it has also been used in the theory of liquids.

We start with a Hamiltonian:

$$H-\mu\hat{N} = \sum_{\mathbf{k}} (\varepsilon_{\mathbf{k}}-\mu)a_{\mathbf{k}}^{+}a_{\mathbf{k}} + \sum_{\mathbf{k}\mathbf{k}'} A_{\mathbf{k}\mathbf{k}'}a_{\mathbf{k}}^{+}a_{\mathbf{k}'} \tag{4.33}$$

(see eqn (1.13)).

Momentum is no longer conserved at each vertex, and so we are forced to consider the Green function:

$$G_{\mathbf{k}\mathbf{k}'}(\tau-\tau') = \langle \mathscr{T} a_{\mathbf{k}}^{\mathrm{H}}(\tau)\tilde{a}_{\mathbf{k}'}^{\mathrm{H}}(\tau')\rangle.$$

The diagrams for G, shown in Fig. 4.8, are deceptively simple, as can be seen

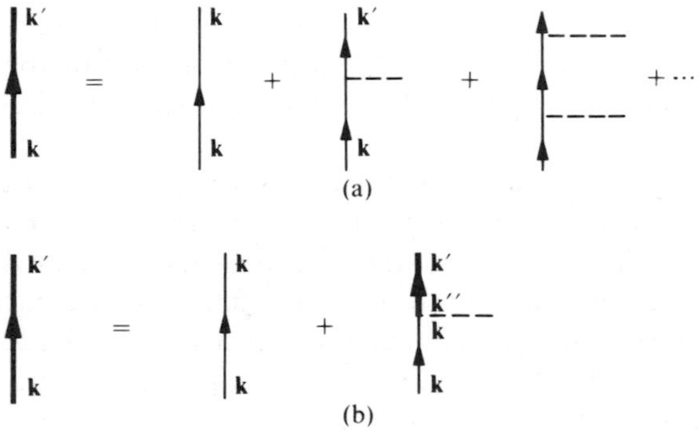

FIG. 4.8.

by writing down the integral equation corresponding to the diagrammatic equation of Fig. 4.8(b) which gives their sum. In ω-space, this takes the form

$$\bar{G}(\mathbf{k},\mathbf{k}',\mathrm{i}\omega_n) = \bar{g}_{\mathbf{k}}(\mathrm{i}\omega_n)\delta_{\mathbf{k},\mathbf{k}'} + \sum_{\mathbf{k}''} \bar{g}_{\mathbf{k}}(\mathrm{i}\omega_n)A_{\mathbf{k}\mathbf{k}''}\bar{G}(\mathbf{k}'',\mathbf{k}',\mathrm{i}\omega_n). \tag{4.34}$$

In some cases the form of $A_{\mathbf{k}\mathbf{k}''}$ is such that this integral equation can be solved exactly (see exercise 4.4.1, below). If we are considering a random distribution of impurities, however, an exact solution is no longer possible. In such a case, we are usually interested in the mean value of a Green function over a random distribution of impurities. We write quantities which have been averaged over this distribution with a curly bracket round them: e.g. $\{\bar{G}(\mathbf{k}, \mathbf{k}', \mathrm{i}\omega_n)\}$. If the impurities are situated at the points $\mathbf{x}_i$,

$$\begin{aligned} A_{\mathbf{kk}'} &= \frac{1}{V}\sum_i \int A(\mathbf{x}-\mathbf{x}_i)\exp[\mathrm{i}(\mathbf{k}-\mathbf{k}').\mathbf{x}]\,\mathrm{d}^3\mathbf{x} \\ &= \frac{1}{V}\sum_i \int A(\mathbf{x}-\mathbf{x}_i)\exp[\mathrm{i}(\mathbf{k}-\mathbf{k}').(\mathbf{x}-\mathbf{x}_i)]\,\mathrm{d}^3\mathbf{x}\exp[\mathrm{i}(\mathbf{k}-\mathbf{k}')\mathbf{x}_i] \\ &\equiv A(\mathbf{k}-\mathbf{k}')\sum_i \exp[\mathrm{i}(\mathbf{k}-\mathbf{k}').\mathbf{x}_i]. \end{aligned}$$

Thus

$$\begin{aligned} \{A_{\mathbf{kk}'}\} &= A(\mathbf{k}-\mathbf{k}')\left\{\sum_i \exp[\mathrm{i}(\mathbf{k}-\mathbf{k}').\mathbf{x}_i]\right\} \\ &= CA(0)\delta_{\mathbf{k},\mathbf{k}'}, \end{aligned} \tag{4.35}$$

since the $\mathbf{x}_i$ are randomly distributed.† C is the number of impurities.

Similarly

$$\begin{aligned} \{A_{\mathbf{k}_1\mathbf{k}_1'}A_{\mathbf{k}_2\mathbf{k}_2'}\} &= A(\mathbf{k}_1-\mathbf{k}_1')A(\mathbf{k}_2-\mathbf{k}_2')\left\{\sum_{ij}\exp(\mathrm{i}[(\mathbf{k}_1-\mathbf{k}_1').\mathbf{x}_i+(\mathbf{k}_2-\mathbf{k}_2').\mathbf{x}_j])\right\} \\ &= A(\mathbf{k}_1-\mathbf{k}_1')A(\mathbf{k}_2-\mathbf{k}_2')\left\{\left(\sum_{i=j}+\sum_{i\neq j}\right)\exp(\cdots)\right\} \\ &= C(C-1)[A(0)]^2\delta_{\mathbf{k}_1,\mathbf{k}_1'}\delta_{\mathbf{k}_2,\mathbf{k}_2'}+C[A(\mathbf{k}_1-\mathbf{k}_1')]^2\delta_{\mathbf{k}_1+\mathbf{k}_2,\mathbf{k}_1'+\mathbf{k}_2'}. \end{aligned} \tag{4.36}$$

In the general term, there will be a δ-function for the momentum labels associated with each independent sum. To proceed further algebraically is unduly tedious and the time has come to turn to diagrams for help. The nth-order term in an expansion of $\bar{G}(\mathbf{k}, \mathbf{k}')$ will be

$$\sum_{\mathbf{k}_2\ldots\mathbf{k}_n} g_{\mathbf{k}}g_{\mathbf{k}_2}\cdots g_{\mathbf{k}_n}g_{\mathbf{k}'}A_{\mathbf{kk}_2}A_{\mathbf{k}_2\mathbf{k}_3}\cdots A_{\mathbf{k}_n\mathbf{k}'} \tag{4.37}$$

and the corresponding term in $\{\bar{G}\}$ will have an average of the products of nAs. The first-order term eqn (4.35) we represent by Fig. 4.9(a); the cross at the end of the dotted line indicates that the average has been taken. In second order, the two terms of eqn (4.36) are represented by Fig. 4.9(b). The As associated with the same δ-function are joined to the same cross. Since the δ-function results from averaging over one $\mathbf{x}_i$, it is sometimes said that the scattering occurs at the same impurity. The third-order terms, and some fourth-order terms are shown in Fig. 4.9(c) and (d). The first term of eqn (4.34)

† $A(0)$ can be chosen to be zero, so that this term vanishes, by including all the diagonal terms of $\sum_{\mathbf{kk}'} A_{\mathbf{kk}'}a_{\mathbf{k}}^+a_{\mathbf{k}'}$ in the first term of the Hamiltonian, thus shifting the unperturbed energies $\varepsilon_{\mathbf{k}}$.

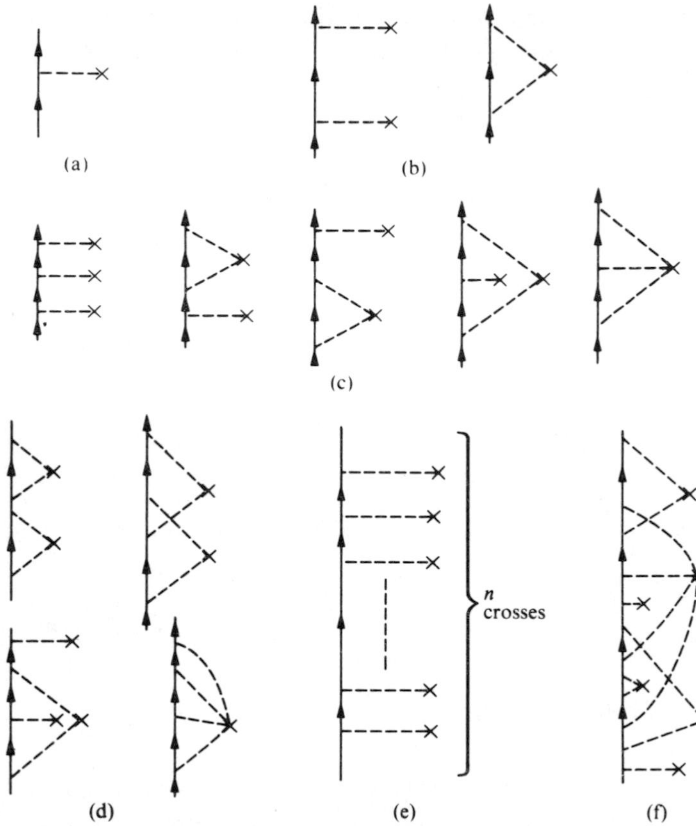

FIG. 4.9.

for the nth order is shown in (e), and a general (twelfth-order) term is shown in (f).

The δ-functions associated with each cross ensure that $\{\bar{G}(\mathbf{k}, \mathbf{k}', i\omega_n)\}$ is zero unless $\mathbf{k} = \mathbf{k}'$. The diagrams now have many similarities to those for Green functions for mutually interacting particles. We may define a proper self-energy part as the sum of all diagrams which cannot be split entirely into two by cutting only one full line. If we call this Σ, we have, in the usual way,†

$$\{\bar{G}(\mathbf{k}, i\omega_n)\} = [i\omega_n - (\varepsilon_{\mathbf{k}} - \mu) - \Sigma(\mathbf{k}, i\omega_n)]^{-1}. \tag{4.38}$$

The simplest diagram for Σ is shown in Fig. 4.10(a), and gives

$$\Sigma^{(a)}(\mathbf{k}, i\omega_n) = CA(0).$$

† This is not strictly true, since, for example, the first term in eqn (4.36) carries a factor $C(C-1)$, rather than C^2. Since we shall have a macroscopic number of impurities, however, we may at this stage neglect this difference.

This merely shifts the energies of the particles by a constant (cf. footnote on p. 123). A simple series of diagrams for Σ is shown in Fig. 4.10(b), and can be thought of as including all scattering events at a single impurity. All these terms give contributions to Σ proportional to C, the number of impurities. The sum of the series $\Sigma^{(b)}$ may be written in terms of the solution to an integral equation:

$$\Sigma^{(b)}(\mathbf{k}, \mathrm{i}\omega_n) = C\Big[A(0) + \sum_{\mathbf{k}_1} A(\mathbf{k}-\mathbf{k}_1)A(\mathbf{k}_1-\mathbf{k})\bar{g}(\mathbf{k}_1, \mathrm{i}\omega_n) + \\ + \sum_{\mathbf{k}_1,\mathbf{k}_2} A(\mathbf{k}-\mathbf{k}_1)A(\mathbf{k}_1-\mathbf{k}_2)A(\mathbf{k}_2-\mathbf{k}) \times \\ \times \bar{g}(\mathbf{k}_1, \mathrm{i}\omega_n)\bar{g}(\mathbf{k}_2, \mathrm{i}\omega_n) + \cdots\Big]$$

$$= Ct(\mathbf{k}, \mathbf{k}, \mathrm{i}\omega_n), \tag{4.39}$$

where

$$t(\mathbf{k}, \mathbf{k}', \mathrm{i}\omega_n) = A(\mathbf{k}-\mathbf{k}') + \sum_{\mathbf{k}_1} A(\mathbf{k}-\mathbf{k}_1)A(\mathbf{k}_1-\mathbf{k}')\bar{g}(\mathbf{k}_1, \mathrm{i}\omega_n) + \cdots$$

$$= A(\mathbf{k}-\mathbf{k}') + \sum_{\mathbf{k}_1} A(\mathbf{k}-\mathbf{k}_1)\bar{g}(\mathbf{k}_1, \mathrm{i}\omega_n)t(\mathbf{k}_1, \mathbf{k}', \mathrm{i}\omega_n). \tag{4.40}$$

This equation has obvious similarities with the t-matrix equation of the previous section.

All terms in Σ depending linearly on the concentration have now been summed. For discussion of how terms of higher order in the concentration can be included, we refer the reader to the literature.

For low concentrations and weak interactions, the first two terms in eqn (4.36) will give a reasonable approximation for Σ. Extending this approximation for Σ to the whole of the complex ω-plane, we have, near the real axis,

$$\Sigma(\mathbf{k}, \omega - \mathrm{i}\delta) = C\Big[A(0) + \frac{V}{(2\pi)^3} P\int |A(\mathbf{k}-\mathbf{k}_1)|^2[\omega - \varepsilon_{\mathbf{k}_1} + \mu]^{-1}\,\mathrm{d}^3\mathbf{k}_1 + \\ + \mathrm{i}\pi\frac{V}{(2\pi)^3}\int \mathrm{d}^3\mathbf{k}_1 |A(\mathbf{k}-\mathbf{k}_1)|^2 \delta(\omega - \varepsilon_{\mathbf{k}_1} + \mu)\Big].$$

Thus we have quasi-particles with energy

$$\varepsilon_{\mathbf{k}} + CA(0) + C\frac{V}{(2\pi)^3} P\int \frac{\mathrm{d}^3\mathbf{k}_1}{\varepsilon_{\mathbf{k}} - \varepsilon_{\mathbf{k}_1}} |A(\mathbf{k}-\mathbf{k}_1)|^2$$

and lifetime

$$\tau = \Big[C\pi\frac{V}{(2\pi)^3}\int \mathrm{d}^3\mathbf{k}_1 |A(\mathbf{k}-\mathbf{k}_1)|^2 \delta\,(\varepsilon_{\mathbf{k}} - \varepsilon_{\mathbf{k}_1})\Big]^{-1}.$$

(a) (b)

FIG. 4.10.

A very simple kinetic theory approach would thus suggest that the conductivity was $ne^2\tau/m$ where n is the density of the electrons and e their charge. We shall return to a discussion of the conductivity in Chapter 7.

Further reading

Edwards (1958) gives an account of the scattering of electrons by random impurities. Soven (1969), using slightly different Green functions from ours, describes an approximation which goes beyond the t-matrix approximation described above.

Exercise

4.4.1. Find the exact solution of eqn (4.34) for the case in which $A_{\mathbf{kk}''}$ is separable as a product of factors: $A_{\mathbf{kk}''} = \alpha(\mathbf{k})\alpha(\mathbf{k}'')$. [Hint: put

$$\gamma_{\mathbf{k}}(\mathrm{i}\omega_n) = \sum_{\mathbf{k}''} \alpha(\mathbf{k}'')\bar{G}(\mathbf{k}'', \mathbf{k}, \mathrm{i}\omega_n)$$

and use eqn (4.34) to write

$$\gamma_{\mathbf{k}}(\mathrm{i}\omega_n) = \sum_{\mathbf{k}'} \alpha(\mathbf{k}')\bar{g}_{\mathbf{k}'}(\mathrm{i}\omega_n)\delta_{\mathbf{k},\mathbf{k}'} + \sum_{\mathbf{k}'} [\alpha(\mathbf{k}')]^2\bar{g}_{\mathbf{k}'}(\mathrm{i}\omega_n)\gamma_{\mathbf{k}}(\mathrm{i}\omega_n).]$$

These equations may be used to discuss the Kronig–Penney model for electrons in a solid.

4.5. The interacting Bose gas below the transition temperature

The Grand Canonical Ensemble may not be applied directly to a Bose gas in which a finite fraction of the total number of particles is condensed into one single-particle state. For the non-interacting gas, for example, the number of particles, N_0, in the zero-momentum state is macroscopic below the transition temperature, and the fluctuations in the total number of particles, measured by

$$[\langle \hat{N}^2 \rangle - \langle \hat{N} \rangle^2] \sim N_0,$$

is of the same order as $\bar{N}$, the total number of particles. (See exercise 2.2.1 and Johnston (1970) for a further discussion of this fact). To avoid this difficulty, we may treat the macroscopically occupied state separately, and regard it as a reservoir of particles for the other states. If this idea is used for the interacting gas, one often also uses the Bogoliubov idea (see exercise 1.2.5) that, since the difference between $a_0^+ a_0$ and $a_0 a_0^+$ is

$$a_0 a_0^+ - a_0^+ a_0 = 1,$$

and since this is very small compared with $\langle a_0^+ a_0 \rangle$ which is of order $\bar{N}$ for the condensed system, a_0^+ and a_0 may be treated as c-numbers, $(N_0)^{\frac{1}{2}}$ say. The thermodynamic potential for the whole system is thus given by

$$\Omega = -\beta^{-1} \ln \left[\sum_{N_0} \exp(\beta\mu N_0) Q'(N_0, \mu) \right], \tag{4.41}$$

where Q' is the grand partition function of the particles with $\mathbf{k} \neq 0$; that is, the grand partition function calculated using the Hamiltonian

$$\begin{aligned} H' = {} & \sum{}' \frac{k^2}{2m} a_{\mathbf{k}}^+ a_{\mathbf{k}} + \tfrac{1}{2} \sum_{\mathbf{k}_1 \mathbf{k}_2\, \mathbf{q}}{}' \, \bar{v}(q) a_{\mathbf{k}_1}^+ a_{\mathbf{k}_2}^+ a_{\mathbf{k}_2 - \mathbf{q}} a_{\mathbf{k}_1 + \mathbf{q}} \; + \\ & + N_0^{\frac{1}{2}} \sum_{\mathbf{k}_1, \mathbf{q}}{}' \, \bar{v}(q) [a_{\mathbf{k}_1}^+ a_{\mathbf{q}}^+ a_{\mathbf{k}_1 + \mathbf{q}} + a_{\mathbf{k}_1}^+ a_{\mathbf{q}} a_{\mathbf{k}_1 - \mathbf{q}}] \; + \\ & + N_0 \sum_{\mathbf{q}}{}' \, [\bar{v}(q) + \bar{v}(0)] a_{\mathbf{q}}^+ a_{\mathbf{q}} + \tfrac{1}{2} \sum_{\mathbf{q}}{}' \, \bar{v}(q) [a_{\mathbf{q}}^+ a_{-\mathbf{q}}^+ + a_{\mathbf{q}} a_{-\mathbf{q}}] + \\ & + \tfrac{1}{2} \bar{v}(0) N_0^2 \end{aligned} \tag{4.42}$$

where primes on the summation symbols indicate that the terms with momentum equal to zero are to be omitted from the sums. If the fluctuations in N_0 are small, the logarithm of the sum in eqn (4.41) is well approximated by the logarithm of the largest term in the sum, and we have:

$$\begin{aligned} \Omega &= -\mu \bar{N}_0 - \beta^{-1} \ln Q'(\bar{N}_0, \mu) \\ &\equiv -\mu \bar{N}_0 + \Omega' \end{aligned} \tag{4.43}$$

where

$$\mu = \left(\frac{\partial \Omega'}{\partial N_0} \right)_{N_0 = \bar{N}_0} \tag{4.44}$$

(See Hugenholtz and Pines (1959), Parry and Turner (1962*b*), Mills (1971) for more details and for justification of this procedure.)

The several different types of vertex contained in the Hamiltonian H' are shown in Fig. 4.11.

We now define two Green functions:

$$G_{\mathbf{k}}(\tau_1, \tau_2) = \langle \mathscr{T} a_{\mathbf{k}}^{\mathrm{H}}(\tau_1) \bar{a}_{\mathbf{k}}^{\mathrm{H}}(\tau_2) \rangle, \tag{4.45}$$

$$\tilde{G}_{\mathbf{k}}(\tau_1, \tau_2) = \langle \mathscr{T} a_{\mathbf{k}}^{\mathrm{H}}(\tau_1) a_{-\mathbf{k}}^{\mathrm{H}}(\tau_2) \rangle. \tag{4.46}$$

In normal systems, for which the Hamiltonian contains only particle-conserving terms, the latter function will be zero. In our case, we may write down the two diagrammatic equations in the **k**, ω representation:

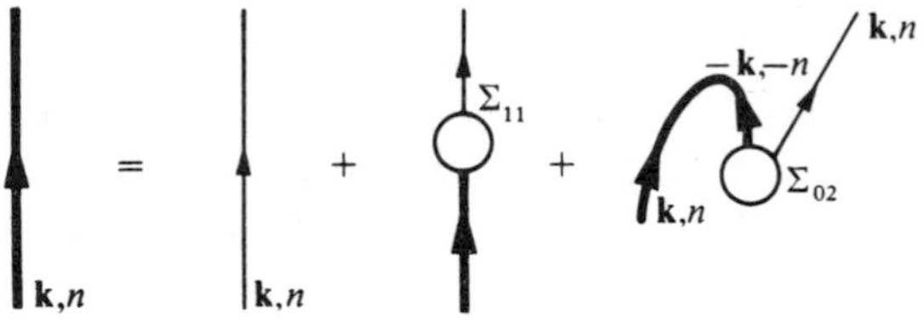

(4.47)

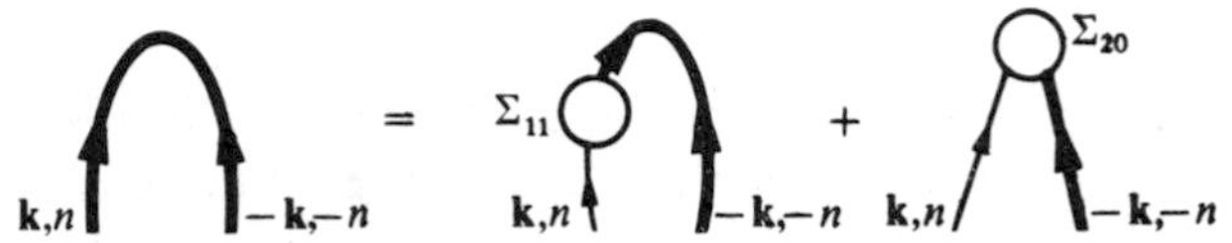

(4.48)

with algebraic counterparts:

$$\bar{G}(\mathbf{k}, \mathrm{i}\omega_n) = \bar{g}(\mathbf{k}, \mathrm{i}\omega_n) + \bar{g}(\mathbf{k}, \mathrm{i}\omega_n)\,\Sigma_{11}(\mathbf{k}, \mathrm{i}\omega_n)\bar{G}(\mathbf{k}, \mathrm{i}\omega_n) + \\ + \bar{g}(\mathbf{k}, \mathrm{i}\omega_n)\Sigma_{02}(\mathbf{k}, \mathrm{i}\omega_n)\,\tilde{G}(\mathbf{k}, \mathrm{i}\omega_n), \tag{4.47a}$$

$$\tilde{G}(\mathbf{k}, \mathrm{i}\omega_n) = \bar{g}(\mathbf{k}, \mathrm{i}\omega_n)\Sigma_{11}(\mathbf{k}, \mathrm{i}\omega_n)\,\tilde{G}(\mathbf{k}, \mathrm{i}\omega_n) + \\ + \bar{g}(\mathbf{k}, \mathrm{i}\omega_n)\Sigma_{20}(\mathbf{k}, \mathrm{i}\omega_n)\,\bar{G}(-\mathbf{k}, -\mathrm{i}\omega_n), \tag{4.48a}$$

where Σ_{11} is the sum of all diagrams with one line entering and one leaving which cannot be separated into two parts by cutting a single full line, and $\Sigma_{20}(\mathbf{k}, \omega)$ and $\Sigma_{02}(\mathbf{k}, \omega)$ are defined similarly for two lines entering and none leaving, and for none entering and two leaving respectively. These equations may be solved, to give

$$\bar{G}(\mathbf{k}, \mathrm{i}\omega_n) = (\mathrm{i}\omega_n + \varepsilon_{\mathbf{k}} - \mu + \Sigma_{11})/\{[\mathrm{i}\omega_n - \tfrac{1}{2}(\Sigma_{11}^{+} - \Sigma_{11}^{-})]^2 - \\ - [\varepsilon_{\mathbf{k}} - \mu + \tfrac{1}{2}(\Sigma_{11}^{+} + \Sigma_{11}^{-})]^2 + \Sigma_{02}\Sigma_{20}\}, \tag{4.49}$$

where

$$\Sigma_{11}^{+} = \Sigma_{11}(\mathbf{k}, \mathrm{i}\omega_n) \text{ and } \Sigma_{11}^{-} = \Sigma_{11}^{-}(-\mathbf{k}, -\mathrm{i}\omega_n).$$

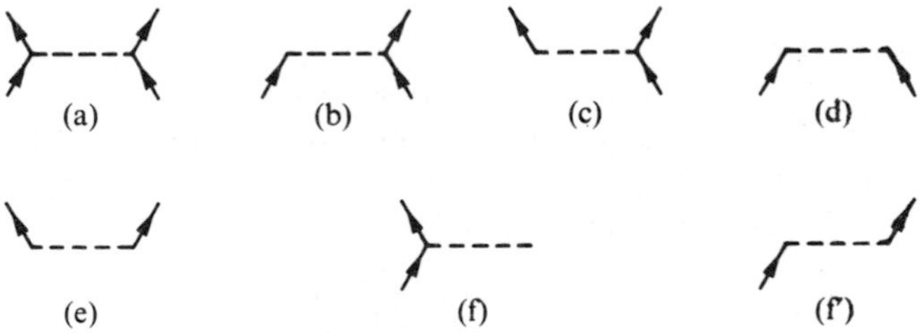

FIG. 4.11. The vertices in H' of eqn. (4.42).

It may be shown (see exercise 4.5.1 below) that eqn (4.44) leads to

$$\mu = \operatorname*{Lt}_{\mathbf{k}\to 0}\,[\Sigma_{11}(\mathbf{k},0)-\Sigma_{02}(\mathbf{k},0)] \tag{4.50}$$

which shows that $\mathrm{Lt}_{\mathbf{k}\to 0}\,\omega_{\mathbf{k}} = 0$, where ω_k is the quasi-particle energy.

The Bogoliubov results (see exercise 1.2.5) may be obtained by taking the lowest-order approximations for Σ_{11} and Σ_{02}:

$$\Sigma_{11}(\mathbf{q},\mathrm{i}\omega_n) = \bar{N}_0[\bar{v}(0)+\bar{v}(q)]$$

$$\Sigma_{02}(\mathbf{q},\mathrm{i}\omega_n) = \Sigma_{20}(\mathbf{q},\mathrm{i}\omega_n) = \bar{N}_0\bar{v}(q) \tag{4.51}$$

The analytic continuation of G to the whole of the complex plane is straightforward in this case:

$$\bar{G}(\mathbf{k},\omega) = [\omega+k^2/2m+\bar{N}_0\bar{v}(k)]/\{\omega^2-[k^2/2m+\bar{N}_0\bar{v}(k)]^2+\bar{N}_0^2[\bar{v}(k)]^2\} \tag{4.52}$$

and we obtain poles at the points on the real axis:

$$\omega_k^2 = k^4/4m^2+\bar{N}_0\bar{v}(k)k^2/m \tag{4.53}$$

corresponding to quasi-particles with these energies. The quasi-particle energies in liquid helium have been obtained experimentally by neutron scattering (e.g. Henshaw and Woods (1961)). They are roughly of the form given by eqn (4.53). In particular, provided $\bar{v}(0) \neq 0$, ω_k given by eqn (4.53) is linear in k for low k. A great deal of work has been done to try to obtain higher-order results than Bogoliubov, so that the theory could be applied in a quantitative way to liquid helium. (See Brandow (1971) and references contained in that paper.) Although success has sometimes been claimed, there is still no satisfactory theory of liquid helium which starts from the interatomic potential, and ends at the quasi-particle energy spectrum, making only well-justified approximations on the way.

Exact results on the low-temperature behaviour of the system have been derived by Hugenholtz and Pines (1959), Gavoret and Nozières (1964), Götze and Wagner (1965), and Kehr (1967). Books by Atkins (1959) and Wilks (1967) review the experimental and theoretical situation.

Exercise

4.5.1. Every diagram of order higher than one for Σ_{11} may be produced from a diagram for Ω' by adding one ingoing line at one of the vertices of type c, e, f or f′, and one outgoing line at one of the vertices of type b, d, f, or f′. If there are n_{a} vertices of type a, etc., in a particular diagram for Ω', show that the additional lines may be added in

$$[n_{\mathrm{b}}+2n_{\mathrm{d}}+n_{\mathrm{f}}+n_{\mathrm{f'}}][n_{\mathrm{c}}+2n_{\mathrm{e}}+n_{\mathrm{f}}+n_{\mathrm{f'}}]$$

ways, to form different diagrams for Σ_{11}. Similarly, show that the number of ways of converting a diagram for Ω' into one for Σ_{02} is

$$n_{\mathrm{b}}(n_{\mathrm{b}}-1+2n_{\mathrm{d}}+n_{\mathrm{f}}+n_{\mathrm{f'}})+2n_{\mathrm{d}}(n_{\mathrm{b}}+1+2n_{\mathrm{d}}-2+n_{\mathrm{f}}+n_{\mathrm{f'}})+$$
$$+(n_{\mathrm{f}}+n_{\mathrm{f'}})(n_{\mathrm{b}}+2n_{\mathrm{d}}+n_{\mathrm{f}}+n_{\mathrm{f'}}-1).$$

Notice that the number of factors of N_0 in such a diagram for Ω' is

$$s = \tfrac{1}{2}(n_{\mathrm{b}}+n_{\mathrm{c}}+2n_{\mathrm{d}}+2n_{\mathrm{e}}+2n_{\mathrm{f}}+2n_{\mathrm{f'}})$$
$$= n_{\mathrm{b}}+2n_{\mathrm{d}}+n_{\mathrm{f}}+n_{\mathrm{f'}} \quad \text{since } n_{\mathrm{b}}+2n_{\mathrm{d}} = n_{\mathrm{c}}+2n_{\mathrm{e}}.$$

Hence, using eqn (4.44) prove eqn (4.50). (The first-order terms in $\Sigma_{11}(0,0)-$ $-\Sigma_{02}(0,0)$ come from differentiating with respect to N_0 the last term in H' in eqn (4.42), which, being merely an additive constant, appears in the same form in Ω'.) Verify the statement that $\mathrm{Lt}_{k\to 0}\,\omega_k = 0$. (See Hugenholtz and Pines (1959).)

This procedure has been extended by Gavoret and Nozières (1964) to show generally that $\omega_k \propto k$ for small k, with the velocity of sound as the constant of proportionality.

5

THE EQUATION OF MOTION METHOD

OUR Green functions were originally introduced in the discussion of the thermodynamic potential. In previous sections we have seen that they have significance in their own right; from their form, one can make deductions about the elementary excitations in our system; through eqn (3.82) they can give thermodynamic properties of the system; they may be used, through eqns (3.55 and 59) to calculate correlation functions which one needs, for example, in the theory of scattering from many-body systems (exercise 3.6.4) and in the theory of linear response and relaxation (Chapter 7). Green functions thus assume a central role in our theory, and it is reasonable to ask if there are ways of evaluating them other than the diagrammatic method described above. It is always useful to have more than one way of evaluating quantities of interest, but an alternative method is particularly useful for magnetic systems, for which a diagrammatic method is rather difficult to develop in some cases (see Chapter 6).

5.1. The equation of motion and the Hartree–Fock approximation

The starting point of the method we now wish to describe is the equation of motion for Green functions. Since, for operators in the modified Heisenberg representation we are using,

$$A(t) = \exp[\mathrm{i}(H-\mu\hat{N})t]A\exp[-\mathrm{i}(H-\mu\hat{N})t], \tag{5.1}$$

we have

$$\mathrm{i}\frac{\mathrm{d}A}{\mathrm{d}t} = A(t)(H-\mu\hat{N})-(H-\mu\hat{N})A(t) \tag{5.2}$$

assuming that the operator A is not explicitly time-dependent. Further

$$\frac{\mathrm{d}}{\mathrm{d}t}\theta(t) = \delta(t). \tag{5.3}$$

Thus, we may write for the retarded, advanced, or causal Green functions, defined in eqns (3.50, 51, 52),

$$\mathrm{i}\frac{\mathrm{d}}{\mathrm{d}t}\langle\langle A(t);B(t')\rangle\rangle_{\mathrm{r,a,c}} = \delta(t-t')\langle[A(t),B(t')]_{\pm}\rangle + \\ + \langle\langle[A(t),H-\mu\hat{N}]_{-};B(t')\rangle\rangle_{\mathrm{r,a,c}}. \tag{5.4}$$

Since the equation of motion is the same for all three functions, we shall often omit the suffices in what follows. If we knew the Green function on the right-hand side of this equation. $\langle\langle [A(t), H-\mu\hat{N}]_-; B(t')\rangle\rangle$, we should have a differential equation for $\langle\langle A; B\rangle\rangle$. Unfortunately, the new Green function is usually more complicated than $\langle\langle A; B\rangle\rangle$ and the only step we seem able to take is to write down the equation of motion of this new Green function, and thereby obtain an even more complicated Green function on the right-hand side of this new equation. Proceeding further in the same way, we shall obtain an infinite set of coupled equations, each probably more complicated than its predecessor.

To illustrate this procedure, we shall write down the first two equations in the chain starting with the single-particle Green function for a system of fermions, or bosons above the transition temperature, with the usual Hamiltonian (eqns (2.7, 8 and 9)). We shall work with the Fourier transform:

$$\langle\langle A(t); B(t')\rangle\rangle = \frac{1}{2\pi}\int_{-\infty}^{\infty} \langle\langle A; B:\omega\rangle\rangle \exp[-\mathrm{i}\omega(t-t')]\,\mathrm{d}\omega. \tag{5.5}$$

Our two equations are then:

$$(\omega-\varepsilon_{\mathbf{k}}+\mu)\langle\langle a_{\mathbf{k}}; a^+_{\mathbf{k}}:\omega\rangle\rangle = 1+\sum_{\mathbf{k}_1,\mathbf{q}} \bar{v}(q)\langle\langle a^+_{\mathbf{k}_1} a_{\mathbf{k}_1+\mathbf{q}} a_{\mathbf{k}-\mathbf{q}}; a^+_{\mathbf{k}}:\omega\rangle\rangle, \tag{5.6}$$

$$\begin{aligned}(\omega-\varepsilon_{\mathbf{k}_1+\mathbf{q}}-\varepsilon_{\mathbf{k}-\mathbf{q}}+\varepsilon_{\mathbf{k}_1}+\mu)&\langle\langle a^+_{\mathbf{k}_1} a_{\mathbf{k}_1+\mathbf{q}} a_{\mathbf{k}-\mathbf{q}}; a^+_{\mathbf{k}}:\omega\rangle\rangle\\
&= \langle a^+_{\mathbf{k}_1} a_{\mathbf{k}_1}\rangle[\delta_{\mathbf{q},0}+\varepsilon\delta_{\mathbf{k},\mathbf{k}_1+\mathbf{q}}]+\\
&\quad+\sum_{\mathbf{k}'_1\mathbf{q}'} \bar{v}(q')\{\langle\langle a^+_{\mathbf{k}_1} a_{\mathbf{k}_1+\mathbf{q}} a^+_{\mathbf{k}'_1} a_{\mathbf{k}'_1-\mathbf{q}'} a_{\mathbf{k}-\mathbf{q}+\mathbf{q}'}; a^+_{\mathbf{k}}:\omega\rangle\rangle+\\
&\quad+\langle\langle a^+_{\mathbf{k}_1} a^+_{\mathbf{k}'_1} a_{\mathbf{k}'_1-\mathbf{q}'} a_{\mathbf{k}_1+\mathbf{q}+\mathbf{q}'} a_{\mathbf{k}-\mathbf{q}}; a^+_{\mathbf{k}}:\omega\rangle\rangle -\\
&\quad-\varepsilon\langle\langle a^+_{\mathbf{k}'_1} a^+_{\mathbf{k}_1+\mathbf{q}'} a_{\mathbf{k}'_1+\mathbf{q}'} a_{\mathbf{k}_1+\mathbf{q}} a_{\mathbf{k}-\mathbf{q}}; a^+_{\mathbf{k}}:\omega\rangle\rangle\},\end{aligned} \tag{5.7}$$

where we have carried out some rearrangement after working out the commutators.

If we proceed in the same way, our equations are clearly going to become more and more complicated. If we are interested in approximations, however, we can make some progress. We notice that the difficult terms in eqn (5.6) carry a factor $\bar{v}(q)$, so that if we were interested in obtaining $\langle\langle a_{\mathbf{k}}; a^+_{\mathbf{k}}:\omega\rangle\rangle$ to first order, we could insert a zero-order expression for $\langle\langle a^+_{\mathbf{k}_1} a_{\mathbf{k}_1+\mathbf{q}} a_{\mathbf{k}-\mathbf{q}}; a^+_{\mathbf{k}}:\omega\rangle\rangle$. There are several reasons why we are rarely interested in an expansion of a Green function, however. One is apparent from eqn (5.6); if we divide that equation by $(\omega-\varepsilon_{\mathbf{k}}+\mu)$ to obtain $\langle\langle a_{\mathbf{k}}; a^+_{\mathbf{k}}:\omega\rangle\rangle$, the last term can be made arbitrarily large even if $\bar{v}(q)$ is small, in the region $\omega \sim \varepsilon_{\mathbf{k}}-\mu$, and this is the region in which we are most interested. Secondly, we have seen above that we are often interested in the position of the poles of a Green function:

this means that we are more interested in an expansion of $\bar{G}^{-1}$, rather than of $\bar{G}$. We may obtain the first-order terms in the denominator of $\bar{G}$ by making the approximation:

$$\langle\langle a^+_{\mathbf{k}_1} a_{\mathbf{k}_1+\mathbf{q}} a_{\mathbf{k}-\mathbf{q}} ; a^+_{\mathbf{k}} : \omega \rangle\rangle \simeq \langle a^+_{\mathbf{k}_1} a_{\mathbf{k}_1} \rangle \langle\langle a_{\mathbf{k}} ; a^+_{\mathbf{k}} : \omega \rangle\rangle [\delta_{\mathbf{q},0} + \varepsilon \delta_{\mathbf{k},\mathbf{k}_1+\mathbf{q}}], \quad (5.8)$$

which we can justify by saying that we are neglecting the fluctuations of $a^+_s a_t$ about its 'mean value' $\langle a^+_s a_t \rangle$, which is zero unless $s = t$. Using this approximation, we obtain

$$\langle\langle a^+_{\mathbf{k}} ; a_{\mathbf{k}} : \omega \rangle\rangle \left[\omega - \varepsilon_{\mathbf{k}} + \mu - \sum_{\mathbf{k}_1} \langle a^+_{\mathbf{k}_1} a_{\mathbf{k}_1} \rangle [\bar{v}(0) + \varepsilon \bar{v}(|\mathbf{k} - \mathbf{k}_1|)] \right] = 1. \quad (5.9)$$

We notice that by this means we have obtained the self-consistent Hartree–Fock result for the quasi-particle energies (compare with eqn (4.4)). To find $\langle a^+_{\mathbf{k}_1} a_{\mathbf{k}_1} \rangle$, we shall have to use eqns (3.55) and (3.59), which give:

$$\begin{aligned} \langle a^+_{\mathbf{k}_1} a_{\mathbf{k}_1} \rangle &= \{\exp[\beta(E_{\mathbf{k}} - \mu)] + 1\}^{-1} \\ E_{\mathbf{k}} &= \varepsilon_{\mathbf{k}} + \sum_{\mathbf{k}_1} \langle a^+_{\mathbf{k}_1} a_{\mathbf{k}_1} \rangle [\bar{v}(0) + \varepsilon \bar{v}(|\mathbf{k} - \mathbf{k}_1|)]. \end{aligned} \quad (5.10)$$

To solve these self-consistent equations for $\langle a^+_{\mathbf{k}} a_{\mathbf{k}} \rangle$ is difficult, as we said in the first section of the previous chapter. To obtain the straightforward, non-self-consistent Hartree–Fock result, we merely replace, $\langle a^+_{\mathbf{k}} a_{\mathbf{k}} \rangle$ by $\langle a^+_{\mathbf{k}} a_{\mathbf{k}} \rangle_0$.

Exercises

5.1.1. Use the equation of motion method to find the single-particle Green function for a system described by the Bogoliubov Hamiltonian of exercise 1.2.5.

5.1.2. Discuss the scattering of neutrons from a boson system described by the Bogoliubov Hamiltonian (see exercise 3.6.4 and Parry and Turner (1962a)).

5.1.3. Bound pairs of electrons with equal and opposite momentum and spin are thought to be very important in superconductors. In this case, the terms $\langle\langle a^+_{-\mathbf{k}} a_{-\mathbf{k}+\mathbf{q}} a_{\mathbf{k}-\mathbf{q}} ; a^+_{\mathbf{k}} : \omega \rangle\rangle$ in eqn (5.9) will be important, as well as the term $\langle\langle a^+_{\mathbf{k}_1} a_{\mathbf{k}_1} a_{\mathbf{k}} ; a^+_{\mathbf{k}} : \omega \rangle\rangle$. (As $\mathbf{k}$ includes a spin label, we define $-\mathbf{k}$ to denote the opposite momentum and opposite spin to $\mathbf{k}$.) Make the approximation

$$\begin{aligned} &\langle\langle a^+_{-\mathbf{k}} a_{-\mathbf{k}+\mathbf{q}} a_{\mathbf{k}-\mathbf{q}} ; a^+_{\mathbf{k}} : \omega \rangle\rangle \\ &\quad = \langle\langle a^+_{-\mathbf{k}} ; a^+_{\mathbf{k}} : \omega \rangle\rangle \langle a_{-\mathbf{k}+\mathbf{q}} a_{\mathbf{k}-\mathbf{q}} \rangle \end{aligned}$$

in the equation of motion for $\langle\langle a_{\mathbf{k}} ; a^+_{\mathbf{k}} : \omega \rangle\rangle$. In a normal system, both factors would be zero, since they are expectation values of products which do not conserve particle number. However, it is convenient to pretend that they are non-zero in order to take into account in a simple way the important term mentioned above, justifying the procedure by adding a term to the Hamiltonian of the form $\mathrm{Lt}_{\alpha\to 0}\alpha\sum_{\mathbf{k}}(a_{\mathbf{k}}^{+}a_{-\mathbf{k}}^{+}+a_{\mathbf{k}}a_{-\mathbf{k}})$ which allows particle non-conservation, but leaves the Hamiltonian unchanged when we take the limit $\alpha\to 0$.

Write down the approximate equation of motion corresponding to eqn (5.9) including these extra terms, and a corresponding equation for $\langle\langle a_{-\mathbf{k}}^{+};a_{\mathbf{k}}^{+}:\omega\rangle\rangle$. (Compare these with the equations of motion in exercise 5.1.1.) Consider the case in which the interaction is zero everywhere except for particles in a small region near the Fermi surface such that

$$\varepsilon_{\mathrm{F}}-\varepsilon_0<\varepsilon_{\mathbf{k}}<\varepsilon_{\mathrm{F}}+\varepsilon_0 \qquad (\alpha)$$

where it is a constant g. (Only pairs near the Fermi surface are thought to be important.) Show that, in this case,

$$\langle\langle a_{\mathbf{k}};a_{\mathbf{k}}^{+}:\omega\rangle\rangle=\left(\omega+\frac{k^2}{2m}-\mu\right)\Big/\left[\omega^2-\left(\frac{k^2}{2m}-\mu\right)^2-\Delta^2\right],$$

where

$$\Delta^2=g^2\sum_{\mathbf{k}_1}{}'\langle a_{\mathbf{k}_1}^{+}a_{-\mathbf{k}_1}^{+}\rangle\sum_{\mathbf{k}_2}{}'\langle a_{\mathbf{k}_2}a_{-\mathbf{k}_2}\rangle,$$

and where the primes on the summation symbols indicate that they are to be taken over the region (α).

We thus have excitations with an energy gap Δ. Find a self-consistent equation for Δ from the expression for $\langle\langle a_{-\mathbf{k}}^{+};a_{\mathbf{k}}^{+}:\omega\rangle\rangle$, and show that there is a transition temperature above which $\Delta=0$, and below which one of the solutions has $\Delta\neq 0$. This can be shown to be the stable solution, justifying our original procedure.

A comprehensive bibliography on superconductivity would be enormously lengthy. The pairing theory is called the BCS theory (Bardeen, Cooper, Schrieffer (1957).) Amongst reviews on the whole subject are books by Rickayzen (1965) and Kuper (1968).

5.2. The electron gas: the random-phase approximation

We shall illustrate the sort of procedure one has to adopt to achieve higher-order results by considering the electron gas again. (Compare section 4.2.) In this case:

$$\bar{v}(q)=4\pi e^2/(V_q^2),\qquad q\neq 0;\qquad \bar{v}(0)=0.$$

We rewrite eqn (5.6) in the form, remembering $\bar{v}(0) = 0$,

$$[\omega - \varepsilon_{\mathbf{k}} + \mu + \sum_{\mathbf{k}_1} \langle n_{\mathbf{k}_1} \rangle \bar{v}(|\mathbf{k} - \mathbf{k}_1|)] \langle\langle a_{\mathbf{k}} ; a_{\mathbf{k}}^+ : \omega \rangle\rangle$$

$$= 1 + \sum_{\mathbf{k}_1, \mathbf{q}}' \bar{v}(q) \langle\langle a_{\mathbf{k}_1}^+ a_{\mathbf{k}_1 + \mathbf{q}} a_{\mathbf{k} - \mathbf{q}} ; a_{\mathbf{k}}^+ : \omega \rangle\rangle -$$

$$- \sum_{\mathbf{k}_1} \bar{v}(|\mathbf{k} - \mathbf{k}_1|) \langle\langle (a_{\mathbf{k}_1}^+ a_{\mathbf{k}_1} - \langle n_{\mathbf{k}_1} \rangle) a_{\mathbf{k}} ; a_{\mathbf{k}}^+ : \omega \rangle\rangle, \tag{5.11}$$

where the prime on the summation symbol indicates that $\mathbf{q} \neq \mathbf{k} - \mathbf{k}_1$. The last term on the right-hand side, a 'fluctuation' term, can be shown to give contributions smaller by a factor of $(V)^{-\frac{1}{2}}$ compared with the other terms, and so we need not consider it any further. For the first term, we use eqn (5.7), for the case $\mathbf{q} \neq \mathbf{k} - \mathbf{k}_1$ or 0, since $\bar{v}(0) = 0$. Following the same principle for approximation in this equation as we used in eqn (5.8), we replace the first higher-order Green function by

$$\langle\langle a_{\mathbf{k}_1}^+ a_{\mathbf{k}_1 + \mathbf{q}} a_{\mathbf{k}_1'}^+ a_{\mathbf{k}_1' - \mathbf{q}'} a_{\mathbf{k} - \mathbf{q} + \mathbf{q}'} ; a_{\mathbf{k}}^+ : \omega \rangle\rangle$$

$$\simeq \delta_{\mathbf{k}_1, \mathbf{k}_1' - \mathbf{q}'} \langle n_{\mathbf{k}_1} \rangle \langle\langle a_{\mathbf{k}_1 + \mathbf{q}} a_{\mathbf{k}_1'}^+ a_{\mathbf{k} - \mathbf{q} + \mathbf{q}'} ; a_{\mathbf{k}}^+ : \omega \rangle\rangle -$$

$$- \delta_{\mathbf{k}_1, \mathbf{k} - \mathbf{q} + \mathbf{q}'} \langle n_{\mathbf{k}_1} \rangle \langle\langle a_{\mathbf{k}_1 + \mathbf{q}} a_{\mathbf{k}_1'}^+ a_{\mathbf{k}_1' - \mathbf{q}'} ; a_{\mathbf{k}}^+ : \omega \rangle\rangle -$$

$$- \delta_{\mathbf{q}' 0} \langle n_{\mathbf{k}_1'} \rangle \langle\langle a_{\mathbf{k}_1}^+ a_{\mathbf{k}_1 + \mathbf{q}} a_{\mathbf{k} - \mathbf{q} + \mathbf{q}'} ; a_{\mathbf{k}}^+ : \omega \rangle\rangle -$$

$$- \delta_{\mathbf{k}_1', \mathbf{k} - \mathbf{q} + \mathbf{q}'} \langle n_{\mathbf{k}_1'} \rangle \langle\langle a_{\mathbf{k}_1}^+ a_{\mathbf{k}_1 + \mathbf{q}} a_{\mathbf{k}_1' - \mathbf{q}'} ; a_{\mathbf{k}}^+ : \omega \rangle\rangle +$$

$$+ (1 - \langle n_{\mathbf{k}_1'} \rangle) \delta_{\mathbf{k}_1', \mathbf{k}_1 + \mathbf{q}} \langle\langle a_{\mathbf{k}_1}^+ a_{\mathbf{k}_1' - \mathbf{q}'} a_{\mathbf{k} - \mathbf{q} + \mathbf{q}'} ; a_{\mathbf{k}}^+ : \omega \rangle\rangle \tag{5.12}$$

with similar approximations for the second and third terms. Eqn (5.7) then becomes

$$\{\omega - \varepsilon_{\mathbf{k}_1 + \mathbf{q}} - \varepsilon_{\mathbf{k} - \mathbf{q}} + \varepsilon_{\mathbf{k}_1} + \mu - (1 - \langle n_{\mathbf{k}_1 + \mathbf{q}} \rangle - \langle n_{\mathbf{k} - \mathbf{q}} \rangle) \bar{v}(|\mathbf{k}_1 - \mathbf{k} + 2\mathbf{q}|) -$$

$$- \sum_{\mathbf{k}_1'} \langle n_{\mathbf{k}_1'} \rangle [\bar{v}(|\mathbf{k}_1' - \mathbf{k}_1|) - \bar{v}(|\mathbf{k}_1 + \mathbf{q} - \mathbf{k}|) -$$

$$- \bar{v}(|\mathbf{k}_1' - \mathbf{k}_1 - \mathbf{q}|)]\} \langle\langle a_{\mathbf{k}_1}^+ a_{\mathbf{k}_1 + \mathbf{q}} a_{\mathbf{k} - \mathbf{q}} ; a_{\mathbf{k}}^+ : \omega \rangle\rangle$$

$$= \sum_{\mathbf{k}_3 \mathbf{k}_4}' [\bar{v}(|\mathbf{k}_1 + \mathbf{q} - \mathbf{k}_3|) - \bar{v}(|\mathbf{k}_1 + \mathbf{q} - \mathbf{k}_4|)] \delta_{\mathbf{k}_1 + \mathbf{k}, \mathbf{k}_3 + \mathbf{k}_4} \times$$

$$\times (1 - \langle n_{\mathbf{k}_1 + \mathbf{q}} \rangle - \langle n_{\mathbf{k} - \mathbf{q}} \rangle) \langle\langle a_{\mathbf{k}_1}^+ a_{\mathbf{k}_3} a_{\mathbf{k}_4} ; a_{\mathbf{k}}^+ : \omega \rangle\rangle +$$

$$+ \sum_{\mathbf{k}_3 \mathbf{k}_4}' [\bar{v}(q) - \bar{v}(|\mathbf{k}_1 + \mathbf{q} - \mathbf{k}_4|)] \delta_{\mathbf{q} + \mathbf{k}_3, \mathbf{k}_4} \times$$

$$\times (\langle n_{\mathbf{k}_1} \rangle - \langle n_{\mathbf{k}_1 + \mathbf{q}} \rangle) \langle\langle a_{\mathbf{k}_3}^+ a_{\mathbf{k}_4} a_{\mathbf{k} - \mathbf{q}} ; a_{\mathbf{k}}^+ : \omega \rangle\rangle +$$

$$+ \sum_{\mathbf{k}_3 \mathbf{k}_4}' \bar{v}(|\mathbf{k} - \mathbf{q} - \mathbf{k}_1|) - \bar{v}(|\mathbf{k} - \mathbf{q} - \mathbf{k}_4|) \delta_{\mathbf{k} - \mathbf{q} + \mathbf{k}_3, \mathbf{k}_1 + \mathbf{k}_4} \times$$

$$\times (\langle n_{\mathbf{k}_1} \rangle - \langle n_{\mathbf{k} - \mathbf{q}} \rangle) \langle\langle a_{\mathbf{k}_3}^+ a_{\mathbf{k}_4} a_{\mathbf{k}_1 + \mathbf{q}} ; a_{\mathbf{k}}^+ : \omega \rangle\rangle. \tag{5.13}$$

We notice that the higher-order Green function occurs in eqn (5.11) multiplied by a potential factor which is singular for $q \to 0$. We shall thus obtain a good approximation by using an expression for $\langle\langle a^{+}_{\mathbf{k}_1} a_{\mathbf{k}_1+\mathbf{q}} a_{\mathbf{k}-\mathbf{q}} ; a^{+}_{\mathbf{k}} : \omega \rangle\rangle$ valid in this region, and by retaining in eqn (5.13) only terms which are singular as $q \to 0$:

$$[\omega - \varepsilon_{\mathbf{k}_1+\mathbf{q}} - \varepsilon_{\mathbf{k}-\mathbf{q}} + \varepsilon_{\mathbf{k}_1} + \mu] \langle\langle a^{+}_{\mathbf{k}_1} a_{\mathbf{k}_1+\mathbf{q}} a_{\mathbf{k}-\mathbf{q}} ; a^{+}_{\mathbf{k}} : \omega \rangle\rangle$$
$$\simeq \sum_{\mathbf{k}_3} (\langle n_{\mathbf{k}_1} \rangle - \langle n_{\mathbf{k}_1+\mathbf{q}} \rangle) \bar{v}(q) \langle\langle a^{+}_{\mathbf{k}_3} a_{\mathbf{k}_3+\mathbf{q}} a_{\mathbf{k}-\mathbf{q}} ; a^{+}_{\mathbf{k}} : \omega \rangle\rangle +$$
$$+ \bar{v}(q)[\langle n_{\mathbf{k}_1} \rangle (1 - \langle n_{\mathbf{k}_1+\mathbf{q}} \rangle - \langle n_{\mathbf{k}-\mathbf{q}} \rangle) + \langle n_{\mathbf{k}_1+\mathbf{q}} \rangle \langle n_{\mathbf{k}-\mathbf{q}} \rangle] \langle\langle a^{+}_{\mathbf{k}} ; a_{\mathbf{k}} : \omega \rangle\rangle. \tag{5.14}$$

By using these approximations, we have in fact discarded the terms which do not contribute to the shielding of the potential as in eqns (4.11) and (4.17). The term in eqn (5.14), $\bar{v}(q) \langle n_{\mathbf{k}_1} \rangle (1 - \langle n_{\mathbf{k}_1+\mathbf{q}} \rangle) \langle\langle a_{\mathbf{k}} ; a^{+}_{\mathbf{k}} ; \omega \rangle\rangle$ does not contribute to this shielding: for low temperatures, the factor $\langle n_{\mathbf{k}_1} \rangle (1 - \langle n_{\mathbf{k}_1+\mathbf{q}} \rangle)$ in zero-order is non-zero for a range of $\mathbf{k}_1$ of the order of $\mathbf{q}$; in this order, this term is thus not singular, and, to be consistent, we should drop it. In this case, eqn (5.14) gives

$$\sum_{\mathbf{k}_1} \langle\langle a^{+}_{\mathbf{k}_1} a_{\mathbf{k}_1+\mathbf{q}} a_{\mathbf{k}-\mathbf{q}} ; a^{+}_{\mathbf{k}} : \omega \rangle\rangle$$
$$= \{\langle n_{\mathbf{k}-\mathbf{q}} \rangle \bar{v}(q) \chi(\mathbf{k}-\mathbf{q}, \mathbf{q}, \omega) / [1 - \bar{v}(q) \chi(\mathbf{k}-\mathbf{q}, \mathbf{q}, \omega)]\} \langle\langle a_{\mathbf{k}} ; a^{+}_{\mathbf{k}} : \omega \rangle\rangle, \tag{5.15}$$

where

$$\chi(\mathbf{k}-\mathbf{q}, \mathbf{q}, \omega) = \sum_{\mathbf{k}_1} (\langle n_{\mathbf{k}_1} \rangle - \langle n_{\mathbf{k}_1+\mathbf{q}} \rangle) / [\omega - \varepsilon_{\mathbf{k}_1+\mathbf{q}} - \varepsilon_{\mathbf{k}-\mathbf{q}} + \varepsilon_{\mathbf{k}_1} + \mu]. \tag{5.16}$$

Thus, substituting this expression into eqn (5.11),

$$\langle\langle a_{\mathbf{k}} ; a^{+}_{\mathbf{k}} : \omega \rangle\rangle$$
$$= \left\{ \omega - \varepsilon_{\mathbf{k}} + \mu + \sum_{\mathbf{k}_1} \langle n_{\mathbf{k}_1} \rangle \bar{v}(|\mathbf{k}_1 - \mathbf{k}|) / [1 - \bar{v}(|\mathbf{k}_1 - \mathbf{k}|) \chi(\mathbf{k}_1, \mathbf{q}, \omega)] \right\}^{-1}. \tag{5.17}$$

We have thus an equation very similar to eqn (4.17) with a shielded Coulomb potential. The $\langle n \rangle$s in eqn (5.16) could, with extreme difficulty, be determined self-consistently from the Green function (5.17), but we obtain the usual shielded potential approximation by using $\langle n \rangle_0$. The problem is considered in more detail by Toigo and Woodruff (1970, 1971).

The disadvantages of the equation of motion method are clearly seen in this example. The 'decoupling' of the Green functions to break off the infinite chain of coupled equations leads to rather poorly understood approximations. Further, the complexity of the calculations is entirely algebraic, whereas in the perturbation method described previously, it was possible to set up a diagrammatic method which helped to reduce the algebra involved. It is rather easy to see in any approximation which diagrams are being neglected, and sometimes to consider under what circumstances the contributions of neglected diagrams will be small. It is much more difficult to assess the importance of the approximations made in the equation of motion method.

It is thus appropriate to consider the application of this method to magnetic systems, where diagrammatic methods are less straightforward to develop.

5.3. The Heisenberg model of a ferromagnet

We shall consider the Heisenberg model of a ferromagnet with a Hamiltonian:

$$H = -B\sum_{\mathbf{i}} S^z_{\mathbf{i}} - \tfrac{1}{2}\sum_{\mathbf{ij}} v_{\mathbf{ij}}\mathbf{S}_{\mathbf{i}}\cdot\mathbf{S}_{\mathbf{j}} \tag{5.18}$$

with the spin operators as defined in exercise 1.4.1. For simplicity, we shall take $v_{\mathbf{ij}} = 0$ unless **i** and **j** are nearest-neighbour sites in which case it is $2J$, a constant. We shall consider the Green function, with ε chosen as $+1$:

$$\langle\langle S^+_{\mathbf{g}}\,;\,S^-_{\mathbf{l}} : \omega\rangle\rangle$$

which satisfies the equation of motion:

$$(\omega - B)\langle\langle S^+_{\mathbf{g}}\,;\,S^-_{\mathbf{l}} : \omega\rangle\rangle = 2\delta_{\mathbf{gl}}\langle S^z_{\mathbf{g}}\rangle + 2J\sum_{\boldsymbol{\delta}}\langle\langle S^z_{\mathbf{g}+\boldsymbol{\delta}}S^+_{\mathbf{g}} - S^z_{\mathbf{g}}S^+_{\mathbf{g}+\boldsymbol{\delta}}\,;\,S^-_{\mathbf{l}} : \omega\rangle\rangle, \tag{5.19}$$

where $\boldsymbol{\delta}$ is a vector connecting any site with a nearest neighbour. We could now write down the equation of motion of $\langle\langle S^z_{\mathbf{g}+\boldsymbol{\delta}}S^+_{\mathbf{g}} - S^z_{\mathbf{g}}S^+_{\mathbf{g}+\boldsymbol{\delta}}\,;\,S^-_{\mathbf{l}} : \omega\rangle\rangle$ which would involve even more complicated Green functions. Instead, we perform the decoupling at this stage, by ignoring the fluctuations of $S^z_{\mathbf{g}}$, replacing it by its average value, $\langle S^z_{\mathbf{g}}\rangle$:

$$\langle\langle S^z_{\mathbf{g}+\boldsymbol{\delta}}S^+_{\mathbf{g}}\,;\,S^-_{\mathbf{l}} : \omega\rangle\rangle \simeq \langle S^z_{\mathbf{g}+\boldsymbol{\delta}}\rangle\langle\langle S^+_{\mathbf{g}}\,;\,S^-_{\mathbf{l}} : \omega\rangle\rangle. \tag{5.20}$$

We are then neglecting correlations between S^z on one site and S^+ and S^- on another. We are thus left with the equation of motion:

$$\left(\omega - B - 2J\sum_{\boldsymbol{\delta}}\langle S^z_{\mathbf{g}+\boldsymbol{\delta}}\rangle\right)\langle\langle S^+_{\mathbf{g}}\,;\,S^-_{\mathbf{l}} : \omega\rangle\rangle = 2\delta_{\mathbf{gl}}\langle S^z_{\mathbf{g}}\rangle - 2J\langle S^z_{\mathbf{g}}\rangle\sum_{\boldsymbol{\delta}}\langle\langle S^+_{\mathbf{g}+\boldsymbol{\delta}}\,;\,S^-_{\mathbf{l}} : \omega\rangle\rangle. \tag{5.21}$$

Because of the translational invariance of the lattice, the average value of $S^z_{\mathbf{i}}$ is independent of **i**. Also, we can Fourier transform the Green function with respect to vectors in the reciprocal lattice:

$$\begin{aligned}\langle\langle S^+_{\mathbf{g}}\,;\,S^-_{\mathbf{l}} : \omega\rangle\rangle &= \frac{1}{N}\sum_{\mathbf{k}}\bar{G}_{\mathbf{k}}(\omega)\exp[i\mathbf{k}\cdot(\mathbf{g}-\mathbf{l})]\\ \bar{G}_{\mathbf{k}}(\omega) &= \sum_{\mathbf{g}}\exp[-i\mathbf{k}\cdot(\mathbf{g}-\mathbf{l})]\langle\langle S^+_{\mathbf{g}}\,;\,S^-_{\mathbf{l}} : \omega\rangle\rangle\end{aligned} \tag{5.22}$$

where N is the total number of spins on the lattice. The equation of motion then takes the form

$$[\omega - B - 2Jn\langle S^z\rangle]\bar{G}_{\mathbf{k}}(\omega) = 2\langle S^z\rangle - 2Jn\langle S^z\rangle\gamma_{\mathbf{k}}\bar{G}_{\mathbf{k}}(\omega) \tag{5.23}$$

where n is the number of nearest neighbours of any site, and where

$$\gamma_{\mathbf{k}} = n^{-1}\sum_{\boldsymbol{\delta}}\exp(i\mathbf{k}\,.\,\boldsymbol{\delta}). \tag{5.24}$$

Thus

$$\bar{G}_{\mathbf{k}}(\omega) = 2\langle S^z\rangle/(\omega - E_{\mathbf{k}}), \tag{5.25}$$

where

$$E_{\mathbf{k}} = B + 2Jn\langle S^z\rangle(1-\gamma_{\mathbf{k}}). \tag{5.26}$$

This approximation thus gives us undamped excitations with energies $E_{\mathbf{k}}$. Unfortunately, to determine $E_{\mathbf{k}}$ we need to know $\langle S^z\rangle$. For spin S, we may write

$$\prod_{r=-S}^{r=S}(S^z - r) = 0 \tag{5.27}$$

where r takes on integral or half-integral values according to whether S is integral or half-integral. (One factor will be zero for each basis vector in the representation in which S^z is diagonal). Thus, for spin $\frac{1}{2}$,

$$(S^z)^2 = \tfrac{1}{4}. \tag{5.28}$$

But from eqns (3.55), (3.59), and (5.25)

$$\langle S^- S^+\rangle = 2\langle S^z\rangle\Phi,$$

where

$$\Phi = N^{-1}\sum_{\mathbf{k}}[\exp(\beta E_{\mathbf{k}}) - 1]^{-1}. \tag{5.29}$$

And for spin $\frac{1}{2}$

$$\langle S^- S^+\rangle = \tfrac{3}{4} - \langle S^z\rangle - \langle (S^z)^2\rangle. \tag{5.30}$$

Thus, from eqns (5.28) and (5.29) for $S = \frac{1}{2}$

$$\langle S^z\rangle_{S=\frac{1}{2}} = 1/[2(1+2\Phi)]. \tag{5.31}$$

Similarly, it may be shown that

$$\langle S^z\rangle_{S=1} = (1+2\Phi)/\{1+3\Phi+3(\Phi)^2\}. \tag{5.32}$$

The results for higher values of S are given by Tahir-Kheli and ter Haar (1962).

Eqns (5.26), (5.31), and (5.32) are a set of simultaneous equations for the quasi-particle energies: we have thus generalized the spin-wave theory of exercise 1.4.1 to non-zero temperatures, and have also obtained an expression for the magnetization, which is proportional to $N\langle S^z\rangle$. Below a certain temperature, T_C, these equations predict spontaneous magnetization, that is, a non-zero $\langle S^z\rangle$ even when the external field, B, is zero. Above T_C, they give an expression for the susceptibility $\langle S^z\rangle/B$. (See the exercises below, and Tahir-Kheli and ter Haar (1962)).

The results of this simple calculation, a theory of ferromagnetism and of paramagnetism for the whole temperature range, is impressive. Once again, however, it is rather difficult to discuss the conditions under which the decoupling approximation used (eqn (5.20)) will be a good one. The low-temperature results may be compared with an exact low-temperature expansion due to Dyson (1956). The dependence of the quasi-particle energy on temperature is given incorrectly by eqn (5.21). Terms in the magnetization proportional to T^0, $T^{3/2}$, $T^{5/2}$, and $T^{7/2}$ are given correctly, but there is a spurious T^3 term in the Green-function expression.

The decoupling procedure for this problem is discussed, for example, by Haas and Jarrett (1964) and by Devlin and Vertogen (1972). Roth (1968) discusses a decoupling scheme for the equation of motion.

In summary, the equation of motion method is an easy way of obtaining low-order approximations to Green functions and correlation functions. It is easy formally to introduce self-consistency into the equations, since the decoupling procedure usually involves taking out the expectation value of an operator (e.g. S^z above) which is determined in terms of the Green function itself. Such self-consistency is often most important in the discussion of collective effects such as ferromagnetism. The method is an extremely valuable tool in cases in which a diagrammatic perturbation theory cannot easily be developed. However, it is difficult to use when one wishes to make controlled approximations and to discuss the range of validity of these approximations.

Exercise

5.3.1. Verify the statement above that spontaneous magnetization occurs in this approximation below a certain temperature, T_C, and determine T_C. Draw a rough graph to show the variation of the susceptibility with temperature.

5.4. Functional derivatives

The equation of motion for the single-particle Green function may be written in a more compact way by using functional derivative techniques, which also lead to an alternative derivation of the diagrammatic expansion for the Green function. We use an amended version of the τ-ordered Green function of eqn (3.16), which is the most convenient for our purposes, written in the coordinate representation:

$$G_1(\mathbf{r}_1, \mathbf{r}_2, \tau_1, \tau_2, U) = \langle \mathscr{T}\{a^{\mathrm{H}}(\mathbf{r}_1, \tau_1)\tilde{a}^{\mathrm{H}}(\mathbf{r}_2, \tau_2)S_U^{(\beta)}\}\rangle / \langle S_U^{(\beta)}\rangle, \tag{5.33}$$

where

$$S_U^{(\beta)} = \mathscr{T}\left\{\exp\left[-\int_0^\beta \mathrm{d}\tau \int \mathrm{d}^3\mathbf{x}\; U(\mathbf{x}, \tau)\tilde{a}^{\mathrm{H}}(\mathbf{x}, \tau)a^{\mathrm{H}}(\mathbf{x}, \tau)\right]\right\} \tag{5.34}$$

and where $U(\mathbf{x}, \tau)$ is a function introduced purely for mathematical convenience. We shall use H in the coordinate representation:

$$H = -(2m)^{-1}\int \mathrm{d}^3\mathbf{x}\; a^+(\mathbf{x})\nabla^2 a(\mathbf{x}) +$$

$$+\tfrac{1}{2}\int v(|\mathbf{x}-\mathbf{x}'|)a^+(\mathbf{x})a^+(\mathbf{x}')a(\mathbf{x}')a(\mathbf{x})\,\mathrm{d}^3\mathbf{x}\,\mathrm{d}^3\mathbf{x}'. \tag{5.35}$$

By comparing the definitions (5.33) and (5.34) with eqns (3.18) and (2.28) one can see that we have effectively added a term

$$\int \mathrm{d}^3\mathbf{x}\; U(\mathbf{x}, \tau)\, a^+(\mathbf{x})a(\mathbf{x})$$

to the Hamiltonian. Using this idea, or differentiating directly, one obtains the equation of motion of G_1:

$$\frac{\partial}{\partial\tau_1}G_1(\mathbf{r}_1, \mathbf{r}_2, \tau_1, \tau_2, U) = \delta(\tau_1-\tau_2)\delta(\mathbf{r}_1-\mathbf{r}_2) +$$

$$+\left[\frac{\nabla_1^2}{2m}+\mu-U(\mathbf{r}_1, \tau_1)\right]G_1(\mathbf{r}_1, \mathbf{r}_2, \tau_1, \tau_2, U)-$$

$$-\int \mathrm{d}^3\mathbf{r}'_1\, v(|\mathbf{r}_1-\mathbf{r}'_1|)G_2(\mathbf{r}'_1, \mathbf{r}_1, \mathbf{r}_2, \tau_1, \tau_1, \tau_2, U), \tag{5.36}$$

where

$$G_2(\mathbf{r}'_1, \mathbf{r}_1, \mathbf{r}_2, \tau'_1, \tau_1, \tau_2, U)$$
$$= \langle \mathscr{T}\{\tilde{a}^{\mathrm{H}}(\mathbf{r}'_1, \tau'_1)a^{\mathrm{H}}(\mathbf{r}'_1, \tau'_1)a^{\mathrm{H}}(\mathbf{r}_1, \tau_1)\tilde{a}^{\mathrm{H}}(\mathbf{r}_2, \tau_2)S_U^{(\beta)}\}\rangle / \langle S_U^{(\beta)}\rangle. \tag{5.37}$$

Because of the τ-dependence of U, G is no longer a function of $\tau_1-\tau_2$, and the periodicity conditions (eqn (3.55)), which may be regarded as boundary conditions on the differential equation (5.36), are no longer satisfied. However, following arguments very similar to those leading to eqn (3.35), one may show

$$G_1(\mathbf{r}_1, \mathbf{r}_2, 0, \tau_2, U) = \varepsilon G_1(\mathbf{r}_1, \mathbf{r}_2, \beta, \tau_2, U). \tag{5.38}$$

Consider a small change in U:

$$U(\mathbf{x}, \tau) \to U(\mathbf{x}, \tau) + \delta U(\mathbf{x}, \tau). \tag{5.39}$$

Because, in eqns (5.33) and (5.34), all the factors appear under the $\mathscr{T}$-operator, we may differentiate without worrying about the order of the operators. Hence,

$$\delta S_U^{(\beta)} = \mathscr{T}\left\{-\int_0^\beta \mathrm{d}\tau \int \mathrm{d}^3\mathbf{x}\, \delta U(\mathbf{x}, \tau)\tilde{a}^{\mathrm{H}}(\mathbf{x}, \tau)a^{\mathrm{H}}(\mathbf{x}, \tau)S_U^{(\beta)}\right\}, \tag{5.40}$$

and, after some algebra,

$$\begin{aligned}\delta G_1(\mathbf{r}_1, \mathbf{r}_2, \tau_1, \tau_2, U) = &-\iint_0^\beta [G_2(\mathbf{x}, \mathbf{r}_1, \mathbf{r}_2, \tau, \tau_1, \tau_2, U) - \\ &-\varepsilon G_1(\mathbf{r}_1, \mathbf{r}_2, \tau_1, \tau_2, U)G_1(\mathbf{x}, \mathbf{x}, \tau, \tau, U)] \times \\ &\times \delta U(\mathbf{x}, \tau)\, \mathrm{d}^3\mathbf{x}\, \mathrm{d}\tau. \end{aligned} \tag{5.41}$$

If we change U only at the point x and time τ, we may write

$$\begin{aligned}\frac{\delta G_1(\mathbf{r}_1, \mathbf{r}_2, \tau_1, \tau_2, U)}{\delta U(\mathbf{x}, \tau)} = &-G_2(\mathbf{x}, \mathbf{r}_1, \mathbf{r}_2, \tau, \tau_1, \tau_2, U) + \\ &+\varepsilon G_1(\mathbf{r}_1, \mathbf{r}_2, \tau_1, \tau_2, U)G_1(\mathbf{x}, \mathbf{x}, \tau, \tau, U). \end{aligned} \tag{5.42}$$

$\delta G/\delta U$ is called the functional derivative (sometimes the variational derivative) of G with respect to U. The equation of motion for the single-particle Green function may now be written:

$$\begin{aligned}\frac{\partial}{\partial \tau_1} G_1(\mathbf{r}_1, \mathbf{r}_2, \tau_1, \tau_2, U) = &\ \delta(\tau_1 - \tau_2)\delta(\mathbf{r}_1 - \mathbf{r}_2) + \\ &+ \left[\frac{\nabla_1^2}{2m} + \mu - U(\mathbf{r}_1, \tau_1)\right] G_1(\mathbf{r}_1, \mathbf{r}_2, \tau_1, \tau_2, U) + \\ &+ \int \mathrm{d}^3\mathbf{r}_1' v(|\mathbf{r}_1 - \mathbf{r}_1'|) \times \\ &\times \left[\frac{\delta}{\delta U(\mathbf{r}_1', \tau_1)} - \varepsilon G_1(\mathbf{r}_1', \mathbf{r}_1', \tau_1, \tau_1, U\right] \times \\ &\times G_1(\mathbf{r}_1, \mathbf{r}_2, \tau_1, \tau_2, U). \end{aligned} \tag{5.43}$$

Unfortunately, no way is known of solving such an equation. It may be used, however, to generate perturbation expansions for G. This is most conveniently done by writing it in the form of an integral equation:

$$G_1(\mathbf{r}_1, \mathbf{r}_2, \tau_1, \tau_2, U) = g(\mathbf{r}_1, \mathbf{r}_2, \tau_1, \tau_2, U) +$$

$$+ \iiint_0^\beta \mathrm{d}^3\mathbf{r}'_1\, \mathrm{d}^3\mathbf{r}'_2\, \mathrm{d}\tau\, g(\mathbf{r}_1, \mathbf{r}'_1, \tau_1, \tau, U) v(|\mathbf{r}'_1 - \mathbf{r}'_2|) \times$$

$$\times \left[-\varepsilon G_1(\mathbf{r}'_2, \mathbf{r}'_2, \tau, \tau, U) + \frac{\delta}{\delta U}(\mathbf{r}'_2, \tau) \right] G_1(\mathbf{r}'_1, \mathbf{r}_2, \tau, \tau_2, U) \tag{5.44}$$

where $g(\mathbf{r}_1, \mathbf{r}_2, \tau_1, \tau_2, U)$ is the single-particle Green function with no inter-particle interaction so that

$$\frac{\partial}{\partial \tau_1} g(\mathbf{r}_1, \mathbf{r}_2, \tau_1, \tau_2, U) = \delta(\tau_1 - \tau_2)\delta(\mathbf{r}_1 - \mathbf{r}_2) +$$

$$+ \left[\frac{\nabla_1^2}{2m} + \mu - U(\mathbf{r}_1, \tau_1) \right] g(\mathbf{r}_1, \mathbf{r}_2, \tau_1, \tau_2, U). \tag{5.45}$$

It is easy to check by direct differentiation that the G given by eqn (5.44) satisfies eqn (5.36) and the boundary condition (5.38).

To obtain the lowest-order terms in an expansion of G in powers of v, we simply solve eqn (5.44) by iteration. Thus, to first order,

$$G_1(\mathbf{r}_1, \mathbf{r}_2, \tau_1, \tau_2, U) = g(\mathbf{r}_1, \mathbf{r}_2, \tau_1, \tau_2, U) +$$

$$+ \iiint_0^\beta \mathrm{d}^3\mathbf{r}'_1\, \mathrm{d}^3\mathbf{r}'_2\, \mathrm{d}\tau\, g(\mathbf{r}_1, \mathbf{r}'_1, \tau_1, \tau, U) v(|\mathbf{r}'_1 - \mathbf{r}'_2|) \times$$

$$\times \left[-\varepsilon g(\mathbf{r}'_2, \mathbf{r}'_2, \tau, \tau, U) + \frac{\delta}{\delta U(\mathbf{r}'_2, \tau)} \right] g(\mathbf{r}'_1, \mathbf{r}_2, \tau, \tau_2, U).$$

But from eqn (5.42) and Wick's theorem

$$\frac{\delta g(\mathbf{r}'_1, \mathbf{r}_2, \tau, \tau_2, U)}{\delta U(\mathbf{r}'_2, \tau)} = -g(\mathbf{r}'_1, \mathbf{r}'_2, \tau, \tau, U) g(\mathbf{r}'_2, \mathbf{r}_2, \tau, \tau_2, U) \tag{5.46}$$

and hence

$$G_1(\mathbf{r}_1, \mathbf{r}_2, \tau_1, \tau_2, U)$$

$$= g(\mathbf{r}_1, \mathbf{r}_2, \tau_1, \tau_2, U) - \iiint_0^\beta \mathrm{d}^3\mathbf{r}'_1\, \mathrm{d}^3\mathbf{r}'_2 \mathrm{d}\tau\, g(\mathbf{r}_1, \mathbf{r}'_1, \tau_1, \tau, U) \times$$

$$\times v(|\mathbf{r}'_1 - \mathbf{r}'_2|) [\varepsilon g(\mathbf{r}'_2, \mathbf{r}'_2, \tau, \tau, U) g(\mathbf{r}'_1, \mathbf{r}_2, \tau, \tau_2, U) +$$

$$+ g(\mathbf{r}'_1, \mathbf{r}'_2, \tau, \tau, U) g(\mathbf{r}'_2, \mathbf{r}_2, \tau, \tau_2, U)] \tag{5.47}$$

or, in diagrammatic language:

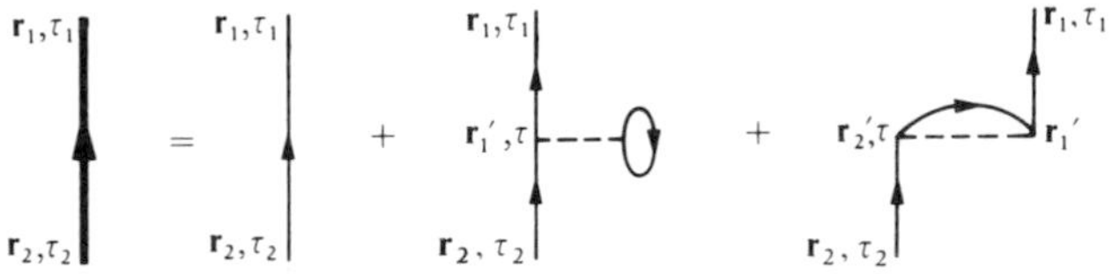

which looks very familiar. If we now put $U = 0$, we have the start of the usual perturbation formula for G. Further terms in this expansion, and the rules for the general term, may now be obtained by a straightforward generalization of this procedure.

We are by now familiar with the fact that it is usually more convenient to consider an expansion for Σ, the proper self-energy, than for G. In the coordinate representation, Σ is defined:

$$\int_0^\beta d\tau \int d^3\mathbf{x}[g^{-1}(\mathbf{r}_1, \mathbf{x}, \tau_1, \tau, U) - \Sigma(\mathbf{r}_1, \mathbf{x}, \tau_1, \tau, U)]G_1(\mathbf{x}, \mathbf{r}_2, \tau, \tau_2, U)$$
$$= \delta(\mathbf{r}_1 - \mathbf{r}_2)\delta(\tau_1 - \tau_2), \tag{5.48}$$

so that

$$G_1^{-1}(\mathbf{r}_1, \mathbf{r}_2, \tau_1, \tau_2, U) = g^{-1}(\mathbf{r}_1, \mathbf{r}_2, \tau_1, \tau_2, U) - \Sigma(\mathbf{r}_1, \mathbf{r}_2, \tau_1, \tau_2, U), \tag{5.49}$$

where g^{-1} and G^{-1} are defined in the operator sense:

$$\int_0^\beta d\tau \int d^3\mathbf{x}\, G_1^{-1}(\mathbf{r}_1, \mathbf{x}, \tau_1, \tau, U) G_1(\mathbf{x}, \mathbf{r}_2, \tau, \tau_2, U) = \delta(\tau_1 - \tau_2)\delta(\mathbf{r}_1 - \mathbf{r}_2). \tag{5.50}$$

Multiplying the fundamental equation (5.43) for G_1 on the right by G_1^{-1}, we have

$$G_1^{-1}(\mathbf{r}_1, \mathbf{r}_2, \tau_1, \tau_2, U)$$
$$= g^{-1}(\mathbf{r}_1, \mathbf{r}_2, \tau_1, \tau_2, U) + \varepsilon \int d^3\mathbf{r}_1'\, G_1(\mathbf{r}_1', \mathbf{r}_1', \tau_1, \tau_1, U) v(|\mathbf{r}_1 - \mathbf{r}_1'|)\delta(\tau_1 - \tau_2) -$$
$$- \int\int\int_0^\beta d^3\mathbf{r}_1'\, d^3\mathbf{r}_2'\, d\tau_2' v(|\mathbf{r}_1 - \mathbf{r}_1'|) \frac{\delta}{\delta U(\mathbf{r}_1', \tau_1)} G_1(\mathbf{r}_1, \mathbf{r}_2', \tau_1, \tau_2', U) \times$$
$$\times G_1^{-1}(\mathbf{r}_2', \mathbf{r}_2, \tau_2', \tau_2, U), \tag{5.51}$$

where we have used

$$g^{-1}(\mathbf{r}_1, \mathbf{r}_2, \tau_1, \tau_2, U) = \left(\frac{\partial}{\partial \tau_1} - \frac{\nabla_1^2}{2m} - \mu + U(\mathbf{r}_1, \tau_1)\right) \delta(\mathbf{r}_1 - \mathbf{r}_2)\delta(\tau_1 - \tau_2), \tag{5.52}$$

which follows from the analogue of eqns (5.43 and 5.50) for g.

But

$$G_1 G_1^{-1} = 1, \qquad G_1 \delta G_1^{-1} + \delta G_1 G_1^{-1} = 0,$$

so that, using eqns (5.48) and (5.46) and $g\delta g^{-1}+\delta g g^{-1}=0$

$$\iint_0^\beta \mathrm{d}^3\mathbf{r}_2'\,\mathrm{d}\tau_2' \frac{\delta}{\delta U(\mathbf{r}_1',\tau_1)} G_1(\mathbf{r}_1,\mathbf{r}_2',\tau_1,\tau_2',U) G_1^{-1}(\mathbf{r}_2',\mathbf{r}_2,\tau_2',\tau_2,U)$$

$$= -\iint_0^\beta \mathrm{d}^3\mathbf{r}_2'\,\mathrm{d}\tau_2'\, G_1(\mathbf{r}_1,\mathbf{r}_2',\tau_1,\tau_2',U) \frac{\delta}{\delta U(\mathbf{r}_1',\tau_1)} G_1^{-1}(\mathbf{r}_2',\mathbf{r}_2,\tau_2',\tau_2,U)$$

$$= -G_1(\mathbf{r}_1,\mathbf{r}_2,\tau_1,\tau_2)\delta(\mathbf{r}_1'-\mathbf{r}_2)\delta(\tau_1-\tau_2)+$$

$$+\iint_0^\beta \mathrm{d}^3\mathbf{r}_2'\,\mathrm{d}\tau_2\, G_1(\mathbf{r}_1,\mathbf{r}_2',\tau_1,\tau_2',U) \frac{\delta}{\delta U(\mathbf{r}_1',\tau_1)} \Sigma(\mathbf{r}_2',\mathbf{r}_2,\tau_2',\tau_2,U). \quad (5.53)$$

Thus, from eqns (5.49), (5.51), and (5.53)

$$\Sigma(\mathbf{r}_1,\mathbf{r}_2,\tau_1,\tau_2,U) = -\varepsilon\int \mathrm{d}^3\mathbf{r}_1'\, v(|\mathbf{r}_1-\mathbf{r}_1'|) G_1(\mathbf{r}_1',\mathbf{r}_1',\tau_1,\tau_1,U)\delta(\tau_1-\tau_2)-$$

$$-v(|\mathbf{r}_1-\mathbf{r}_2|)\delta(\tau_1-\tau_2)G_1(\mathbf{r}_1,\mathbf{r}_2,\tau_1,\tau_1,U)+$$

$$\iiint_0^\beta \mathrm{d}^3\mathbf{r}_1'\,\mathrm{d}^3\mathbf{r}_2'\,\mathrm{d}\tau_2'\, v(|\mathbf{r}_1-\mathbf{r}_1'|) G_1(\mathbf{r}_1,\mathbf{r}_2',\tau_1,\tau_2',U)\times$$

$$\times\frac{\delta}{\delta U(\mathbf{r}_1',\tau_1)}\Sigma(\mathbf{r}_2',\mathbf{r}_2,\tau_2',\tau_2,U). \quad (5.54)$$

This equation is now of a form suitable for an iterative expansion, giving Σ as a power series in v. To lowest order

$$\Sigma_{(1)}(\mathbf{r}_1,\mathbf{r}_2,\tau_1,\tau_2,U) = \Big[-\varepsilon\int \mathrm{d}^3\mathbf{r}_1'\, v(|\mathbf{r}_1-\mathbf{r}_1'|)\, G_1(\mathbf{r}_1',\mathbf{r}_1',\tau_1,\tau_1,U)-$$

$$-v(|\mathbf{r}_1-\mathbf{r}_2|)G_1(\mathbf{r}_1,\mathbf{r}_2,\tau_1,\tau_2,U)\Big]\delta(\tau_1-\tau_2) \quad (5.55)$$

as we expect. $\Sigma_{(1)}$, when inserted into eqn (5.54), will then yield the usual second-order terms. The diagrammatic expansion may be developed as before, U being put equal to zero once the expansion has been made.

In fact, an iterative solution eqn (5.54) leads to an expansion of Σ in powers of v, each term being a function of G_1, the full Green function, rather than g. It therefore leads to self-consistent equations, which, as we have seen, are usually difficult to solve but which often lead to results which are physically more meaningful than a straightforward power expansion.

Exercise

5.4.1. Show that the second-order terms in the self-consistent expansion of Σ in powers of the potential agree with those obtained by the rules of Chapter 3.

Further reading

The equation of motion method is described by Zubarev (1960) and, in more detail, by Bonch-Bruevich and Tyablikov (1962). Tyablikov and Bonch-Bruevich (1962) develop a perturbation series from the equation of motion. A linked-cluster theorem based on the equations of motion is given by Kobe (1966). Functional derivative techniques are fully described in *Quantum Statistical Mechanics*, by Kadanoff and Baym (1962).

6

MAGNETISM: THE DRONE–FERMION REPRESENTATION

6.1. The drone–fermion representation

THE diagrammatic methods developed in previous chapters are not directly applicable to Hamiltonians involving spin operators, such as the Heisenberg Hamiltonian (eqn (5.18)). The development of the diagrammatic expansion rests on Wick's theorem, which cannot be used directly for spin operators, since the commutator of two spin operators is itself a spin operator (see exercise 1.4.1). The last term on the right-hand side of eqn (2.29) thus has a new operator, $[a, b]$ (not a c-number) in it, and the proof can no longer be carried through. We have seen in the previous chapter that the equation of motion method may be applied to such systems, but the difficulties and uncertainties of the decoupling procedures have led to a search for possible ways of putting magnetic problems in a form suitable for the application of the standard perturbation approach.

A method for replacing the spin operators by boson operators has been mentioned in exercise 1.4.1, but this method has the disadvantage that great care has to be taken to ensure that one considers only the finite number of physical spin states amongst the infinite number of states generated by the boson operators in the transformed problem. Other methods are very briefly described in the reference section at the end of the chapter. We wish to describe in more detail a method developed by Spencer (1968*a*) which puts the spin problem for $S = \frac{1}{2}$ and $S = 1$, in such a form that the methods developed in Chapters 2 and 3 may be applied with very little modification.

(a) Spin $\frac{1}{2}$.

Spencer used the 'drone-fermion' representation of the spin operators introduced by Mattis (1965). For spin-$\frac{1}{2}$ operators localized at site positions $j, k, l, \ldots$, we put, using the notation of exercise 1.4.1,

$$S_j^z = c_j^+ c_j - \tfrac{1}{2}; \qquad S_j^+ = c_j^+ \phi_j; \qquad \phi_j = d_j + d_j^+. \tag{6.1}$$

The cs and ds are chosen to satisfy anti-commutation relations:

$$\begin{aligned} [c_j, c_l^+]_+ &= [d_j, d_l^+]_+ = \delta_{jl}, \\ [c_j, c_l]_+ &= [d_j, d_l]_+ = 0; \end{aligned} \tag{6.2}$$

with all cs commuting with all ds. One may now easily verify that the usual relations for spin operators are satisfied:

$$\mathbf{S}_j \cdot \mathbf{S}_j = \tfrac{3}{4}; \qquad [S_i^z, S_j^\pm] = \pm\delta_{ij} S_i^\pm; \qquad [S_i^+, S_j^-] = 2\delta_{ij} S_i^z \tag{6.3}$$

so that the cs and ds are a possible representation of the spin operators. The 'particles' created by the c^+s and d^+s are called the fermions and the drones of the system respectively. Unfortunately, we have too many states for each site. This may easily be seen by examining the possible states for one site in the c–d representation if we define, in the usual way (dropping site labels for the moment):

$$\begin{aligned} c|0,0\rangle &= 0; \qquad d|0,0\rangle = 0 \\ c^+|0,0\rangle &= |1,0\rangle; \quad d^+|0,0\rangle = |0,1\rangle \quad \text{etc.} \end{aligned} \tag{6.4}$$

We find:

$$\begin{aligned} S^z|0,0\rangle &= -\tfrac{1}{2}|0,0\rangle; S^z|0,1\rangle = -\tfrac{1}{2}|0,1\rangle \\ S^z|1,0\rangle &= \tfrac{1}{2}|1,0\rangle; S^z|1,1\rangle = \tfrac{1}{2}|1,1\rangle \end{aligned} \tag{6.5}$$

Fortunately, there are just two c–d states for each physical state. Thus, for each site, and for any operator A,

$$\mathrm{Tr}_S\{A\} = \tfrac{1}{2}\,\mathrm{Tr}_{cd}\{A\}, \tag{6.6}$$

where Tr_S indicates that the trace is to be taken over the physical (spin) states, and Tr_{cd} that it is to be taken over the c–d states. Thus, for the whole system with N sites:

$$\mathrm{Tr}_s\{A\} = \mathrm{Tr}_{cd}\{A\}/2^N. \tag{6.7}$$

An expectation value, which is the quotient of two traces, is therefore the same in the two representations. The anticommutation relations obeyed by the c and d operators are such that the usual proof of Wick's theorem may be used, so that our diagrammatic techniques may be taken over unchanged in this case. We give an example of such work below.

(b) Spin 1.

The drone–fermion representation may rather easily be generalized to higher spin values by associating with each localized spin operator $\mathbf{S}_j$, $2S$ spin-$\frac{1}{2}$ operators $\mathbf{S}_{i\alpha}$, each of which is represented in terms of drone-fermion operators, as above, so that:

$$\begin{aligned} &\mathbf{S}_i = \sum_\alpha \mathbf{S}_{i\alpha}; \\ S_{i\alpha}^z = c_{i\alpha}^+ c_{i\alpha} - \tfrac{1}{2}; \qquad &S_{i\alpha}^+ = c_{i\alpha}^+ \phi_{i\alpha}; \qquad \phi_{i\alpha} = d_{i\alpha} + d_{i\alpha}^+ \end{aligned} \tag{6.8}$$

with the commutation rules:

$$[c_{i\alpha}, c_{j\beta}^+]_+ = [d_{i\alpha}, d_{j\beta}^+]_+ = \delta_{ij}\delta_{\alpha\beta}, \text{ etc.} \tag{6.9}$$

The commutation rules for the spin operators are given correctly, but unfortunately, traces in the two systems are no longer simply related. We write, for spin 1 ($\alpha = 1, 2$):

$$
\begin{aligned}
c_\alpha|0,0,0,0\rangle &= d_\alpha|0,0,0,0\rangle = 0\\
c_1^+|0,0,0,0\rangle &= |1,0,0,0\rangle\\
c_2^+|0,0,0,0\rangle &= |0,0,1,0\rangle\\
d_1^+|0,0,0,0\rangle &= |0,1,0,0\rangle\\
d_2^+|0,0,0,0\rangle &= |0,0,0,1\rangle \quad \text{etc.}
\end{aligned}
\tag{6.10}
$$

where we have again dropped site labels, so that the subscripts refer to the value of α. The S^z values for the c–d states are now:

$$
\begin{aligned}
S^z = -1:&\quad |0,0,0,0\rangle, |0,0,0,1\rangle, |0,1,0,0\rangle, |0,1,0,1\rangle\\
S^z = 0:&\quad |1,0,0,0\rangle, |1,0,0,1\rangle, |1,1,0,0\rangle, |1,1,0,1\rangle\\
&\quad |0,0,1,0\rangle, |0,0,1,1\rangle, |0,1,1,0\rangle, |0,1,1,1\rangle\\
S^z = 1:&\quad |1,0,1,0\rangle, |1,0,1,1\rangle, |1,1,1,0\rangle, |1,1,1,1\rangle
\end{aligned}
\tag{6.11}
$$

so that there are four c–d states for each physical state $S^z = \pm 1$, and eight for $S^z = 0$.

The fact that these overweighting factors are not equal is not as big a drawback as it at first appears. We notice that, if A and B are any operators (still considering only one site),

$$
\mathrm{Tr}_S(AS^zB) = \mathrm{Tr}_S(BAS^z) = \tfrac{1}{4}\,\mathrm{Tr}_{cd}(BAS^z) = \tfrac{1}{4}\,\mathrm{Tr}_{cd}(AS^zB). \tag{6.12}
$$

The first and last steps follow from the cyclic invariance of the trace, and the second from the fact that the awkward terms in the trace, for which $S^z = 0$, are projected out by the S^z operator.

We may use eqn (6.12) to relate traces of arbitrary spin operators in the two representations. Eqn (6.12) tells us that to obtain the trace over the S states from the trace over the c–d states of any product of spin operators containing S^z as a factor, we merely divide by four. Secondly, consider the expectation value, in the noninteracting ensemble, of a product of m S^+ operators and n S^- operators. We shall take as our zero-order Hamiltonian H_0 those terms which are of the form $\omega_0 S^z$, so that in developing our particular expansion we have traces of the form

$$
T = \mathrm{Tr}\{\exp(-\beta\omega_0 S^z)S^+S^+S^+ \ldots S^-S^-S^- \ldots\} \tag{6.13}
$$

where we have assumed initially that the S^+ operators lie to the left of all S^- operators. (In fact, $n = m$ for non-zero traces, but this fact need not concern us here.) We commute the first S^+ operator through each of the other operators, leaving a sum of products of $m-1$ S^+ operators, $n-1$ S^- operators and one S^z operator, since

$$[S^+, S^-] = 2S^z.$$

Thus

$$\begin{aligned} T = {} & 2\mathrm{Tr}\{\exp(-\beta\omega_0 S^z)\overleftrightarrow{S^+ \dots S^+}^{m-1} S^z \overleftrightarrow{S^- \dots S^-}^{n-1}\} + \\ & +2\,\mathrm{Tr}\{\exp(-\beta\omega_0 S^z)\overleftrightarrow{S^+ \dots S^+}^{m-1} S^- S^z \overleftrightarrow{S^- \dots S^-}^{n-2}\} + \dots + \\ & +\mathrm{Tr}\{\exp(-\beta\omega_0 S^z)\overleftrightarrow{S^+ \dots S^+}^{m-1} \overleftrightarrow{S^- \dots S^-}^{n} S^+\}. \end{aligned} \tag{6.14}$$

Using the cyclic invariance of the trace, and eqn (vi) of Appendix B, the final term may be written as $T\exp(\beta\omega_0)$. Thus

$$T \times [1-\exp(\beta\omega_0)] = \mathrm{Tr}\{\exp(-\beta\omega_0 S^z)S^+ \dots S^+ S^z S^- \dots S^-\} + \cdots \tag{6.15}$$

Each term on the right-hand side of this equation contains a factor of S^z, which projects out the awkward terms in the trace, as in eqn (6.12). But a product of m S^+ operators and n S^- operators in an arbitrary order differs from the product considered in eqn (6.13) only by terms containing factors of S^z, to which we may again apply eqn (6.12). Thus traces of such products over the physical states may also be obtained from those over the c–d states by dividing by four.

From eqn (2.27) we see that the expectation values we shall need in developing our perturbation expansion will all be of one of the forms considered above. However, when we take expectation values, we have also to evaluate the denominator:

$$\mathrm{Tr}_S\{\exp(-\beta\omega_0 S^z)\} = \exp(-\beta\omega_0)+1+\exp(\beta\omega_0) \tag{6.16}$$

whereas

$$\mathrm{Tr}_{cd}\{\exp(-\beta\omega_0 S^z)\} = 4[\exp(-\beta\omega_0)+2+\exp(\beta\omega_0)]. \tag{6.17}$$

Thus we have, finally, for expectation values in the noninteracting ensemble of all products of operators we shall need,

$$\langle AB \dots\rangle_{0,S} = Y\langle AB \dots\rangle_{0,cd} \tag{6.18}$$

for one site, where

$$\begin{aligned} Y &= [\exp(-\beta\omega_0)+2+\exp(\beta\omega_0)]/[\exp(-\beta\omega_0)+1+\exp(\beta\omega_0)] \\ &= [1+\exp(\beta\omega_0)][1+\exp(-\beta\omega_0)]/[1+\exp(-\beta\omega_0)+\exp(\beta\omega_0)], \end{aligned} \tag{6.19}$$

and where $\langle \dots \rangle_{0,cd}$ indicates an expectation value in the non-interacting ensemble (with $H_0 = \omega_0(c_1^+ c_1 + c_2^+ c_2 - 1)$) taken over the c–d states, and $\langle \dots \rangle_{0,S}$ is the expectation value over the non-interacting ensemble taken over the physical (spin) states.

By a simple extension of the argument above, one may prove that the corresponding result for a product of spin operators referring to t different sites is

$$\langle AB\ldots\rangle_{0,S} = Y^t\langle AB\ldots\rangle_{0,cd}. \tag{6.20}$$

For systems of higher spin, no trick has yet been found which relates expectation values taken over the c–d ensemble to those for the physical ensemble.

6.2. The Heisenberg ferromagnet

(a) Spin $\frac{1}{2}$.

It is convenient to transform from site labels to momentum labels in the usual way:

$$c_i^+ = N^{-\frac{1}{2}}\sum_{\mathbf{k}} \exp[\mathrm{i}\mathbf{k}\,.\,\mathbf{r}_i]c_{\mathbf{k}}^+ \tag{6.21}$$

with similar equations for the ϕs and ds, and with

$$v(\mathbf{r}_{ij}) = N^{-1}\sum_{\mathbf{k}} \bar{v}(\mathbf{k})\exp[\mathrm{i}\mathbf{k}\,.\,\mathbf{r}_{ij}]. \tag{6.22}$$

The Hamiltonian, given in the S-language in eqn (5.18), then takes the form (putting the external field in the negative z-direction for convenience):

$$\begin{aligned} H &= H_0 + H_1, \\ H_0 &= -\tfrac{1}{2}N\left(B+\frac{\bar{v}(0)}{4}\right) + \sum_{\mathbf{k}}\left(\frac{\bar{v}(0)}{2}+B\right)c_{\mathbf{k}}^+c_{\mathbf{k}}, \\ H_1 &= -\tfrac{1}{2}N^{-1}\sum_{\mathbf{k},\mathbf{k}',\mathbf{q}} \bar{v}(\mathbf{q})[c_{\mathbf{k}}^+c_{\mathbf{k}'}^+c_{\mathbf{k}'-\mathbf{q}}c_{\mathbf{k}+\mathbf{q}} + c_{\mathbf{k}'}^+c_{\mathbf{k}}\phi_{\mathbf{k}'+\mathbf{q}}^+\phi_{\mathbf{k}+\mathbf{q}}], \end{aligned} \tag{6.23}$$

where we have used the fact that $v(\mathbf{r}_{ij}=0)=0$, so that $\sum_{\mathbf{k}}\bar{v}(\mathbf{k}) = 0$. H_0 represents the unperturbed Hamiltonian for 'c-ons', with energy $\bar{v}(0)/2+B$, and 'd-ons' with zero energy respectively. We note that both are independent of $\mathbf{k}$. The constant term in H_0 may be omitted if we redefine our zero of energy. H_1 represents interactions between the fermions, and between the fermions and drones. The first term in H_1 is sometimes called the longitudinal interaction, because, in the S-representation, it involves only the S^z operators, and the second term, the transverse interaction.

The unperturbed propagators are

$$C_{\mathbf{k}\mathbf{k}'}^0(\tau) \equiv \langle\mathscr{T}\{c_{\mathbf{k}}(\tau_1)c_{\mathbf{k}'}^+(\tau_2)\}\rangle \equiv \delta_{\mathbf{k},\mathbf{k}'}C_{\mathbf{k}}^0(\tau), \tag{6.24}$$

where $\tau = \tau_1-\tau_2$, with Fourier transform

$$\bar{C}_{\mathbf{k}}^0(\mathrm{i}\omega_l) = (\mathrm{i}\omega_l-\varepsilon_0)^{-1}; \qquad \omega_l = (2l+1)\pi/\beta; \qquad \varepsilon_0 = B+\tfrac{1}{2}\bar{v}(0);$$

and

$$D_\mathbf{k}^0(\tau) \equiv \langle \mathscr{T}\{\phi_\mathbf{k}(\tau_1)\phi_\mathbf{k}(\tau_2)\}\rangle_0, \tag{6.25}$$

with Fourier transform $\bar{D}_\mathbf{k}^0(i\omega_l) = 2(i\omega_l)^{-1}$. It is convenient not to separate the ϕs into ds (eqn (4.1)), since the ds appear in the interaction Hamiltonian only in the ϕ combination. This, together with the fact that the drones have zero energy, gives $\bar{D}_\mathbf{k}^0$ a rather unfamiliar appearance. Since neither the number of c-ons nor of d-ons is fixed, there is no $\exp[\beta(\mu_d\hat{N}_d + \mu_c\hat{N}_c)]$ factor in the statistical operator $\rho^{(\beta)}$, and the usual μ terms are missing from the free propagators.†

We represent $\bar{C}_\mathbf{k}^0(i\omega_l)$ by a full line with an arrow: $\xrightarrow{\mathbf{k},l}$, and $\bar{D}_\mathbf{k}^0(i\omega_l)$ by a wavy line with an arrow: $\overset{\mathbf{k},l}{\sim\!\!\sim\!\!\sim}$. Since $\phi_\mathbf{k} = d_\mathbf{k} + d_{-\mathbf{k}}^+$ we do not need the arrow to indicate at which end the d-on is created or destroyed, but merely to indicate in which direction the momentum is being carried. The two terms in H_1 are shown in Fig. 6.1. The diagrammatic expansion for the

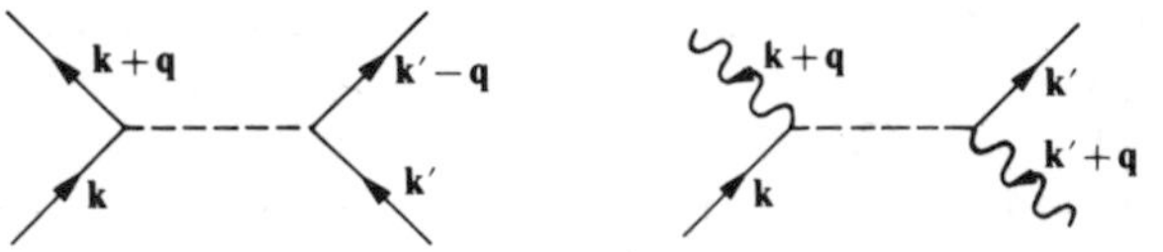

$$-(2N)^{-1}\bar{v}(\mathbf{q})c_{\mathbf{k}+\mathbf{q}}^+ c_{\mathbf{k}'-\mathbf{q}}^+ c_{\mathbf{k}'}\cdot c_\mathbf{k} \qquad\qquad -(2N)^{-1}\bar{v}(\mathbf{q})c_{\mathbf{k}'+\mathbf{q}}^+ \phi_{\mathbf{k}'+\mathbf{q}}^+ \phi_{\mathbf{k}'+\mathbf{q}}c_{\mathbf{k}'}.$$

Fig. 6.1.

free energy, or for any Green function, now follows in the usual way. Thus, for example, for the full propagator $\bar{C}_\mathbf{k}(i\omega_l)$, the Fourier coefficient of $\langle \mathscr{T}\{c_\mathbf{k}(\tau_1)c_\mathbf{k}^+(\tau_2)\}\rangle$, we have:

$$\bar{C}_\mathbf{k}(i\omega_l) \equiv {\uparrow}\, \mathbf{k}, l = {\uparrow}\mathbf{k}, l + \;\bigcirc\; \Sigma(\mathbf{k}, i\omega_l)$$

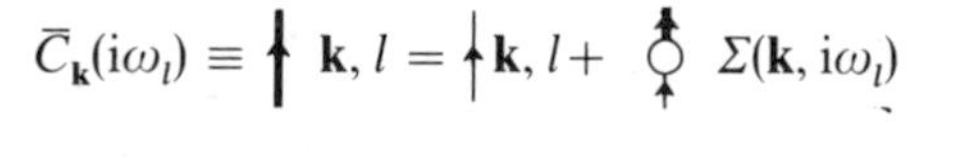

where Σ, the self-energy, includes terms like:

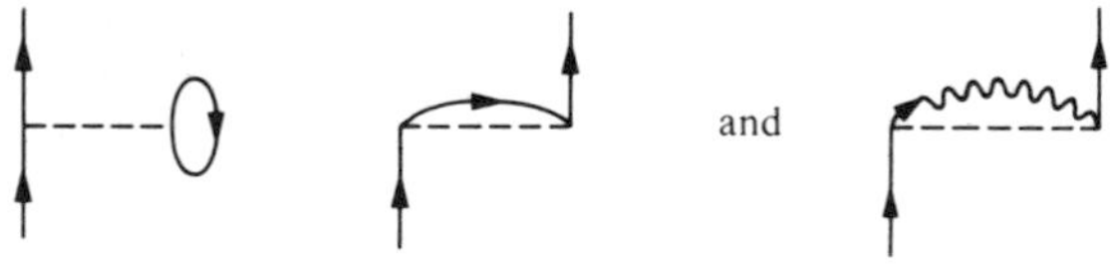

from the longitudinal interaction   and   from the transverse interaction

It is possible to regroup the terms in the perturbation expansion so that one obtains a series in powers of $1/n$, where n is the number of spins linked to a given spin through the interaction, or, in the case that only nearest-

† Note, however, that the cs and ϕs nevertheless appear in the Hamiltonian in pairs only: this ensures that eqn (2.27), for example, still holds.

neighbour interactions are important, n is the number of nearest neighbours. In the τ-representation, each vertex has a factor $\bar{v}(\mathbf{k})$, and a label τ_i which is to be integrated over from 0 to β. To obtain an order-of-magnitude estimate of the contribution of any diagram, we approximate the vertex contribution by $\beta\bar{u}$, where $\bar{u}$ is some average interaction. The C^0 and D^0 propagators are independent of k; thus the summations over k will involve only the k-dependence of the $\bar{v}$s. Since $\bar{v}(\mathbf{q}) = \sum_i v(\mathbf{r}_{ij}) \exp(-i\mathbf{q}\,.\,\mathbf{r}_{ij})$, $\bar{v}(\mathbf{q})$ is proportional to n. However, if we associate each $\bar{u}$ with a factor n, putting $\bar{u} = \gamma n$, each sum over a momentum label appearing in the argument of a $\bar{v}$ must then be considered to carry a factor n^{-1}, since, for example,

$$\begin{aligned}\sum_{\mathbf{q}} [\bar{v}(\mathbf{q})]^2 &= \sum_{\mathbf{q}} \sum_{ii'} v(\mathbf{r}_{ij}) v(\mathbf{r}_{i'j'}) \exp[-i\mathbf{q}\,.\,(\mathbf{r}_{ij}+\mathbf{r}_{i'j'})] \\ &= \sum_{ii'} v(\mathbf{r}_{ij}) v(\mathbf{r}_{i'j'}) \delta(\mathbf{r}_{ij}+\mathbf{r}_{i'j'}) \\ &= \sum_{i} v(\mathbf{r}_{ij}) v(-\mathbf{r}_{ij}),\end{aligned}$$

which is proportional to n and not n^2. Thus a diagram which has V vertices, and L v-dependent $\mathbf{k}$ summations, will have (V-L) independent site summations, and will have an n, β-dependence of the form $(\beta\gamma)^V n^{V-L}$. But, from elementary treatments of the ferromagnet, we know that the Curie temperature, T_C, is of the order of $\bar{u}/\kappa$, so that the diagram's contribution may be written $(T_C/T)^V n^{-L}$. Thus, if we are interested in the system at temperatures on the scale of the Curie temperature, the n-dependence of the diagram should be taken to be n^{-L}. The n-dependence of several diagrams for Σ is therefore as shown in Fig. 6.2.

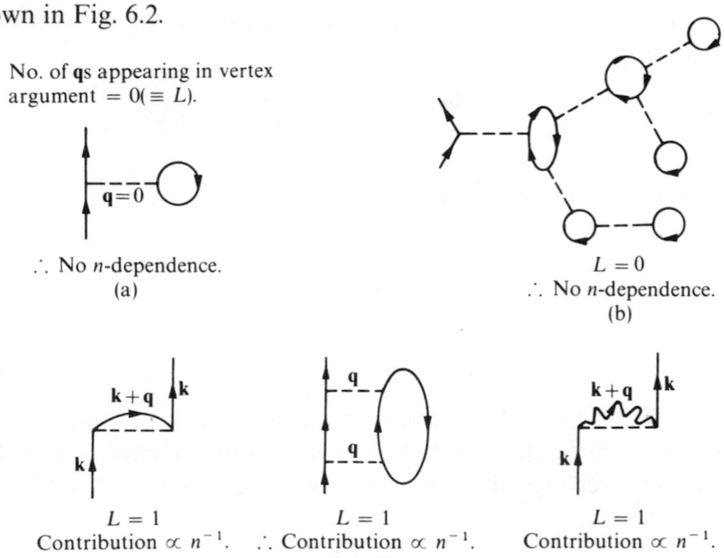

FIG. 6.2.

The dominant terms for large n are therefore just those summed by the Hartree self-consistent approximation:

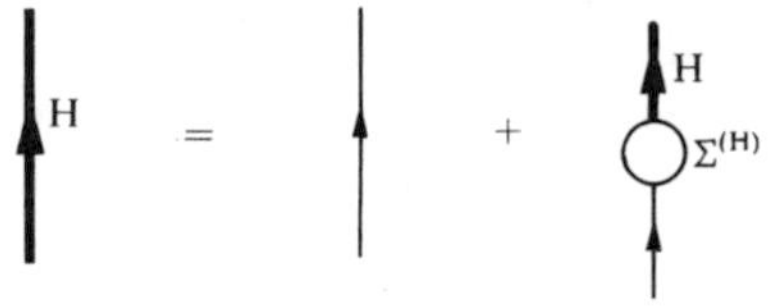

(6.26)

(6.27)

i.e.

$$\Sigma^{(\mathrm{H})}(\mathbf{k}, \mathrm{i}\omega_l) = -\frac{\bar{v}(0)}{N\beta} \underset{\substack{\tau\to 0\\ \tau<0}}{\mathrm{Lt}} \sum_{\mathbf{k}',l'} \bar{C}^{(\mathrm{H})}_{\mathbf{k}'}(\mathrm{i}\omega_{l'}) \exp[-\mathrm{i}\omega_{l'}(\tau)]$$

$$= -\frac{\bar{v}(0)}{N}(\tfrac{1}{2}N - M^{(\mathrm{H})})$$

where

$$M^{(\mathrm{H})} = \frac{N}{2} - \underset{\substack{\tau\to 0\\ \tau<0}}{\mathrm{Lt}} \sum_{\mathbf{k},l} \frac{1}{\beta} \bar{C}^{(\mathrm{H})}_{\mathbf{k}}(\mathrm{i}\omega_l) \exp(-\mathrm{i}\omega_l \tau). \tag{6.28}$$

Since the magnetization of the system is proportional to $-\langle S^z \rangle$, and

$$S^z = \sum_i (c_i^+ c_i - \tfrac{1}{2}) = \sum_{\mathbf{k}} c_{\mathbf{k}}^+ c_{\mathbf{k}} - \tfrac{1}{2}N,$$

$$-\langle S^z \rangle = \frac{N}{2} + \underset{\substack{\tau\to 0\\ \tau<0}}{\mathrm{Lt}} \sum_{\mathbf{k}} C_{\mathbf{k}}(\tau) = M^{(\mathrm{H})}, \tag{6.29}$$

$M^{(\mathrm{H})}$ is essentially the magnetization of the system in this approximation. Further

$$M^{(\mathrm{H})} = \frac{N}{2} - \underset{\substack{\tau\to 0\\ \tau<0}}{\mathrm{Lt}} \sum_{\mathbf{k}l} \frac{1}{\beta} \left[\mathrm{i}\omega_l - B - \frac{\bar{v}(0)}{N} M^{(\mathrm{H})} \right]^{-1} \exp(-\mathrm{i}\omega_l \tau)$$

$$= \frac{N}{2} - N \left\{ \exp\left[\beta \left(B + \frac{\bar{v}(0)}{N} M^{(\mathrm{H})} \right) \right] + 1 \right\}^{-1}$$

$$= \frac{N}{2} \tanh\left[\tfrac{1}{2}\beta \left(B + \frac{\bar{v}(0)}{N} M^{(\mathrm{H})} \right) \right]. \tag{6.30}$$

This is the usual equation for the magnetization in molecular field theory, which we have thus shown to be correct in the limit of very large n. The form of the solution of eqn (6.30) is given in most elementary texts on solid-state physics. It gives a non-zero value for the magnetization even for zero external field for temperatures less than $T_C = \bar{v}(0)/(4\kappa)$.

The comparative ease with which we have been able to make a well-controlled approximation in this method of dealing with the ferromagnet should be contrasted with the somewhat haphazard way in which approximations were made in the equation of motion method.

(b) Spin 1.

The corresponding results for $S = 1$ are more difficult to obtain because of the factors of Y which have to be included. This difficulty is illustrated in second order by the (improper) diagram shown in Fig. 6.3(a). This contains the factors

$$\bar{v}(0)\bar{v}(0) = \sum_j v(\mathbf{r}_{ij}) \sum_{j'} v(\mathbf{r}_{ij'})$$

$$= \left(\sum_{j \neq j'} + \sum_{j = j'} \right) v(\mathbf{r}_{ij}) v(\mathbf{r}_{ij'}).$$

The first term on the right-hand side has two site-label indices, and thus the expectation value of the operators associated with it carries a factor of Y^2 if we work in the c–d representation, whereas the second involves only one site index, and thus carries a factor Y. It is therefore easiest to work with site labels rather than momentum labels in the early stages of the calculation.

The interaction part of the Hamiltonian may be written:

$$H_1 = H_1^0 + H_1^T + H_1^L,$$

$$H_1^0 = \tfrac{1}{2} \sum_{ij\alpha\alpha'} v(\mathbf{r}_{ij}) c_{i\alpha}^{+} c_{i\alpha},$$

$$H_1^T + H_1^L = -\tfrac{1}{2} \sum_{ij\alpha\alpha'} v(\mathbf{r}_{ij}) c_{i\alpha}^{+} \phi_{i\alpha} \phi_{j\alpha'} c_{j\alpha'} - \tfrac{1}{2} \sum_{ij\alpha\alpha'} v(\mathbf{r}_{ij}) c_{i\alpha}^{+} c_{i\alpha} c_{j\alpha'}^{+} c_{j\alpha'}, \tag{6.31}$$

where $\alpha, \alpha' = 1, 2$. The term H_1^0 cannot be incorporated with the non-interacting term H_0, as it could be in the spin-$\frac{1}{2}$ problem, since it has a double site summation, and, in nth order, carries a factory of Y^r, where r depends on the number of site coincidences.

If we first disregard the difficulty with respect to the second-order diagram in Fig. 6.3(b) (drawn in the site representation), we shall have to add a correction term shown in (c) with a factor $(Y - Y^2) \sum_j [v(\mathbf{r}_{ij})]^2$. However, if we are calculating the magnetization to order n^0, we may neglect this correction term, since it is clearly of order $1/n$ times the corresponding diagram with independent site summations. This will also be the case for all correction diagrams, and so, to this order in $1/n$, we may ignore the

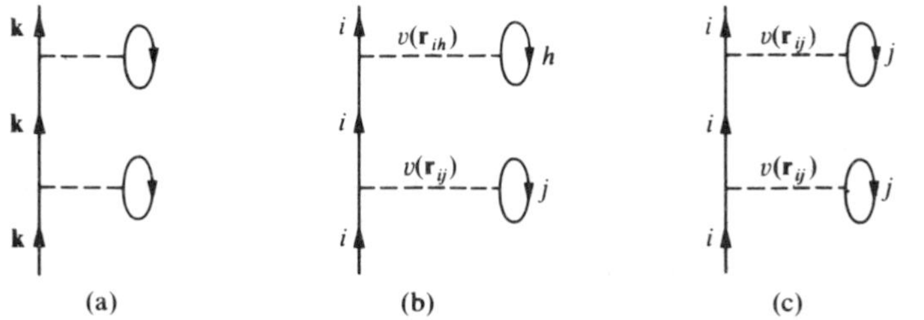

FIG. 6.3.

coincidence problem. We thus have, for the Hartree self-consistent approximation (remembering the extra H^0 term):

$$\bar{C}^{(\mathrm{H})}_{ij\alpha\alpha'}(\mathrm{i}\omega_l) = \delta_{\alpha\alpha'}\delta_{ij}[\mathrm{i}\omega_l - B - \Sigma^{(\mathrm{H})}_{i\alpha}]^{-1} \tag{6.32}$$

$$\begin{aligned}\Sigma^{(\mathrm{H})}_{i\alpha} &= -\operatorname*{Lt}_{\substack{\tau\to 0\\ \tau<0}} Y\beta^{-1}\sum_{j\alpha' l} v(\mathbf{r}_{ij})[\bar{C}^{(\mathrm{H})}_{j\alpha'}(\mathrm{i}\omega_l)\exp(-\mathrm{i}\omega_l\tau)]\\ &= \bar{v}(0)N^{-1}M^{(\mathrm{H})},\end{aligned} \tag{6.33}$$

where

$$M^{(\mathrm{H})} = YN\tanh[\tfrac{1}{2}\beta(B+\bar{v}(0)N^{-1}M^{(\mathrm{H})})] \tag{6.34}$$

and where the trace in the definition of $\bar{C}^{(\mathrm{H})}_j(\mathrm{i}\omega_l)$ has, for convenience, been taken over the c–d ensemble. Thus, in order to obtain $\langle S^z\rangle$ from $C(0)$, we need a further factor of Y. This means that $M^{(\mathrm{H})}$ as defined above is the magnetization in this approximation.

Higher-order terms in n^{-1} may be calculated in a straightforward manner for $S = \frac{1}{2}$; for $S = 1$, this is still possible, but the coincidence problem now has to be dealt with. We refer the reader to Spencer's paper for details.

Exercise

6.2.1. Compare the results of this chapter with those of section 5.3 and exercise 1.4.1.

Further reading

Chapter 4 of Kittel (1963) contains a review of spin-wave theory. Mattis (1965) gives a full account of work up to 1965, including much of the material in the references below, and introduces the drone–fermion representation.

The transformation of spin operators to boson or fermion operators is discussed by Schwinger (1952), Holstein and Primakoff (1940) and Kennan (1966) (the drone–fermion representation).

Other forms of Wick's theorem for the spin problem are given by Davis (1960), Mills, Kenan and Korringa (1960), Yolin (1965), Abrikosov (1965), Yang-li Wang, Shtrikman, and Callen (1966), Doniach (1966), Giovanni,

Peter and Koidé (1966), Lewis and Stinchcombe (1967). These methods lead to somewhat unconventional diagrammatic expansions, and the linked-cluster theorem cannot always be proved.

Stinchcombe, Horwitz, Englert and Brout (1963) discuss expansions in inverse powers of the number of nearest neighbours, using the method of semi-invariants. See also Vaks, Larkin and Pekin (1967).

The work described in this chapter is taken from Spencer (1968*a*), who applies the same method to the theory of *s–d* scattering in dilute magnetic alloys with spin-half impurities (Spencer, 1968*b*).

7

LINEAR RESPONSE AND TRANSPORT PROCESSES

7.1. Linear-response theory

WHEN considering the properties of many-body systems, we are often interested in the response of the system to an external disturbance. A simple example is provided by the susceptibility of a system containing particles with a magnetic moment: we apply a magnetic field to the system, and wish to calculate the resulting magnetic moment. This has been done for a simple case in the previous chapter. Problems in the theory of transport present more complicated examples, but the fundamental idea is much the same. If we are considering the electrical conductivity, for example, we are interested in the current resulting from the application of an electric field.

Elementary kinetic theory provides one way of discussing transport processes: we introduce a mean free path or a relaxation time, and, by physical arguments, relate this to transport coefficients, such as electrical conductivity, viscosity and mobility. At a higher degree of sophistication, the Boltzmann equation may be used, and, for the first half of this century, the kinetic theory of transport processes was concerned with the justification of this equation, or with calculations of transport coefficients based on it. Much of this work is described in Chapman and Cowling (1952). Unfortunately, the Boltzmann equation is applicable only to dilute systems. Further, the usual derivation of the quantum-mechanical version of the equation (Peierls, 1934, 1955) depends on the use of the repeated random-phase approximation, which cannot be given a general justification.

In 1956, Kubo (see also Kubo (1958)) used linear response theory to give *exact* expressions for transport coefficients in terms of correlation functions for the *equilibrium* system. As we have seen in previous chapters, to evaluate such correlation functions for any particular system, approximations have to be made. But the existence of an exact formulation of the problem helps one to consider the effects of such approximations, and useful results can be obtained for some strong-coupling problems. Further, some exact results can be obtained.

We shall first give a general discussion of linear-response theory. We confine ourselves to the problem of evaluating the expectation value of some operator, A, to first order in the applied disturbance. This will give us, for example, an expression for the conductivity, since, for most systems, the current, $\mathbf{j} = \sigma\mathbf{E}$, where $\mathbf{E}$ is the applied field, and σ is a constant for low fields. There is also interest in non-linear responses, as a result, for example, of the present-day possibility of producing intense electromagnetic fields. The techniques used for this are similar to, but more complex than, those used in linear response theory, described below, and we refer the reader to the literature (e.g. Peterson, 1967).

We consider the response of our system to an external field described by an additional term, $H_1(t)$, in the Hamiltonian:

$$H' - \mu\hat{N} = H - \mu\hat{N} + H_1(t) \tag{7.1}$$

For example, in the case of an electric field, constant in space, but varying periodically with time, $H_1(t)$ could be taken to be

$$H_1(t) = \mathrm{e}\sum_i \mathbf{E}_0 \,.\, \mathbf{r}_i \cos \omega t. \tag{7.2}$$

The mean value of a dynamical variable with operator A is

$$\langle A \rangle_t = \mathrm{Tr}\{\rho^{(\beta)}(t)A\}, \tag{7.3}$$

where $\rho(t)$ is the density matrix which is now time-dependent, because of the time-dependence of the Hamiltonian. (Note the different normalization from that used in Chapter 2). The equation of motion for ρ is (see Appendix F)

$$\mathrm{i}\frac{\mathrm{d}\rho}{\mathrm{d}t} = [H', \rho]. \tag{7.4}$$

To determine the linear response of the system, we have to solve eqn (7.4) to first order in H_1. We consider the case in which H_1 is switched on at time $t = 0$, before which the system is in thermal equilibrium, so that:

$$H_1(t) = 0 \qquad t < 0, \tag{7.5}$$

$$\rho^{(\beta)}(0) = \exp[-\beta(H - \mu\hat{N})]/\mathrm{Tr}\{\exp[-\beta(H - \mu\hat{N})]\}. \tag{7.6}$$

We write

$$\rho^{(\beta)}(t) = \rho_0(t) + \sum_{n=1} \rho_n(t), \tag{7.7}$$

where ρ_n is of nth order in H_1. Thus, from eqn (7.4):

$$\mathrm{i}\frac{\mathrm{d}\rho_0}{\mathrm{d}t} = [H - \mu\hat{N}, \rho_0(t)]$$

$$\mathrm{i}\frac{\mathrm{d}\rho_n}{\mathrm{d}t} = [H - \mu\hat{N}, \rho_n(t)] + [H_1(t), \rho_{n-1}(t)], \qquad n \geqslant 1. \tag{7.8}$$

From the boundary conditions at $t = 0$:

$$\rho_0(0) = \exp[-\beta(H - \mu\hat{N})]/\mathrm{Tr}\{\exp[-\beta(H - \mu\hat{N})]\}$$

$$\rho_n(0) = 0 \qquad n \geq 1. \tag{7.9}$$

The solutions

$$\rho_0(t) = \exp[-\beta(H-\mu\hat{N})]/\mathrm{Tr}\{\exp[-\beta(H-\mu\hat{N})]\}, \tag{7.10}$$

$$\rho_1(t) = -\mathrm{i}\int_0^t \mathrm{d}t' \exp[-\mathrm{i}(H-\mu\hat{N})(t-t')][H_1(t'), \rho_0]\exp[\mathrm{i}(H-\mu\hat{N})(t-t')], \tag{7.11}$$

may be seen to satisfy the differential eqns (7.8), by direct differentiation, and the boundary conditions eqn (7.9).

The expectation value at time t of an operator A, $\langle A\rangle_t$, to first order in H_1, is thus

$$\langle A\rangle_t = \mathrm{Tr}\{\rho_0 A\} - \mathrm{i}\int_0^t \mathrm{d}t'\,\mathrm{Tr}\{\exp[-\mathrm{i}(H-\mu\hat{N})(t-t')\times$$
$$\times[H_1(t'), \rho_0]\exp[\mathrm{i}(H-\mu\hat{N})(t-t')]A\}. \tag{7.12}$$

This expression may be considerably simplified. Our aim is to write it as an expectation value $\langle X\rangle = \mathrm{Tr}\{\rho_0 X\}$ of some operator X in the equilibrium ensemble. We expand the commutator, and use the cyclic invariance of the trace, to write

$$\langle A\rangle_t = \langle A\rangle - \mathrm{i}\int_0^t \mathrm{d}t'\,\mathrm{Tr}\{\rho_0\exp[\mathrm{i}(H-\mu\hat{N})(t-t')]A\,\times$$
$$\times\exp[-\mathrm{i}(H-\mu\hat{N})(t-t')]H_1(t')\}+$$
$$+\mathrm{i}\int_0^t \mathrm{d}t'\,\mathrm{Tr}\{\rho_0\{H_1(t')\exp[\mathrm{i}(H-\mu\hat{N})(t-t')]A\,\times$$
$$\times\exp[-\mathrm{i}(H-\mu\hat{N})(t-t')]\}. \tag{7.13}$$

But ρ_0 and $H-\mu\hat{N}$ commute, as they are both diagonal in a representation in which $H-\mu\hat{N}$ is diagonal. Again using cyclic invariance, we may write

$$\langle A\rangle_t = \langle A\rangle - \mathrm{i}\int_0^t \mathrm{d}t'\,\mathrm{Tr}\{\rho_0\exp[\mathrm{i}(H-\mu\hat{N})t]A\exp[-\mathrm{i}(H-\mu\hat{N})t]\times$$
$$\times\exp[\mathrm{i}(H-\mu\hat{N})t']H_1(t')\exp[-\mathrm{i}(H-\mu\hat{N})t']\}+$$
$$+\mathrm{i}\int_0^t \mathrm{d}t'\,\mathrm{Tr}\{\rho_0\exp[\mathrm{i}(H-\mu\hat{N})t']H_1(t')\exp[-\mathrm{i}(H-\mu\hat{N})t']\times$$
$$\times\exp[\mathrm{i}(H-\mu\hat{N})t]A\exp[-\mathrm{i}(H-\mu\hat{N})t]\}$$
$$= \langle A\rangle - \mathrm{i}\int_0^t \mathrm{d}t'\langle[A^{\mathrm{H}}(t), H_1^{\mathrm{H}}(t')]\rangle, \tag{7.14}$$

where $B^{\mathrm{H}}(t)$ is the operator $B(t)$ in the Heisenberg representation:

$$B^{\mathrm{H}}(t) = \exp[\mathrm{i}(H-\mu\hat{N})t]\,B(t)\exp[-\mathrm{i}(H-\mu\hat{N})t].$$

Thus we have written the linear response of the system in terms of the equilibrium ensemble expectation value of products of operators, and the techniques developed in previous chapters may be used to evaluate these expectation values. In particular, eqn (7.14) may be written in terms of the retarded Green function of A and H_1 (eqn (3.50)):

$$\langle A\rangle_t = \langle A\rangle + \int_0^\infty \mathrm{d}t'\langle\langle A; H_1(t')\rangle\rangle_\mathrm{r} \tag{7.15}$$

with the choice $\varepsilon = 1$ in the definition of the Green function.

If the perturbation is such that, after a sufficiently long time, the mean value of A will not depend on the time at which the perturbation was switched on, the lower limit of the integral in eqn (7.15) may be extended to $-\infty$. This will normally be the case, as we can see by considering the response of the system to an electric field, with A as the current or polarization vector. The connection between this assumption and the introduction of irreversibility into a theory of transport is discussed by Chester (1963). In order to avoid the transient effects of a sudden switching on of the perturbation, it is sometimes necessary to include a factor $\exp[\eta(t'-t)]$ in the integral, and take the limit as η tends to zero, which means that the perturbation is switched on slowly.

We shall assume for simplicity that $\langle A\rangle$ is zero; this can always be arranged if A has no explicit time-dependence by subtracting a constant from the operator A. With these changes, eqn (7.15) takes the form

$$\langle A\rangle_t = \operatorname*{Lt}_{\eta\to 0} \int_{-\infty}^{\infty} \mathrm{d}t'\langle\langle A; H(t')\rangle\rangle_\mathrm{r}\exp[\eta(t'-t)]. \tag{7.16}$$

This formula is an *exact* expression for the *linear* response of a physical observable A to a perturbation H'. Causality is built into the equation by the fact that the Green function is zero if $t' > t$: the mean value of the operator A at time t cannot be affected by the applied field for times greater than t. In field theory, causality is often used to give dispersion relations, and a dispersion relation may be derived from eqn (7.16) by using the analytic properties of the retarded Green function.

Consider, for example, the case in which the applied perturbation can be expressed as a Fourier integral in the form:

$$H_1(t') = B\int_{-\infty}^{\infty} \exp(-\mathrm{i}\omega t')f(\omega)\,d\omega, \tag{7.17}$$

where B is a Hermitian operator, which, in the Schrödinger representation, does not depend on the time. Consider the effect of just one component of H_1, and write:

$$H_1^{(\omega)}(t') = B\exp(-\mathrm{i}\omega t')f(\omega)$$

$$\langle A\rangle_t^{(\omega)} = f(\omega)\int_{-\infty}^{\infty} \mathrm{d}t' \langle\!\langle A^{\mathrm{H}}(t); B^{\mathrm{H}}(t')\rangle\!\rangle_{\mathrm{r}} \exp(-\mathrm{i}\omega t').$$

(Remember that, in the Green function, B is in the Heisenberg representation and thus depends on the time.) By a trivial change of variables,

$$\begin{aligned}\langle A\rangle_t^{(\omega)} &= f(\omega)\exp(-\mathrm{i}\omega t)\int_{-\infty}^{\infty} \mathrm{d}t'' \langle\!\langle A^{\mathrm{H}}(t), B^{\mathrm{H}}(t-t'')\rangle\!\rangle_{\mathrm{r}} \exp(\mathrm{i}\omega t'')\\ &= f(\omega)\exp(-\mathrm{i}\omega t)\mathscr{G}_{\mathrm{r}}^{AB}(\omega).\end{aligned} \tag{7.18}$$

$\langle A\rangle_t^{(\omega)}/[Bf(\omega)\exp(-\mathrm{i}\omega t)]$ is the quotient of the response and the field, of frequency ω, both at time t. We may therefore call it the generalized susceptibility $\chi(\omega)$. Splitting it into its real and imaginary parts $\chi'(\omega)$ and $\chi''(\omega)$ and using eqns (3.70 and 71):

$$\chi(\omega) = \chi'(\omega)+\mathrm{i}\chi''(\omega)$$

where:

$$\chi'(\omega) = \frac{P}{\pi}\int_{-\infty}^{\infty} \frac{\chi''(\omega')}{\omega'-\omega}\,\mathrm{d}\omega' \tag{7.19}$$

$$\chi''(\omega) = -\frac{P}{\pi}\int_{-\infty}^{\infty} \frac{\chi'(\omega')}{\omega'-\omega}\,\mathrm{d}\omega'. \tag{7.20}$$

These are generalized forms of the Kramers–Kronig (Kronig (1926), Kramers (1927)) relations. For the case of the current response to an applied alternating electric field, for example, they relate the resistive and capacitative parts of the conductivity.

Exercise

7.1.1. An impurity is at rest in a many-body system. At time $t = 0$, it starts to move. Derive a formal expression for the rate of transfer of momentum to the system at $t = 0$.

(Write the Hamiltonian as:

$$H = H_0 + H_1 t + \tfrac{1}{2}H_2 t^2 + \cdots$$

and expand the density matrix similarly:

$$\rho = \rho_0 + \rho_1 t + \tfrac{1}{2}\rho_2 t^2 + \cdots.$$

At $t = 0$, the system is in equilibrium, so that $[H_0, \rho_0] = 0$. If P is the momentum operator for the system, show that:

$$\langle \mathbf{P} \rangle_t = -\mathrm{i}\frac{t^2}{2}\,\mathrm{Tr}\{\rho_0[\mathbf{P}, H_1]\} + \text{terms of higher order in } t$$

$$= -\mathrm{i}\frac{t^2}{2}\langle[\mathbf{P}, H_1]\rangle_{t=0} + \text{terms of higher order in } t.)$$

A system of phonons in a liquid in the presence of a moving scattering centre may sometimes be described by the Hamiltonian:

$$H = \sum_{\mathbf{k}} \omega_{\mathbf{k}} a_{\mathbf{k}}^{+} a_{\mathbf{k}} + \sum_{\mathbf{k}} [f_{\mathbf{k}}(t) a_{\mathbf{k}} + f_{\mathbf{k}}^{*}(t) a_{\mathbf{k}}^{+}].$$

Find $\langle \mathbf{P} \rangle_t$, correct to terms of second-order in t.

7.2. Dielectric response

If we introduce a test charge density, $\phi_{\mathbf{q}} \exp(\mathrm{i}\mathbf{q}\,.\,\mathbf{r}).\ \exp(-\mathrm{i}\omega t)$, into a system of particles of charge e, the electric displacement will be given by Poisson's equation:

$$\mathrm{Div}\, \mathbf{D}_{\mathbf{q}} = 4\pi\phi_{\mathbf{q}} \exp(\mathrm{i}\mathbf{q}\,.\,\mathbf{r}) \exp(-\mathrm{i}\omega t).$$

The electric field will be given by:

$$\mathrm{Div}\, \mathbf{E}_{\mathbf{q}} = 4\pi\ (\text{Test charge} + \text{Induced charge})$$

$$= 4\pi\phi_{\mathbf{q}} \exp(\mathrm{i}\mathbf{q}\,.\,\mathbf{r}) \exp(-\mathrm{i}\omega t) + 4\pi e\langle\rho(\mathbf{r})\rangle_{\mathbf{q}}$$

where $\rho(\mathbf{r})$ is the particle density in our system. In the case that the system is isotropic, $\mathbf{D}$, $\mathbf{E}$, and $\langle\rho(\mathbf{r})\rangle_{\mathbf{q}}$ will vary as $\exp(\mathrm{i}\mathbf{q}\,.\,\mathbf{r})$, and these equations may be written:

$$\mathrm{i}\mathbf{q}\,.\,\mathbf{D}_{\mathbf{q}} = 4\pi\phi_{\mathbf{q}} \exp(\mathrm{i}\mathbf{q}\,.\,\mathbf{r}) \exp(-\mathrm{i}\omega t)$$

$$\mathrm{i}\mathbf{q}\,.\,\mathbf{E}_{\mathbf{q}} = 4\pi\phi_{\mathbf{q}} \exp(\mathrm{i}\mathbf{q}\,.\,\mathbf{r}) \exp(-\mathrm{i}\omega t) + 4\pi\frac{e}{V}\langle\rho_{\mathbf{q}}\rangle \exp(\mathrm{i}\mathbf{q}\,.\,\mathbf{r}).$$

Thus, defining a (possibly complex) dielectric function for wave-number $\mathbf{q}$ and frequency ω by

$$\mathbf{D}_{\mathbf{q}} = \varepsilon(\mathbf{q}, \omega)\mathbf{E}_{\mathbf{q}},$$

we have

$$[\varepsilon(\mathbf{q}, \omega)]^{-1} = 1 + \frac{e}{V}\langle\rho_{\mathbf{q}}\rangle \exp(\mathrm{i}\mathbf{q}\,.\,\mathbf{r})/[\phi_{\mathbf{q}} \exp(\mathrm{i}\mathbf{q}\,.\,\mathbf{r}) \exp(-\mathrm{i}\omega t)]. \quad (7.21)$$

The second term on the right-hand side of this equation is exactly of the form of a generalized susceptibility† introduced above. Putting $\varepsilon(\mathbf{q}, \omega) = n^2(\mathbf{q}, \omega)$, where n is the complex refractive index, and noting that in most cases $n \approx 1$, we have, from eqns (7.19) and (7.20), the Kramers–Kronig relations in their original form:

$$\operatorname{Re}\{n(\mathbf{k}, \omega)\} = 1 + \frac{P}{\pi} \int_{-\infty}^{\infty} d\omega' \operatorname{Im}\{n(\mathbf{k}, \omega')\}/(\omega' - \omega),$$

$$\operatorname{Im}\{n(\mathbf{k}, \omega)\} = -\frac{P}{\pi} \int_{-\infty}^{\infty} d\omega' \operatorname{Re}\{n(\mathbf{k}, \omega') - 1\}/(\omega' - \omega). \tag{7.22}$$

In order to find the form of the Green function we need in eqn (7.18), we must determine the extra term which appears in our Hamiltonian as a result of the test charge. If the latter has a density in real space of $\phi(\mathbf{r})$, the additional potential energy, and hence the extra term in the Hamiltonian, will be

$$e \int \rho(\mathbf{r})\phi(\mathbf{r}')/|\mathbf{r} - \mathbf{r}'| \, d^3\mathbf{r} \, d^3\mathbf{r}' = 4\pi e \int d^3\mathbf{q} \, \rho_{-\mathbf{q}}\phi_{\mathbf{q}}/q^2. \tag{7.23}$$

where $\phi_{\mathbf{q}}$ is the Fourier transform of $\phi(\mathbf{r})$. In the case above, this has only one component, $\phi_{\mathbf{q}}$, and thus

$$[\varepsilon(\mathbf{q}, \omega)]^{-1} = 1 + [4\pi e^2/(q^2 V)] \langle\langle \rho_{\mathbf{q}}; \rho_{-\mathbf{q}} : \omega \rangle\rangle_{\mathrm{r}}. \tag{7.24}$$

This equation relates the dielectric constant to the density–density correlation functions.

We have determined (approximately) the density–density Green function at the infinite set of points along the imaginary axis in section 4.2, since

$$\rho_{\mathbf{q}} = \sum_{\mathbf{k}} a_{\mathbf{k}}^{+} a_{\mathbf{k}+\mathbf{q}}.$$

From eqn (4.11) we have, extending $\bar{K}$ to the whole of the complex plane,

$$\bar{K}(\mathbf{q}, \omega) \simeq \Pi^{(a)}(\mathbf{q}, \omega)/[1 - \bar{v}(q)\Pi^{(a)}(\mathbf{q}, \omega)] \tag{7.25}$$

where $\Pi^{(a)}$ is given by eqn (4.14). Thus, from eqns (4.8) and (7.24),

$$\varepsilon(\mathbf{q}, \omega) = 1 - \bar{v}(q)\Pi^{(a)}(\mathbf{q}, \omega) \tag{7.26}$$

which should be compared with eqn (4.18).

If we put a unit point charge, constant in time, at the origin, we have

$$\phi_{\mathbf{q}} = 1.$$

The induced charge density is thus, from eqn (7.21),

$$e \int \langle \rho_{\mathbf{q}} \rangle \exp(-i\mathbf{q} \cdot \mathbf{r}) \, d^3\mathbf{q} \sim e\lambda^2 r^{-1} \exp(-\lambda r) \quad \text{for large } r,$$

† Note that this is not the susceptibility as normally defined for a dielectric, for which $\varepsilon = 1 + \chi$.

(with the notation of section 4.2) while the potential behaves, for large r, as $(e/r)\exp(-\lambda r)$, a typical shielded Coulomb potential. We have thus accounted for the shielding of the point charge at the origin by the charges in the many-body system.

Exercise

7.2.1. Use eqns (3.94), (3.58), (3.59), (3.64) and (7.24) to show that

$$[\varepsilon(\mathbf{q}, \omega)]^{-1} = 1+\frac{4\pi e^2}{Vq^2}\operatorname*{Lt}_{\gamma\to 0}\int_{-\infty}^{\infty} \mathrm{d}E'\, S(\mathbf{q}, -E')[\exp(\beta E')-1]/(\omega-E'+\mathrm{i}\gamma)$$

$$= 1+\frac{4\pi e^2}{Vq^2}\operatorname*{Lt}_{\gamma\to 0}\int_{-\infty}^{\infty} \mathrm{d}E'\, S(\mathbf{q}, E')[(\omega-E'+\mathrm{i}\gamma)^{-1}-(\omega+E'+\mathrm{i}\gamma)^{-1}],$$

since $\exp(\beta E')S(\mathbf{q}, -E') = S(\mathbf{q}, E')$ as

$$\langle\rho_{\mathbf{q}}(t)\rho_{-\mathbf{q}}(0)\rangle = \langle\rho_{\mathbf{q}}(0)\rho_{-\mathbf{q}}(-t)\rangle = \langle\rho_{-\mathbf{q}}(0)\rho_{\mathbf{q}}(-t)\rangle.$$

Put $\varepsilon(\mathbf{q}, \omega) = \varepsilon_1(\mathbf{q}, \omega)+\mathrm{i}\varepsilon_2(\mathbf{q}, \omega)$, where ε_1 and ε_2 are both real, and show that

$$\varepsilon_2(\mathbf{q}, \omega)/|\varepsilon(\mathbf{q}, \omega)|^2 = \frac{4\pi^2 e^2}{Vq^2}[S(\mathbf{q}, \omega)-S(\mathbf{q}, -\omega)].$$

Use eqn (3.97) to show that

$$\int_0^{\infty} \mathrm{d}\omega\, \omega\varepsilon_2(\mathbf{q}, \omega)/|\varepsilon(\mathbf{q}, \omega)|^2 = 2\pi^2 ne^2/m.$$

Show further that

$$\operatorname*{Lt}_{\omega\to\infty} \varepsilon(\mathbf{q}, \omega) = 1-\omega_0^2/\omega^2$$

where $\omega_0^2 = 4\pi ne^2/m$.

7.3. Electrical conductivity

In a similar way, eqn (7.18) may be used to evaluate the conductivity of a system of a large number of particles of mass m carrying a charge e. We apply an electric field, $\mathbf{E}_0\exp(-\mathrm{i}\omega t)$, alternating in time but constant in space, to the system: for our purposes, this is most easily expressed in terms of the scalar and vector potentials:

$$\phi = 0, \qquad \mathbf{A} = -\mathrm{i}\mathbf{E}_0 c \exp(-\mathrm{i}\omega t)/\omega. \tag{7.27}$$

To first order in **A**, the additional term in the Hamiltonian is

$$H_1 = -e\mathbf{A}\,.\,\mathbf{P}/mc \tag{7.28}$$

where **P** is the total momentum operator for the whole system. The current may be written in terms of the momentum operators, using

$$\dot{\mathbf{x}}_i = -\mathrm{i}[\mathbf{x}_i, H - \mu\hat{N} + H_1] = \frac{\mathbf{p}_i}{m} - \frac{e\mathbf{A}}{mc}$$

for the coordinate, $\mathbf{x}_i$, and momentum $\mathbf{p}_i$ of the ith particle. Thus, if the field is applied in the νth direction, the μth component of the current density is given by

$$\langle j_{\mu\nu}(t)\rangle = (\,\mathrm{i}\delta_{\mu\nu}\frac{\bar{N}e^2}{m\omega}E_0 \exp(-\mathrm{i}\omega t) + \mathrm{i}e^2 E_0 \exp(-\mathrm{i}\omega t)/(\omega m^2) \times$$
$$\times \langle\!\langle P_\mu ; P_\nu : \omega\rangle\!\rangle_{\mathrm{r}}/V,$$

where $\bar{N}$ is the total number of particles in the system. Thus, the conductivity tensor is

$$\sigma_{\mu\nu} = (\mathrm{i}\delta_{\mu\nu}\bar{N}e^2/m\omega + \mathrm{i}(e^2/m^2\omega)\langle\!\langle P_\mu ; P_\nu : \omega\rangle\!\rangle_{\mathrm{r}})/V. \tag{7.29}$$

Both the real and imaginary (resistive and capacitative) parts of the conductivity are included in this expression.

The first term corresponds to the conductivity of a system of non-interacting particles. To see that there is an equal and opposite contribution from the second term as $\omega \to 0$, for a system of interacting particles, so that the conductivity remains finite, we use the exact formula for the retarded Green function, eqn (3.64):

$$\mathscr{G}_{\mathrm{r}}^{BC}(\omega) = \frac{1}{2\pi} \underset{\gamma\to 0}{\mathrm{Lt}} \int_{-\infty}^{\infty} \mathrm{d}\omega'(\exp(\beta\omega') - \varepsilon)J^{CB}(\omega')/(\omega - \omega' + \mathrm{i}\gamma)$$

$$\therefore \quad \underset{\omega\to 0}{\mathrm{Lt}}\, \mathscr{G}_{\mathrm{r}}^{BC}(\omega) = \frac{P}{2\pi}\int \mathrm{d}\omega'[\exp(\beta\omega') - \varepsilon]J^{CB}(\omega')]/(-\omega') -$$
$$-\frac{\mathrm{i}}{2}\int \mathrm{d}\omega'[\exp(\beta\omega') - \varepsilon]J^{CB}(\omega')\delta(-\omega'). \tag{7.30}$$

The second term is zero, since $J^{CB}(0)$ is finite, and, from eqn (7.15), we choose $\varepsilon = +1$. Using eqn (3.58) for J, we may rewrite the first term:

$$\begin{aligned}\operatorname*{Lt}_{\omega\to 0}\ \mathscr{G}_{\mathrm{r}}^{BC}(\omega) &= \frac{1}{Q}\sum_{\substack{vv'\\ NN'}}\{\exp[-\beta(E_{N'v'}-\mu N')]\times\\
&\quad\times\langle v'N'|B|vN\rangle\langle vN|C|v'N'\rangle/[E_{N'v'}-E_{Nv}+\mu(N-N')]+\\
&\quad+\exp[-\beta(E_{Nv}-\mu N)]\times\\
&\quad\times\langle vN|C|v'N'\rangle\langle v'N'|B|vN\rangle/[E_{Nv}-E_{N'v'}+\mu(N'-N)]\}\\
&= \frac{2}{Q}\sum_{\substack{v\neq v'\\ N}}\exp[-\beta(E_{Nv'}-\mu N)]\times\\
&\quad\times\langle v'N|P_{\mu}|vN\rangle\langle vN|P_{\mu}|v'N\rangle/[E_{Nv'}-E_{Nv}]\\
&\qquad\text{in the case}\quad B = P_{\mu};\, C = P_{\mu}.\end{aligned}\tag{7.31}$$

(The terms with $v = v'$ in the sum have to be omitted, since the principal value of the integral occurs in eqn (7.30).) In order to evaluate this sum, we note that $\sum_{j\neq i} B_{ij}B_{ji}/(E_i-E_j)$ is the second-order shift in ith energy level resulting from a perturbation B applied to the system. A zero magnetic field may be described by the potentials

$$\phi = 0,\qquad \mathbf{A} = \text{constant}.$$

Thus, the Hamiltonian,

$$H_1 = \frac{1}{2m}\sum_s(\mathbf{p}_s)^2+V(\{x_s\}),$$

has the same eigenvalues as the Hamiltonian

$$H_2 = \frac{1}{2m}\sum_s(\mathbf{p}_s-e\mathbf{A}/c)^2+V(\{x_s\})$$

In particular, the second-order shift in the energies in H_2, with the **A**-dependent terms regarded as a perturbation, must be zero. Hence

$$\sum_{v\neq v'}\langle v|P_{\mu}|v'\rangle\langle v'|P_{\mu}|v\rangle/(E_v-E_{v'})+\frac{mN}{2} = 0.$$

Since this is true for any level v', it is true of the statistical mean in eqn (7.31). Hence

$$\operatorname*{Lt}_{\omega\to 0}\ \mathscr{G}_{\mathrm{r}}^{P_{\mu}P_{\mu}}(\omega) = -m\overline{N}$$

and the terms in ω^{-1} in eqn (7.29) subtract out in the limit as $\omega\to 0$, leaving only finite terms, which are evaluated below. For a system of noninteracting particles, however, only the diagonal matrix elements in eqn (7.32) are non-zero, and these are to be excluded in the sum, which is therefore zero.†

Similarly, from eqns (3.70) and (3.71)

$$\operatorname*{Lt}_{\omega\to\infty} \mathscr{G}_{\mathrm{r}}^{BC}(\omega) \to 0 \quad \text{at least as } \omega^{-1}.$$

Thus, the second term in eqn (7.29) vanishes more quickly than the first; and, for high enough frequencies, the system will behave as a system of free charges.

Exercise

7.3.1. From the conservation equation:

$$\operatorname{div} \langle \mathbf{J} \rangle = -e\frac{\partial\langle \rho \rangle}{\partial t}$$

show that the current $\langle \mathbf{J} \rangle_{\mathbf{q}\omega}$ resulting from a field which produces an induced charge $e\langle \rho \rangle_{\mathbf{q}\omega}$ both varying as $\exp(\mathrm{i}\mathbf{q}\,.\,\mathbf{r})\exp(-\mathrm{i}\omega t)$, is given by

$$\mathbf{q}\,.\,\langle \mathbf{J} \rangle_{\mathbf{q},\omega} = e\omega\langle \rho \rangle_{\mathbf{q}\omega}.$$

Using eqn (7.21) show that for the longitudinal part of the conductivity, $\sigma(\mathbf{q}, \omega)$,

$$\varepsilon(\mathbf{q}, \omega) = 1 + 4\pi i\sigma(\mathbf{q}, \omega)/\omega.$$

Use the results of exercise (7.2.1) to show that

$$\int_0^\infty \mathrm{d}\omega\, \sigma_1(\mathbf{q}, \omega) = \pi n e^2/2m,$$

where

$$\sigma(\mathbf{q}, \omega) = \sigma_1(\mathbf{q}, \omega) + \mathrm{i}\,\sigma_2(\mathbf{q}, \omega)$$

where σ_1 and σ_2 are real.

7.4. The conductivity of charges scattered by random impurities

We shall next, as an example, evaluate approximately the conductivity for a system of free charges scattered by randomly distributed impurities,

† The second-order terms in the eigenvalues of the Hamiltonian in the proof above must still be zero, but in this case it is very important to take properly into account the degeneracy of the levels, which we have ignored, since to include them would add terms which are negligible in the limit of large volume.

described in section 4.4. We need the Green function†

$$F(\omega) = \langle\langle \sum_{\mathbf{k}} k_\mu a_{\mathbf{k}}^+ a_{\mathbf{k}} ; \sum_{\mathbf{k}'} k'_\nu a_{\mathbf{k}'}^+ a_{\mathbf{k}'} : \omega \rangle\rangle_r$$

$$= \sum_{\mathbf{k},\mathbf{k}'} k_\mu k_\nu \langle\langle a_{\mathbf{k}}^+ a_{\mathbf{k}} ; a_{\mathbf{k}'}^+ a_{\mathbf{k}'} : \omega \rangle\rangle_r \equiv \sum_{\mathbf{k}\mathbf{k}'} k_\mu k_\nu F(\mathbf{k}, \mathbf{k}', \omega). \qquad (7.32)$$

In order to evaluate F, we first work with the thermodynamic Green function in the usual way, evaluating $F(\mathbf{k}, \mathbf{k}'; \mathrm{i}\omega_n)$ at the infinite set of points along the imaginary axis. We then continue F to the whole of the complex plane. It may easily be seen that in this case there are Fourier coefficients only at the points $\mathrm{i}\omega_n = 2\pi n\mathrm{i}/\beta$. The diagrams for $F(\mathrm{i}\omega_n)$ are of several types, shown in Fig. 7.1, generated by joining the four lines in (a) in all possible ways, with any number of interaction vertices in between. The diagrams in (b), (c), and (d) will only contribute for $n = 0$ and hence will give a δ-function at the origin of the ω-plane: as we shall be interested only in $F(\omega + \mathrm{i}\gamma)$, for the retarded Green function, we may ignore these terms.‡

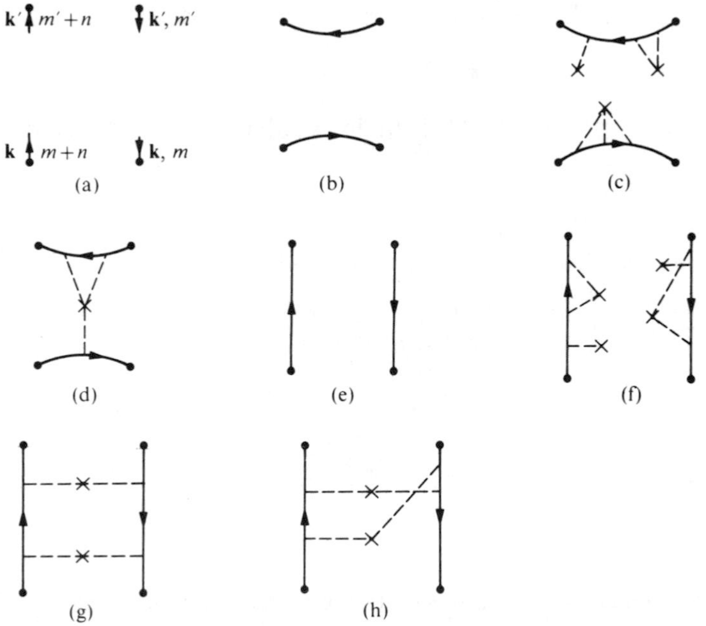

FIG. 7.1.

† Here, and in what follows, we omit the curly brackets introduced in Section 4.4 to denote averages.

‡ Another way of looking at this is to notice that these contributions are τ-independent, and the corresponding terms in $\mathscr{G}_r(t)$ are t-independent. But as $t \to \infty$ $\mathscr{G}_r^{AB}(t) \to -i[\langle A(t)\rangle \langle B(0)\rangle - \langle B(0)\rangle \langle A(t)\rangle]$ since any correlations between A and B will have vanished for large enough times. Hence $\mathscr{G}_r \to 0$, and the weight of the δ-function is zero.

We shall first evaluate the contributions from terms such as (e) and (f), which have no interaction lines crossing from one full line to the other. In this case

$$F_1(\mathbf{k}, \mathbf{k}', \mathrm{i}\omega_n) = -\varepsilon\delta_{\mathbf{k}\mathbf{k}'}\beta^{-1}\sum_m [\mathrm{i}\omega_m - \varepsilon_{\mathbf{k}} + \mu - \Sigma(\mathbf{k}, \mathrm{i}\omega_m)]^{-1} \times$$
$$\times [\mathrm{i}(\omega_m + \omega_n) - \varepsilon_{\mathbf{k}} + \mu - \Sigma(\mathbf{k}, \mathrm{i}\omega_m + \mathrm{i}\omega_n)]^{-1} \qquad (7.33)$$

that is, F is given simply by the product of two single-particle Green functions, which were evaluated in section 4.4. The sum in eqn (7.33) may be evaluated in the usual way by converting to an integral, over ω', say. We shall consider the case in which the charged particles are fermions, so that $\omega_m = (2m+1)\pi/\beta$.

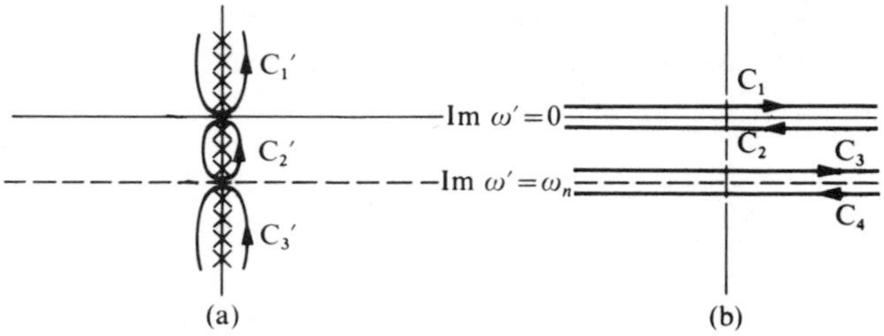

FIG. 7.2.

The singularities of the first factor in the sum are on the line Im $\omega' = 0$ and of the second, on Im $\omega' = -\omega_n$. Thus, with the contours shown in Fig. 7.2, for the case $\omega_n > 0$,

$$F_1(\mathbf{k}, \mathbf{k}', \mathrm{i}\omega_n) = -\delta_{\mathbf{k},\mathbf{k}'}\beta^{-1}\frac{\beta}{2\pi\mathrm{i}}\int\limits_{C_1'+C_2'+C_3'} \mathrm{d}\omega'[\exp(\beta\omega')+1]^{-1} \times$$
$$\times [\omega' - \varepsilon_{\mathbf{k}} + \mu + \mathrm{i}\omega_n - \Sigma(\mathbf{k}, \mathrm{i}\omega_n + \omega')]^{-1}[\omega' - \varepsilon_{\mathbf{k}} + \mu - \Sigma(\mathbf{k}, \omega')]^{-1}. \qquad (7.34)$$

The integral may alternatively be performed over the contour $C_1 + C_2 + C_3 + C_4$, since the integrand vanishes on the circle at infinity at least as $(\omega')^2$.

We shall make the simplification that

$$\Sigma(\mathbf{k}, \omega) = \Delta_{\mathbf{k}} - \mathrm{i}\Gamma, \qquad \mathrm{Im}\,\omega > 0,$$
$$\Sigma(\mathbf{k}, \omega) = \Delta_{\mathbf{k}} + \mathrm{i}\Gamma, \qquad \mathrm{Im}\,\omega < 0,$$

where Γ is independent of $\mathbf{k}$, and both $\Delta_{\mathbf{k}}$ and Γ are independent of ω. $\Delta_{\mathbf{k}}$ may then be absorbed into $\varepsilon_{\mathbf{k}}$, and, from now on, we shall assume that this has been done. In this case

$$F_1(\mathbf{k},\mathbf{k}',\mathrm{i}\omega_n)=\delta_{\mathbf{k},\mathbf{k}'}\frac{\mathrm{i}}{2\pi}\int_{-\infty}^{\infty}\mathrm{d}\omega'[\exp(\beta\omega')+1]^{-1}\times$$
$$\times\{(\omega'-\varepsilon_{\mathbf{k}}+\mu+\mathrm{i}\omega_n+\mathrm{i}\Gamma)^{-1}(\omega'-\varepsilon_{\mathbf{k}}+\mu+\mathrm{i}\Gamma)^{-1}-$$
$$-(\omega'-\varepsilon_{\mathbf{k}}+\mu+\mathrm{i}\omega_n+\mathrm{i}\Gamma)^{-1}(\omega'-\varepsilon_{\mathbf{k}}+\mu-\mathrm{i}\Gamma)^{-1}+$$
$$+(\omega'-\varepsilon_{\mathbf{k}}+\mu+\mathrm{i}\Gamma)^{-1}(\omega'-\varepsilon_{\mathbf{k}}+\mu-\mathrm{i}\omega_n-\mathrm{i}\Gamma)^{-1}-$$
$$-(\omega'-\varepsilon_{\mathbf{k}}+\mu-\mathrm{i}\Gamma)^{-1}(\omega'-\mathrm{i}\omega_n-\varepsilon_{\mathbf{k}}+\mu-\mathrm{i}\Gamma)^{-1}\}. \tag{7.35}$$

Thus, continuing F_1 into the half of the ω-plane for which $\mathrm{Im}\,\omega>0$, in particular to just above the real axis so as to obtain the retarded Green function, we have

$$\underset{\gamma\to 0}{\mathrm{Lt}}\ F_1(\mathbf{k},\mathbf{k}',\omega+\mathrm{i}\gamma)$$
$$=\delta_{\mathbf{k},\mathbf{k}'}\frac{\mathrm{i}}{2\pi}\int_{-\infty}^{\infty}\mathrm{d}\omega'\left\{(h^+ +\mathrm{i}\Gamma)^{-1}(h^- +\mathrm{i}\Gamma)^{-1}\left[\exp\left\{\beta\left(\omega'-\frac{\omega}{2}\right)\right\}+1\right]^{-1}+\right.$$
$$+(h^+ +\mathrm{i}\Gamma)^{-1}(h^- -\mathrm{i}\Gamma)^{-1}\left\{\left[\exp\left\{\beta\left(\omega'+\frac{\omega}{2}\right)\right\}+1\right]^{-1}-\right.$$
$$\left.-\left[\exp\left\{\beta\left(\omega'-\frac{\omega}{2}\right)\right\}+1\right]^{-1}\right\}-$$
$$\left.-(h^+ -\mathrm{i}\Gamma)^{-1}(h^- -\mathrm{i}\Gamma)^{-1}\left[\exp\left\{\beta\left(\omega'+\frac{\omega}{2}\right)\right\}+1\right]^{-1}\right\} \tag{7.36}$$

where $h^\pm=\omega'\pm\omega/2-\varepsilon_{\mathbf{k}}+\mu$, and where we have made slight changes of variable in the integrals. To evaluate the dissipative term in the D.C. conductivity, σ_0, we need

$$\underset{\omega\to 0}{\mathrm{Lt}}\ \frac{1}{\omega}\ \underset{\gamma\to 0}{\mathrm{Lt}}\ \mathrm{Im}\,F_1(\mathbf{k},\mathbf{k}',\omega+\mathrm{i}\gamma)$$
$$=-\delta_{\mathbf{k},\mathbf{k}'}\frac{\beta}{2\pi}\int_{-\infty}^{\infty}\mathrm{d}\omega'\exp(\beta\omega')[\exp(\beta\omega')+1]^{-2}\{[(\omega'-\varepsilon_{\mathbf{k}}+\mu)^2+\Gamma^2]^{-1}-$$
$$-[(\omega'-\varepsilon_{\mathbf{k}}+\mu)^2-\Gamma^2][(\omega'-\varepsilon_{\mathbf{k}}+\mu)^2+\Gamma^2]^{-2}\}. \tag{7.37}$$

The final factor in the integrand of eqn (7.37) is large only in the region

$\omega' = \varepsilon_{\mathbf{k}} - \mu$, so that

$$\underset{\omega\to 0}{\mathrm{Lt}}\ \omega^{-1}\ \underset{\gamma\to 0}{\mathrm{Lt}}\ \mathrm{Im}\, F_1(\mathbf{k}, \mathbf{k}', \omega + \mathrm{i}\gamma)$$

$$= -\delta_{\mathbf{k},\mathbf{k}'}\frac{\beta}{2\Gamma}\{\exp[\beta(\varepsilon_{\mathbf{k}} - \mu)] + 1\}^{-1}[1 - \{\exp[\beta(\varepsilon_{\mathbf{k}} - \mu)] + 1\}^{-1}] \tag{7.38}$$

for $\beta\Gamma \ll 1$. Thus

$$\sigma_0 = \frac{e^2\beta}{2m^2\Gamma}\sum_{\mathbf{k}}(k_z)^2\langle n_{\mathbf{k}}\rangle_0(1 - \langle n_{\mathbf{k}}\rangle_0)/V. \tag{7.39}$$

For $\beta\mu \gg 1$, $\langle n_{\mathbf{k}}\rangle_0(1 - \langle n_{\mathbf{k}}\rangle_0) \simeq \beta^{-1}\delta(\varepsilon_{\mathbf{k}} - \mu)$ and so

$$\sigma_0 = ne^2/2m\Gamma; \qquad (n = \bar{N}/V). \tag{7.40}$$

This is very similar to the usual result for the D.C. conductivity from kinetic theory, except that, from section 4.4,

$$\Gamma = C\pi\,\frac{V}{(2\pi)^3}\int \mathrm{d}^3\mathbf{k}_1|A(\mathbf{k} - \mathbf{k}_1)|^2\delta(\varepsilon_{\mathbf{k}_1} - \mu), \tag{7.41}$$

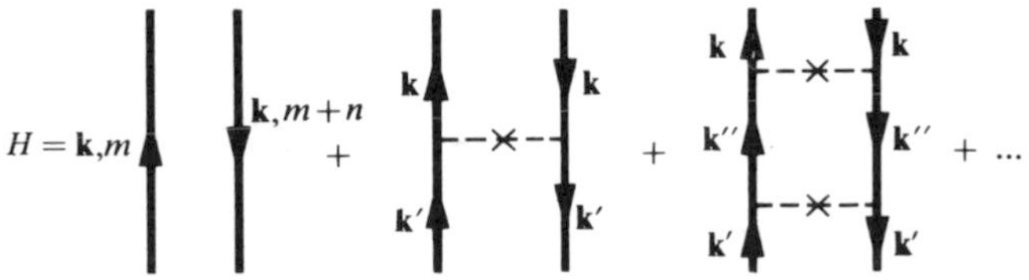

FIG. 7.3.

whereas in the normal expression for the conductivity the integrand for Γ has an extra factor $(1 - \cos\theta)$. This factor can be obtained by summing the series of diagrams shown in Fig. 7.3. Call

$$\begin{aligned} H(\mathbf{k}, \mathbf{k}', \mathrm{i}\omega_n, \mathrm{i}\omega_m) = {} & \bar{G}(\mathbf{k}, \mathrm{i}\omega_m)\bar{G}(\mathbf{k}, \mathrm{i}(\omega_m + \omega_n))[\delta_{\mathbf{k},\mathbf{k}'} + C|A(\mathbf{k} - \mathbf{k}')|^2 \times \\ & \times \bar{G}(\mathbf{k}', \mathrm{i}\omega_m)\bar{G}(\mathbf{k}', \mathrm{i}(\omega_m + \omega_n))] + \\ & + C^2\sum_{\mathbf{k}''}|A(\mathbf{k} - \mathbf{k}'')|^2|A(\mathbf{k}'' - \mathbf{k}')|^2\bar{G}(\mathbf{k}'', \mathrm{i}\omega_m) \times \\ & \times \bar{G}(\mathbf{k}'', \mathrm{i}\omega_m + \mathrm{i}\omega_n)\bar{G}(\mathbf{k}', \mathrm{i}\omega_m)\bar{G}(\mathbf{k}', \mathrm{i}\omega_m + \mathrm{i}\omega_n) + \cdots, \end{aligned} \tag{7.42}$$

where we have used the full Green functions, so that the full lines include any number of interactions with no dotted lines crossing from one full line to the other. The series may be summed:

$$H(\mathbf{k}, \mathbf{k}', \mathrm{i}\omega_m, \mathrm{i}\omega_n) = \bar{G}(\mathbf{k}, \mathrm{i}\omega_n)\bar{G}(\mathbf{k}, \mathrm{i}\omega_m + \mathrm{i}\omega_n) \times$$

$$\times \left[\delta_{\mathbf{k},\mathbf{k}'} + C\sum_{\mathbf{k}''} |A(\mathbf{k}-\mathbf{k}'')|^2 H(\mathbf{k}'', \mathbf{k}', \mathrm{i}\omega_m, \mathrm{i}\omega_n)\right].$$

In general this equation cannot be solved. However, since $k'_\nu = \mathbf{k}' \,.\, \boldsymbol{\nu}$, where $\boldsymbol{\nu}$ is a unit vector in the $\boldsymbol{\nu}$th direction, the current depends only on

$$\sum_{\mathbf{k}'} \mathbf{k}' H(\mathbf{k}, \mathbf{k}', \mathrm{i}\omega_m, \mathrm{i}\omega_n),$$

which, for an isotropic system must be of the form $\mathbf{k}\bar{H}(\mathbf{k}, \mathrm{i}\omega_m, \mathrm{i}\omega_n)$. (Remember that we are to evaluate the Green function in the absence of the electric field.) Thus, suppressing the ω labels

$$\mathbf{k}\bar{H}(\mathbf{k}) = \mathbf{k}\bar{G}(\mathbf{k})\bar{G}(\mathbf{k}) \left[1 + C\sum_{\mathbf{k}''} |A(\mathbf{k}-\mathbf{k}'')|^2 \mathbf{k}'' \,.\, \mathbf{k}k^{-2}\bar{H}(\mathbf{k}'')\right]$$

$$\therefore \quad \bar{H}(\mathbf{k}) = \bar{G}(\mathbf{k})\bar{G}(\mathbf{k}) \left[1 + C\sum_{\mathbf{k}''} |A(\mathbf{k}-\mathbf{k}'')|^2 \mathbf{k}'' \,.\, \mathbf{k}k^{-2}\bar{H}(\mathbf{k}'')\right]. \qquad (7.43)$$

If we define:

$$L(\mathbf{k}) = C\sum_{\mathbf{k}''} |A(\mathbf{k}-\mathbf{k}'')|^2 \mathbf{k}'' \,.\, \mathbf{k}k^{-2}\bar{H}(\mathbf{k}''),$$

eqn (7.43) gives:

$$L(\mathbf{k}) = C\sum_{\mathbf{k}''} |A(\mathbf{k}-\mathbf{k}'')|^2 \mathbf{k}'' \,.\, \mathbf{k}k^{-2}[1 + L(\mathbf{k}'')]\bar{G}(\mathbf{k}'')\bar{G}(\mathbf{k}''). \qquad (7.44)$$

We have seen above that the product $G(\mathbf{k}, \mathrm{i}\omega_m)G(\mathbf{k}, \mathrm{i}\omega_m + \mathrm{i}\omega_n)$ when summed over ω_m contains a factor $\langle n_{\mathbf{k}}\rangle_0(1 - \langle n_{\mathbf{k}}\rangle_0)$ which is small away from the Fermi surface. This corresponds to the intuitive idea that only electrons near the Fermi surface will contribute to the conductivity. Thus we need $L(\mathbf{k})$ only for k near k_{F}, and, once the sum over ω_m is taken, only values of k'' near k_{F} in eqn (7.44) will contribute. Thus we may ignore the $\mathbf{k}$-dependence of L, and write:

$$L(1+L)^{-1} = C\sum_{\mathbf{k}''} |A(\mathbf{k}-\mathbf{k}'')|^2 \cos\theta G(\mathbf{k}'')G(\mathbf{k}''),$$

where θ is the angle between $\mathbf{k}$ and $\mathbf{k}''$. With the same approximation as above for the single-particle Green function,

$$\frac{L}{1+L} = C\sum_{\mathbf{k}'',m} |A(\mathbf{k}''-\mathbf{k})|^2 \cos\theta[\mathrm{i}\omega_m - \varepsilon_{\mathbf{k}} + \mu \pm \mathrm{i}\Gamma]^{-1}[\mathrm{i}\omega_m + \mathrm{i}\omega_n - \varepsilon_{\mathbf{k}} + \mu \pm \mathrm{i}\Gamma]^{-1}$$

$$= \int_{-\infty}^{\infty} \mathrm{d}z\, C\sum_{\mathbf{k}'',m} |A(\mathbf{k}''-\mathbf{k})|^2 \cos\theta\, \delta(z-\varepsilon_{\mathbf{k}}+\mu)[\mathrm{i}\omega_m - z \pm \mathrm{i}\Gamma]^{-1} \times$$

$$\times [\mathrm{i}(\omega_m+\omega_n) - z \pm \mathrm{i}\Gamma]^{-1}, \tag{7.45}$$

where the sign of the Γ term depends on the signs of ω_m and of $(\omega_m+\omega_n)$. But in the work above, we treated $C\sum_{\mathbf{k}''}|A(\mathbf{k}-\mathbf{k}'')|^2\delta(z-\varepsilon_{\mathbf{k}''}+\mu) = \Gamma\pi^{-1}$ as a constant, independent of z, and so we now treat $C\sum_{\mathbf{k}''}|A(\mathbf{k}-\mathbf{k}'')|^2 \times \times \cos\theta\delta(z-\varepsilon_{\mathbf{k}''}+\mu) \equiv \Gamma'\pi^{-1}$ as a constant. In that case

$$\frac{L}{1+L} = \sum_m \frac{\Gamma'}{\pi} \int_{-\infty}^{\infty} \mathrm{d}z[\mathrm{i}\omega_m - z \pm \mathrm{i}\Gamma]^{-1}[\mathrm{i}\omega_m + \mathrm{i}\omega_n - z \pm \mathrm{i}\Gamma]^{-1}.$$

Thus, preparing to perform the sums over ω_m in the same way as above, for $\omega_n > 0$, on the contours of C_1 and C_4 (Fig. 7.2), the signs of the Γ terms are such that $L/(1+L)$ is zero, and on C_2 and C_3,

$$\frac{L}{1+L} = \frac{\Gamma'}{\pi}\,\frac{2\pi\mathrm{i}}{\mathrm{i}\omega_n + 2\mathrm{i}\Gamma}$$

$$L = 2\Gamma'/[\omega_n + 2(\Gamma-\Gamma')].$$

Thus

$$\bar{H}(\mathbf{k}) = \bar{G}(\mathbf{k})\bar{G}(\mathbf{k})[\omega_n+2\Gamma]/[\omega_n+2\Gamma_1],$$

where

$$\Gamma_1 = \pi C\sum_{\mathbf{k}''}|A(\mathbf{k}-\mathbf{k}'')|^2(1-\cos\theta)\delta(\varepsilon_{\mathbf{k}''}-\mu).$$

Since

$$\operatorname*{Lt}_{\omega\to 0}\, \omega^{-1}\bar{H}(\mathbf{k}) = \operatorname*{Lt}_{\omega\to 0}\, \omega^{-1}\bar{G}(\mathbf{k})\bar{G}(\mathbf{k})\Gamma/\Gamma_1,$$

the D.C. conductivity at zero frequency is

$$\sigma_0 = ne^2/2m\Gamma_1, \tag{7.46}$$

which is the result derived in the conventional way from Boltzmann's equation.

The terms we have neglected are of several types, shown in Fig. 7.4. Those in (a) may be included by replacing A by a scattering matrix t, as in section 4.4. Those of type (b) may be shown to be small if $\beta\mu \ll 1$. Those of type (c), similar to diagrams which lead to the Lamb shift in electrodynamics, may also be shown to be small if the interaction is weak.

It may seem that in this work one is using a sledge-hammer to crack a hazel nut; it must be admitted that there are easier ways of obtaining the kinetic theory result. However, there are considerable formal advantages in starting with an exact formulation of the problem in eqn (7.29), rather than, for example, the Boltzmann equation, which may contain hidden approximations. Although one will almost certainly have to make approximations later, one will be able to discuss the conditions under which they apply. Further, equations such as (7.29) may be used to derive results which are exact in some sense, for example, those in exercises 7.2.1 and 7.3.1, and to treat cases in which the Boltzmann equation is not applicable.

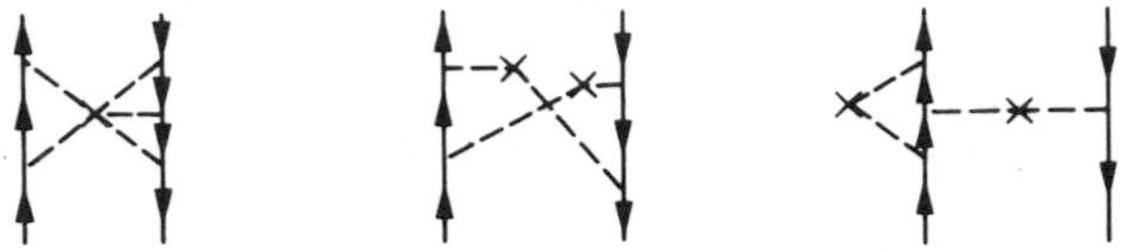

(a) (b) (c)

FIG. 7.4.

Further reading

Chester (1963) reviews linear-response theory and its relationship to Master Equations. Kadanoff and Baym (1962) review the relationship between the equations of motion of Green functions, and the Boltzmann equation and stress the importance of using only approximations which obey certain conservation laws. Chester and Thellung (1961) derive the Wiedemann–Franz law under certain assumptions. The impurity-resistance calculation described in this chapter is considered in more detail by, for example, Edwards (1958), Velicky (1969), and in a series of papers by Langer (1960, 1961, 1962a, 1962b). The mobility of polarons is considered in a manner rather similar to that in this chapter by Mahan (1966). Montroll (1959, 1960) and Gillis (1968) discuss ways of extending this method to transport processes, such as diffusion or viscosity, the cause of which cannot be described by an additional term in the Hamiltonian.

8

MANY-BODY SYSTEMS AT ZERO TEMPERATURE

8.1. Diagrammatic perturbation theory at zero temperature

ALTHOUGH the absolute zero of temperature is not experimentally attainable, one often wishes to consider the properties of many-body systems at this temperature. Such calculations give an idea of some of the properties of such systems at low temperatures. Further, the excitation energies in the system may be so large that effects depending on the temperature do not concern us. This will almost always be the case in nuclear-matter calculations (only almost always, since calculations on the properties of large numbers of elementary particles in the stars are now being performed), and usually the case in atomic physics. Further, since the Fermi energy is, for many systems of experimental interest, much greater than Boltzmann's constant times room temperature, we can learn quite a lot from zero-temperature calculations about correlations, elementary excitations, and similar properties of such systems which will be of use to us at quite high temperatures.

The most obvious way to obtain the limit of the work described in the previous chapters is to use the rules given there and then to take the limit at the end of the calculation. This is not always as simple as it sounds, however. It is customary to take the Volume $\to \infty$ limit in many-body calculations, and it is not at all clear initially whether we should take the $V \to \infty$ or the $T \to 0$ limit first. This is discussed below. It is also possible to make certain simplifications of the rules. We consider the case of fermions and notice that

$$\underset{\beta\to\infty}{\mathrm{Lt}}\ \beta^{-1}\sum_n f(\mathrm{i}\omega_n) = \frac{1}{2\pi\mathrm{i}}\underset{\Delta\omega_n\to 0}{\mathrm{Lt}}\sum_n f(\mathrm{i}\omega_n)\Delta(\mathrm{i}\omega_n)$$

$$= \frac{1}{2\pi\mathrm{i}}\int_{-\mathrm{i}\infty}^{\mathrm{i}\infty} f(z)\,\mathrm{d}z, \tag{8.1}$$

where $\Delta(\mathrm{i}\omega_n) = 2\pi\mathrm{i}/\beta$, since $\omega_n = (2n+1)\pi\mathrm{i}/\beta$.

In order to reach the usual forms of the rules for zero-temperature perturbation theory, we need to convert these integrals along the imaginary axis to integrals along the real axis. The integrals obtained via eqn (8.1) from an expansion of Ω may be replaced by integrals along the real axis and the infinite semi-circle in the upper half plane provided one change is made in

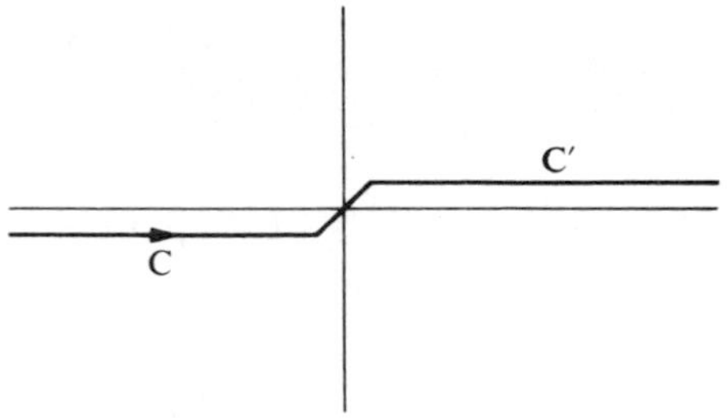

FIG. 8.1.

the definition of the non-interacting Green function. The integrand should be multiplied by $\exp(-itz)$ if this is necessary to make the integral around the semicircle zero. Call the closed contour C_1. It is easy to show that this change of the ranges of integration leaves the final integral unchanged for the simple diagram of Fig. 8.2(a) (see exercise 8.1.1, which the reader should attempt before reading further). To see this in the general case, we consider two types of ω_n sum: those for which $i\omega_n$ appears more than once in the denominator, such as the sum over ω_n in Fig. 8.2(a), where $i\omega_n$ appears in two denominators because of the ω_n-conservation rule at vertices; and those for which $i\omega_n$

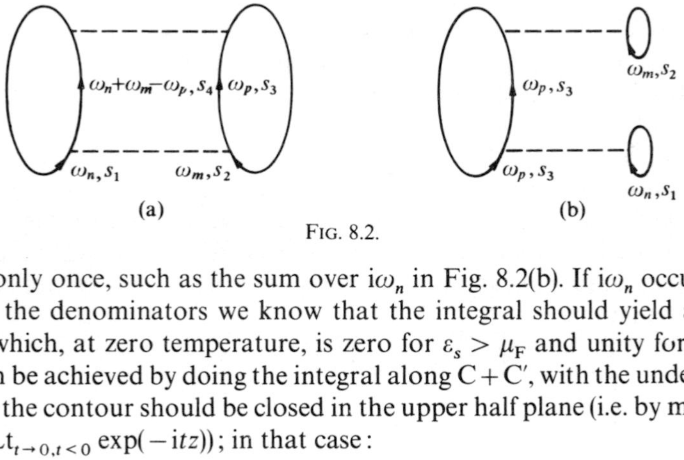

FIG. 8.2.

occurs only once, such as the sum over $i\omega_n$ in Fig. 8.2(b). If $i\omega_n$ occurs only once in the denominators we know that the integral should yield a factor $\langle n_s \rangle_0$, which, at zero temperature, is zero for $\varepsilon_s > \mu_F$ and unity for $\varepsilon_s < \mu$. This can be achieved by doing the integral along $C+C'$, with the understanding that the contour should be closed in the upper half plane (i.e. by multiplying by $\mathrm{Lt}_{t\to 0, t<0} \exp(-itz)$); in that case:

$$\frac{1}{2\pi i}\int_{C+C'} (z-\varepsilon_s+\mu)^{-1} = 0 \text{ if } \varepsilon_s-\mu > 0, \text{ since no pole is enclosed, and}$$

$$\frac{1}{2\pi i}\int_{C+C'} (z-\varepsilon_s+\mu)^{-1} = 1 \text{ if } \varepsilon_s-\mu < 0, \text{ since a simple pole is enclosed.}$$

The same results may be obtained by redefining the zero-order Green function as

$$\bar{g}(s, z) = \mathrm{Lt}_{\gamma\to 0} [z-\varepsilon_s+\mu+i\gamma \,\mathrm{sgn}(\varepsilon_s-\mu)]^{-1}, \tag{8.2}$$

where

$$\operatorname{sgn}(x) = -1 \quad \text{for } x < 0$$
$$= +1 \quad \text{for } x > 0,$$

and performing the integrals along the real z-axis.

If $i\omega_n$ occurs more than once in the denominators the contour in eqn (8.1) may be closed by an infinite semicircle on the left. Call this closed contour C_2. Certain simple† poles will then be enclosed in our contours. The number of poles enclosed in the contour for the z_i integral will depend on the number of times z_i appears in denominators, and also on the signs of the $\varepsilon_s - \mu$ terms in these denominators; but provided that, in doing the integrals along C_1, we use the new form of $\bar{g}(s, z)$ of eqn (8.2), for each pole enclosed by C_1 there will be a corresponding pole enclosed by C_2. After all the integrals have been performed, we shall be left with products of terms all of the form $(\varepsilon_{s_i} \pm \varepsilon_{s_j} + \varepsilon_{s_k} \pm \cdots)^{-1}$, whether we have done the integrals round C_1 or C_2. It is now easy to see that these products will be the same in the two formulations of the integrals.

As $T \to 0$, $\Omega \to E - \mu\bar{N}$, where E is the ground state energy of the system. Thus, we have the rules:

For the nth-order term in an expansion of $E - \mu\bar{N}$: Draw all different nth-order connected diagrams, giving each line a momentum (or state) label and ω-label. The factors making up the contribution of a particular diagram to $E - \mu\bar{N}$ are:

(i) $\Gamma^0_{s_1 s_2 s_3 s_4}$ for each vertex square with lines (s_3, ω_3), (s_4, ω_4) entering and (s_1, ω_1), (s_2, ω_2) leaving, where $\omega_1 + \omega_2 = \omega_3 + \omega_4$.

(ii) $\bar{g}(s, \omega)$ for each full line with labels s and ω. (The minus sign may be omitted, since we are dealing with fermions, and there must be an even number of fermion lines in each diagram.)

(iii) $(-1)^{n+1}$.

(iv) $(-1)^D$ where D is the number of closed fermion loops in the diagram.

(v) $(\frac{1}{2})^{n_e}$, where n_e is the number of pairs of equivalent arms in the diagram.

(vi) $1/p$, where p is the coordination number of the diagram.

(vii) $(2\pi i)^{-1}$ for each independent ω-integral.

Finally, sum over all the s-labels and integrate over all independent ω-labels from $-\infty$ to ∞.

The second-order diagrams of Fig. 2.18 (with ω-labels instead of ω_n-labels), thus have the contributions:

$$\text{(a)} \sum_{s_1 s_2 s_3 s_4} \Gamma^0_{s_1 s_2 s_3 s_4} \Gamma^0_{s_3 s_4 s_1 s_2} \left(\frac{1}{2\pi i}\right)^3 \iiint_{-\infty}^{\infty} d\omega_1\, d\omega_2\, d\omega_3\, \bar{g}(s_1, \omega_1)\bar{g}(s_2, \omega_2) \times$$
$$\times\, \bar{g}(s_3, \omega_3)\bar{g}(s_4, \omega_1 + \omega_2 - \omega_3)(-1)^3(-1)^2 \times \tfrac{1}{4} \times \tfrac{1}{2}, \qquad (8.3)$$

† The special case of sums such as that over ω_p in Fig. 8.2(b) is discussed in section 8.4.

$$\text{(b)} \sum_{s_1 s_2 s_3 s_4} \Gamma^0_{s_1 s_2 s_3 s_2} \Gamma^0_{s_3 s_4 s_1 s_4} \left(\frac{1}{2\pi \mathrm{i}}\right)^3 \iiint\limits_{-\infty}^{\infty} \mathrm{d}\omega_1 \, \mathrm{d}\omega_2 \, \mathrm{d}\omega_4 \, \bar{g}(s_1, \omega_1) \bar{g}(s_2, \omega_2) \times$$

$$\times \bar{g}(s_3, \omega_1) \bar{g}(s_4, \omega_4)(-1)^3(-1)^3 \tfrac{1}{2}. \tag{8.4}$$

Since

$$\frac{1}{2\pi} \int_{-\infty}^{\infty} \mathrm{e}^{-\mathrm{i}\omega(t-t')} \bar{g}(s, \omega) \equiv g(s, t-t')$$

$$= 0, t-t' > 0, \varepsilon_s - \mu < 0$$

$$= 0, t-t' \leqslant 0, \varepsilon_s - \mu > 0$$

$$= \mathrm{i} \exp[-i(\varepsilon_s - \mu)(t-t')], t-t' \leqslant 0, \varepsilon_s - \mu < 0$$

$$= -\mathrm{i} \exp[-\mathrm{i}(\varepsilon_s - \mu)(t-t')], t-t' > 0, \varepsilon_s - \mu > 0 \tag{8.5}$$

(compare with eqn (3.49)) and

$$\int_{-\infty}^{\infty} \exp[\mathrm{i}(\omega_1 + \omega_2 - \omega_3 - \omega_4)t] \, \mathrm{d}t = 2\pi\delta(\omega_1 + \omega_2 - \omega_3 - \omega_4), \tag{8.6†}$$

we may express these rules in a time-dependent form. The lines no longer have ω-labels, and each vertex has a t-label. The rules are then:

(i) $\Gamma^0_{s_1 s_2 s_3 s_4}$ for each vertex square with lines s_1, s_2 leaving and s_3, s_4 entering.
(ii) $g(s, t-t')$, for each full line with label s, going from t' to t.
(iii) $(\mathrm{i})^{n+1}$ (using the fact that there are $n+1$ independent ωs in an nth-order diagram).
(iv)–(vi) as above.

Finally, sum over all the s-labels, and integrate over all but one of the t-labels from $-\infty$ to ∞.

We may write down similar rules for $G^{AB}(\mathrm{i}\omega)$, which will now be defined, in the limit $T \to 0$, for all points along the imaginary axis. Analytic continuation gives us $\bar{G}^{AB}(E)$ in the usual way, for points near the real axis. The rules for $\bar{G}^{AB}(E)$ are:

Draw all connected diagrams with the appropriate number (for the operators A and B) of ingoing and outgoing lines. The factors making up the contribution of a particular nth-order diagram are:

(i) $\Gamma^0_{s_1 s_2 s_3 s_4}$ for each vertex square with lines (s_1, ω_1), (s_2, ω_2) leaving and (s_3, ω_3), (s_4, ω_4) entering; further, $\omega_1 + \omega_2 = \omega_3 + \omega_4$.

† This relation should be interpreted in the sense

$$\operatorname*{Lt}_{\alpha \to 0} \left[\int_{-\infty}^{0} \mathrm{e}^{\mathrm{i}\omega t} \, \mathrm{e}^{\alpha t} \, \mathrm{d}t + \int_{0}^{\infty} \mathrm{e}^{\mathrm{i}\omega t} \, \mathrm{e}^{-\alpha t} \, \mathrm{d}t \right] = 2\pi\delta(\omega)$$

(ii) $\bar{g}(s, \omega)$ for each full line with labels s and ω.
(iii) $(-1)^n$.
(iv), (v), (vii) as above.

Finally, sum over all the state labels of the internal lines, and integrate over the independent ω-labels of the internal lines from $-\infty$ to ∞.

At zero temperature, there is a direct link between $\bar{G}^{AB}(\omega)$ and the causal, or time-ordered Green function $\bar{\mathscr{G}}_c^{AB}(\omega)$. From eqn (3.54), we see that, at zero temperature, all the poles in the first term of $\bar{G}^{AB}(\omega)$ will lie to the right of the imaginary axis, and all those in the second term will lie to the left, since only one term, for which $E_{N\nu}-\mu N$ is a minimum, will contribute significantly to the sum over ν and N. Thus, at zero temperature,

$$\bar{\mathscr{G}}_c^{AB}(E) = \operatorname*{Lt}_{\gamma\to 0, \gamma>0} \bar{G}^{AB}(E+\mathrm{i}\gamma \operatorname{sgn} E).$$

Therefore, in a manner very similar to that followed above, we may transform to a t-representation, and obtain a diagrammatic representation of $\mathscr{G}_c^{AB}(t_2-t_1)$. The lines no longer have ω-labels, and each vertex has a t-label. The points at which the external lines enter and leave the diagram also have t-labels, (t_1 and t_2). The rules are then:

(i)–(v) as above (p. 178) replacing i^{n+1} in rule (iii) by i^n. Finally, sum over all the s-labels and integrate over all the internal t-labels from $-\infty$ to ∞.

Exercises

8.1.1. Consider the z integrals obtained via eqn (8.1) for the diagrams in Fig. 8.2(a):

$$\left(\frac{1}{2\pi\mathrm{i}}\right)^3 \iiint \frac{\mathrm{d}z_1}{z_1-\varepsilon_{s_1}+\mu}\,\frac{\mathrm{d}z_2}{z_2-\varepsilon_{s_2}+\mu}\,\frac{\mathrm{d}z_3}{z_3-\varepsilon_{s_3}+\mu}\,\frac{1}{z_1+z_2-z_3-\varepsilon_{s_4}+\mu}$$

(i) when they are to be taken from $-\mathrm{i}\infty$ to $\mathrm{i}\infty$,
(ii) when they are to be taken along the real axis with $(z-\varepsilon_s+\mu)^{-1}$ replaced by $\bar{g}(s, z)$ of eqn (8.2).

Close the contours by infinite semi-circles, in the left half-plane for (i) and the upper half-plane for (ii).

Consider first the three cases:

(a) $\varepsilon_{s_1} > \mu$, $\varepsilon_{s_4} > \mu$,
(b) $\varepsilon_{s_1} < \mu$, $\varepsilon_{s_2} > \mu$,
(c) $\varepsilon_{s_1} > \mu$, $\varepsilon_{s_2} > \mu$, $\varepsilon_{s_3} < \mu$, $\varepsilon_{s_4} < \mu$.

Show that both integrals are zero in cases (a) and (b), and give a physical explanation of this. Show that, for (c), both integrals give $-(\varepsilon_{s_3}+\varepsilon_{s_4}-\varepsilon_{s_1}-$ $-\varepsilon_{s_2})^{-1}$, and compare the contribution to Ω with conventional second-order perturbation theory. What cases other than (c) give non-zero contributions?

8.1.2. The interested reader should repeat some of the exercises of Chapters 2, 3, and 4 for the zero-temperature case. Notice that, amongst other simplifications, the self-consistent Hartree–Fock equations of Chapter 4 are now much easier to solve for fermions, since they now become:

$$\Sigma^{\mathrm{SCHF}}(\mathbf{k}) = \sum_{\mathbf{k}_1|\mathbf{k}_1|<\mathbf{k}_F} [\bar{v}(0) - \bar{v}(|\mathbf{k}-\mathbf{k}_1|)]$$

$$N = \sum_{\mathbf{k}_1|\mathbf{k}_1|<\mathbf{k}_F} 1 \; .$$

The self-consistent Hartree–Fock approximation at zero-temperature is described by Woo and Jha (1971).

8.2. The Goldstone rules for $E - \mu\bar{N}$†

The time-independent rules for $E-\mu\bar{N}$ may be put in a different form, analogous to the t-independent form of section 2.9. To do this, we start with the time-dependent rules on p. 178 and again split up the time range of each integration, counting the contributions of diagrams such as Figs. 8.3 (a, b, and c) separately. Thus, in the t-dependent form of the expansion of $E-\mu\bar{N}$, in

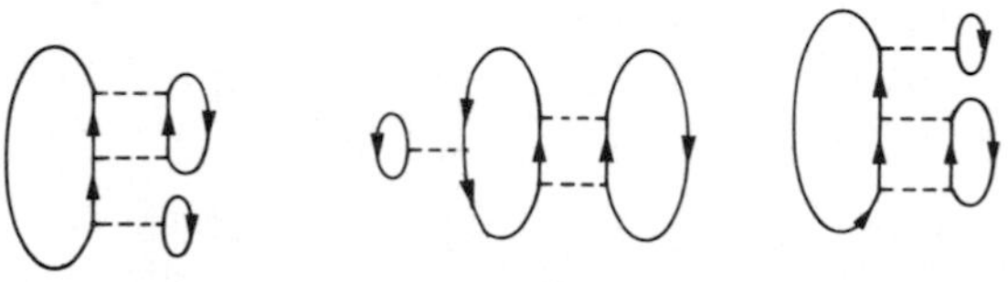

FIG. 8.3.

an nth-order diagram, we have $n-1$ t-integrals to perform with the restriction $\infty > t_n > t_{n-1} > t_{n-2} > \cdots > t_1 > -\infty$. The t-dependence of a particular diagram will be $\exp(\mathrm{i}t_j\Delta E_j)$, where, if lines with labels s_a, s_b, enter, and lines with labels s_c, s_d leave the vertex labelled t_j:

$$E_j = \varepsilon_{s_c} + \varepsilon_{s_d} - \varepsilon_{s_a} - \varepsilon_{s_b}.$$

We now make the transformation to new variables:

$$T_n = t_n; \qquad T_j = t_j - t_{j+1},\, j < n: \qquad \text{i.e. } t_j = \sum_j^n T_j. \tag{8.7}$$

We choose t_n as the t-label over which we do not integrate. The integrals over the other Ts are subject to the restrictions

$$-\infty \leqslant T_j \leqslant 0,$$

and the T_j-dependence of a particular diagram is $\exp(\mathrm{i}T\Delta\mathscr{E})$ where $\Delta\mathscr{E}$ is

† Goldstone (1957).

the sum of the kinetic energies of all the ascending (particle) lines immediately above the jth vertex, minus the sum of the kinetic energies of all the descending (hole) lines.† The integration over T_j thus gives a factor $(\mathrm{i}\Delta\mathscr{E}_j)^{-1}$ (see the footnote on p. 178). The rules in this representation are thus the same as those on p. 177 with rules (ii), (iii) and (vii) replaced by:

(ii) $(-\Delta\mathscr{E}_j)^{-1}$, where $\mathscr{E}_j$ is defined above, for the interval between each vertex.

(iia) a factor $\theta^{+}(\varepsilon_s-\mu)$ for each ascending line and $-\theta^{-}(\varepsilon_s-\mu)$ for each descending line, where:

$$\begin{aligned}\theta^{+}(\varepsilon_s-\mu) &= 1 \qquad \varepsilon_s > \mu\\ &= 0 \qquad \varepsilon_s < \mu,\\ \theta^{-}(\varepsilon_s-\mu) &= 1-\theta^{+}(\varepsilon_s-\mu).\end{aligned}$$

We should emphasize again, as in section 2.8, that although the rules are rather similar to the previous rules, the diagrams are different. Similar rules for the evaluation of Green functions may easily be written down.

Exercises

8.2.1. Use the Goldstone rules to evaluate the first- and second-order corrections to $E-\mu\bar{N}$.

8.2.2. The proof on the discontinuity at the Fermi surface (section 3.9) may now be completed. Refer to the last few pages of Luttinger (1961).

A beautiful example of the use of the t-dependent form of these rules is provided by Nozières and De Dominicis (1969), who find an exact solution for the form of the singularity in the x-ray absorption spectrum in metals.

8.3. The canonical ensemble

We have so far worked entirely with the Grand Canonical Ensemble. μ has been regarded as an independent variable throughout the calculation, to be determined after the potential Ω has been calculated, by using the equation

$$\bar{N} = -\left(\frac{\partial\Omega}{\partial\mu}\right)_{T,V}. \tag{8.8}$$

This procedure must similarly be followed at zero temperature in the work described above. As we pointed out in Chapter 2, the proof of Wick's theorem in the form in which we have used it depends on our working in the Grand Ensemble. However, if we are considering the properties of a system of a

† The diagrams are drawn with t increasing upwards.

relatively small number of particles of the order of ten or a hundred, the Grand Ensemble is not really appropriate, since the fluctuations in the number of particles about the mean is not small compared with the mean. Thus, our present formalism cannot cope with the electrons in an atom, for example. Further, eqn (8.8) is often difficult to solve for μ, and this is inconvenient if we need the explicit $\bar{N}$-dependence of some function.

At zero temperature, it is possible to avoid these difficulties by working with a fixed number of particles, that is, by working in the zero-temperature limit of the Canonical Ensemble. To do this, we repeat the work in the first section of Chapter 2, working always with canonical averages, rather than grand canonical. Thus, for example, we define

$$q = \mathrm{Tr}\{\exp(-\beta H)\},$$
$$F = -\beta^{-1}\ln q, \tag{8.9}$$

and, as shown in the elementary texts on statistical mechanics, F is the free energy of the system:

$$F = E - TS; \qquad p = -\left(\frac{\partial F}{\partial V}\right)_T; \qquad S = -\left(\frac{\partial F}{\partial T}\right)_V, \tag{8.10}$$

where E is the internal energy. We then have

$$q = \langle S^{(\beta)}\rangle_0 \tag{8.11}$$

where the average is now to be taken over the non-interacting canonical ensemble, and $S^{(\beta)}$ is given by eqn (2.27) with

$$H_1(\tau) = \exp(H_0\tau)H_1\exp(-H_0\tau). \tag{8.12}$$

In the zero-temperature limit of this work, we shall thus need the expectation value of products of operators over the ground state of the non-interacting system, since at zero temperature all other terms in the sum for the partition function will be infinitely small compared with the term for the ground state. In order to develop a diagrammatic expansion, we need a form of Wick's theorem for these expectation values. The theorem gives a result very similar to the non-zero-temperature theorem with which we are already familiar; the expectation value, in the non-interacting ground state of the τ-ordered product of creation and annihilation operators is equal to the sum of all possible products of all paired contractions multiplied by $(-1)^C$, where C is the number of interchanges of fermion creation or annihilation operators needed to bring together the contracted operators. The contraction of a pair of operators is defined as:

$$\{\bar{a}_{s_1}(\tau_1)\bar{a}_{s_2}(\tau_2)\}'' = 0 = \{a_{s_1}(\tau_1)a_{s_2}(\tau_2)\}''$$
$$\{a_{s_1}(\tau_1)\bar{a}_{s_2}(\tau_2)\}'' = \delta_{s_1,s_2}\exp[(\tau_2-\tau_1)\varepsilon_{s_1}](1-\langle n_{s_1}\rangle_0) \qquad \tau_1 > \tau_2$$
$$= -\delta_{s_1,s_2}\exp[(\tau_2-\tau_1)\varepsilon_{s_1}]\langle n_{s_1}\rangle_0 \qquad \tau_1 \leqslant \tau_2 \tag{8.13}$$

where now $\langle n_s \rangle_0 = 1$, $\varepsilon_s < \mu^0$; $\langle n_s \rangle_0 = 0$, $\varepsilon_0 > \mu^0$ and μ^0 is the Fermi energy for the unperturbed system. The proof of the theorem is given in Appendix G.

The subsequent development is almost identical to that in Chapter 2. The linked-cluster theorem gives us an expression for the zero-temperature limit of the free energy, which is just the ground-state energy, in terms of connected diagrams. Thus, in the τ-independent form, the rules for the ground-state energy are identical to those on p. 177, except for the definition of the non-interacting Green function, which is now

$$\bar{g}(s, \omega) = [\omega - \varepsilon_s + \mathrm{i}\gamma \operatorname{sgn}[\varepsilon_s - \mu^0)]^{-1}, \tag{8.14}$$

and for the fact that we are now evaluating E, rather than $E - \mu N$. Sometimes, it is convenient to measure the unperturbed energies from the Fermi energy for the unperturbed system, in which case the zero-order Green function looks even more familiar:

$$\bar{g}(s, \omega) = [\omega - (\varepsilon_s - \mu^0) + \mathrm{i}\gamma \operatorname{sgn}(\varepsilon_s - \mu^0)]^{-1}. \tag{8.15}$$

The time-dependent and Goldstone form of the rules now follow easily.

The more usual derivation of the time-dependent form of the rules, starting from the adiabatic hypothesis, rather than the canonical ensemble is described by, for example, Nozières (1963). See also Thouless (1961).

Exercise

8.3.1. The Goldstone form of the rules for the zero-temperature case are often written:

$$E - E_0 = \sum_{n,C} \left\langle \Phi_0 \middle| H_1 \left(\frac{1}{E_0 - H_0} H_1 \right)^n \middle| \Phi_0 \right\rangle,$$

where the sum over C indicates that only connected diagrams are to be included and where Φ_0 is the unperturbed ground state. Show that this form leads to the same set of rules for E as the procedure above.

This is the form of the rules most frequently used in the calculation of the properties of atoms and molecules. An introduction to such calculations is provided by Kelly (1963). For further reading, see, for example, Kelly (1969), Miller and Kelly (1971). For a review of nuclear matter calculations, see Thouless (1964).

8.3.2. The time-development operator,

$$U(t) = \exp(-\mathrm{i}Ht),$$

satisfies the equation of motion

$$\mathrm{i}\frac{\mathrm{d}U}{\mathrm{d}t} = HU.$$

Compare this with eqn (2.16), and, using the zero-temperature form of Wick's theorem, develop a diagrammatic analysis of

$$P(t) = \langle\Phi_0|\exp(-\mathrm{i}H_0t)U(t)|\Phi_0\rangle.$$

Note that this is identical with the development of Chapter 2 under the transformation $\tau \to it$.

$U(t)$ may be expressed in terms of the resolvent operator:

$$R(z) = (H-z)^{-1}$$

$$U(t) = -(2\pi\mathrm{i})^{-1}\int_{\mathrm{C}} \mathrm{d}z\, R(z)\exp(-\mathrm{i}zt)$$

where C is a contour enclosing the real axis, described counterclockwise. Write

$$R(z) = (H_0+H_1-z)^{-1}$$

and derive a perturbation expansion for $R(z)$ and hence a diagrammatic expansion for $\Gamma(z) = \langle\Phi_0|R(z)|\Phi_0\rangle$. How could $\Gamma(z)$ be used to find (a) the ground state energy; (b) the density of states at energy E? (See, for example, Hugenholtz (1965).)

By making a transformation similar to eqn (8.7) deduce the rules for E given in the previous example.

(See e.g. Thouless (1961), Goldstone (1957).)

8.4. Anomalous diagrams

In writing down eqn (8.1) we have implicitly assumed that the energy levels in our system are all different from the Fermi energy, μ. If one of them coincides with the Fermi energy, the integral in eqn (8.1) can include terms which are undefined. Consider, for example, the contribution of the diagram shown in Figure 8.2(b). This contains, at non-zero temperature, the ω_p sum,

$$\sum_p [\mathrm{i}\omega_p-(\varepsilon_{s_3}-\mu)]^{-2}$$

and if we follow the procedure suggested by eqn (8.1), there is a singularity on the contour of integration for $\varepsilon_{s_3} = \mu$ which makes the integral meaningless. If we return to our old method of performing these sums, we obtain

$$\frac{1}{\beta}\sum_p [\mathrm{i}\omega_p-(\varepsilon_{s_3}-\mu)]^{-2} = (2\pi\mathrm{i})^{-1}\int_C [\omega-(\varepsilon_{s_3}-\mu)]^{-2}[\exp(\beta\omega)+1]^{-1}\,\mathrm{d}\omega$$

$$= \left\{\frac{\partial}{\partial\omega}[\exp(\beta\omega)+1]^{-1}\right\}_{\omega=\varepsilon_{s_3}-\mu}$$

where the contour C is defined in Appendix D. In the limit $T \to 0$, this derivative tends to $\delta(\omega) = \delta(\varepsilon_{s_3} - \mu)$, and we obtain a non-zero contribution if $\varepsilon_{s_3} = \mu$. On the other hand, if we follow the procedure suggested in section 8.3, we do not obtain a contribution; the contribution includes factors $g_s(\tau_1 - \tau_2)g_s(\tau_2 - \tau_1)$, one of which must always be zero, since ε_s is either above or below the Fermi surface (or the level with label s is either occupied or unoccupied in the non-interacting ground state). Diagrams which cause this type of difficulty, containing both ascending and descending lines with the same momentum label, are called 'anomalous diagrams'.

If, then, we are considering discrete levels, there is no problem to worry us, since the Fermi energy will normally lie between two single-particle levels; if our energy levels are continuous, however, we shall obtain a finite contribution from the delta-function. Since, for normal systems, we obtain a continuous distribution of energy levels only when we take the limit Volume $\to \infty$, we are at present involved with the subtle question of whether we should first take the $T \to 0$ limit, and then the $V \to \infty$ limit, or vice-versa. If we are to do the former, anomalous diagrams do not contribute at zero temperature; if the latter, they do.

An appeal to nature gives no help: real systems are neither infinite nor at zero temperature; we are using both limits as approximations. We are given a clue about the order in which we should take the limits by considering the equation

$$\bar{N} = -\frac{\partial \Omega}{\partial \mu}. \tag{8.16}$$

In the form of the rules for calculating Ω given on p. 54, μ appears only through Ω_0 and the functions f^+ and f^-. But

$$\operatorname*{Lt}_{T \to 0} \frac{\partial}{\partial \mu} f_s^+ = \delta(\varepsilon_s - \mu)$$

so that in this case, if we take the $V \to \infty$ limit after the $T \to 0$ limit, we have

$$\bar{N} = -\frac{\partial \Omega}{\partial \mu} = -\frac{\partial \Omega_0}{\partial \mu},$$

and hence

$$\mu = \mu^0. \tag{8.17}$$

This is an absurd result, since we know that the chemical potential of an interacting system is not always equal to that of the corresponding non-interacting system. Thus, in this case we must take the $V \to \infty$ limit before the $T \to 0$ limit; then the delta-functions can contribute to the thermodynamic potential, and $\mu \neq \mu^0$.

It is conjectured that what has gone wrong with the discussion above in the case that the $T \to 0$ limit is taken first is that the radius of convergence of the perturbation series as a function of the strength of the interaction, λ, is very small. If V is large and finite, the true energy levels of the interacting system, plotted as functions of λ, have sharp bends, so as not to cross each other. If we take the $T \to 0$ limit, we focus our attention on one of these levels, the lowest, and, as V gets very large, the bends become very sharp and get very close to $\lambda = 0$, and the radius of convergence tends to zero. We then obtain unreliable results such as eqn (8.17). The terms resulting in the non-crossing of the levels, and hence the non-convergence of the series, are proportional to V^{-1}, and thus vanish in the limit $V \to \infty$. If, however, we ignore them, having taken the $T \to 0$ limit first, we focus our attention on one level and follow it as a function of λ. Since it can now cross other levels we cannot be sure that we shall always be considering the ground state (see Fig. 8.4). This suggests that we should first take the $V \to \infty$ limit, to help the convergence of our series, and then take the $T \to 0$ limit, to obtain information about the ground state. This agrees with the clue given by eqn (8.16). (Kohn and Luttinger (1960).)

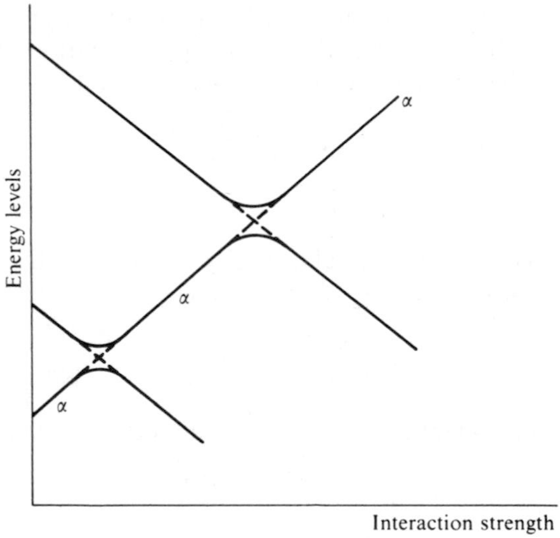

FIG. 8.4. —— With the terms preventing level-crossings. ---- Level-crossings allowed. If one ignores the terms preventing level-crossing and first takes the $T = 0$ limit, one is studying the state α.

In this case, it appears that the rules in section 8.3 are useless, as in deriving them, we have first taken the $T \to 0$ limit. There are, however, several cases in which they may be used. In atomic systems, the lowest levels are discrete, and the Fermi energy lies in this energy range, and thus between discrete levels. In this case, no problem arises. A similar situation occurs in calculations for finite nuclei. What is more surprising is that the two orders of taking the limits lead to identical results if the single-particle energies $\varepsilon_{\mathbf{q}}$ are functions only of the magnitude of the momentum $|\mathbf{q}|$ and the potential energy is spherically symmetric. There are two differences between the two sets of rules: in the first, μ is used and anomalous diagrams should be included; in the second μ^0 is used and anomalous diagrams are not included. It may be shown that in the case mentioned above, the effect of these differences exactly cancel out. This is shown to second order by Kohn and Luttinger (1960) and to all orders by Luttinger and Ward (1960).

It is advisable to remember, however, that this work indicates that for systems for which the $V \to \infty$ leads to a continuous distribution of energy levels in the neighbourhood of the Fermi energy, we have to take particular care with the anomalous diagrams, and with eqn (8.16).

The treatment of bosons at $T = 0$ is not a straightforward extension of the method for fermions, owing to the macroscopic occupation of the zero-momentum state. The difficulties and their resolution are discussed by Hugenholtz and Pines (1959) and by Mills (1971).

9

THE VARIATIONAL PRINCIPLE AND PAIR-WAVE APPROXIMATION

9.1. The variational principle

THE variational principle is often of use in discussing systems of small numbers of particles; for example in many text-books on elementary quantum mechanics, one finds a variational treatment of the helium atom. Useful results can also be obtained in this way for many-body systems, and in this chapter, we shall describe some of the techniques used in this approach and discuss their advantages and disadvantages.

We first define the one- and two-particle density matrices $\gamma(\mathbf{r}|\mathbf{r}')$ and $\Gamma(\mathbf{r}_1, \mathbf{r}_2|\mathbf{r}'_1, \mathbf{r}'_2)$ in the case that the system of N identical particles is in a definite state described by the wave-function $\Psi(\mathbf{r}_1, \mathbf{r}_2, \ldots, \mathbf{r}_N)$:

$$\gamma(\mathbf{r}_1|\mathbf{r}'_1) = \frac{N\int \Psi^*(\mathbf{r}_1, \mathbf{r}_2 \ldots \mathbf{r}_N)\Psi(\mathbf{r}'_1, \mathbf{r}_2 \ldots \mathbf{r}_N)\, \mathrm{d}^3\mathbf{r}_2 \ldots \mathrm{d}^3\mathbf{r}_N}{\int |\Psi(\mathbf{r}_1, \mathbf{r}_2 \ldots \mathbf{r}_N)|^2\, \mathrm{d}^3\mathbf{r}_1 \ldots \mathrm{d}^3\mathbf{r}_N} \tag{9.1}$$

$$\Gamma(\mathbf{r}_1, \mathbf{r}_2|\mathbf{r}'_1, \mathbf{r}'_2) = \frac{N(N-1)\int \Psi^*(\mathbf{r}_1, \mathbf{r}_2, \mathbf{r}_3 \ldots \mathbf{r}_N)\Psi(\mathbf{r}'_1, \mathbf{r}'_2, \mathbf{r}_3 \ldots \mathbf{r}_N)\, \mathrm{d}^3\mathbf{r}_3 \ldots \mathrm{d}^3\mathbf{r}_N}{\int |\Psi(\mathbf{r}_1 \ldots \mathbf{r}_N)|^2\, \mathrm{d}^3\mathbf{r}_1 \ldots \mathrm{d}^3\mathbf{r}_N} \tag{9.2}$$

(We have not taken Ψ to be normalized.)

The average number of particles in a single-particle state with wave-function $\phi_\alpha(\mathbf{r})$ is

$$\langle n_\alpha \rangle = \int \gamma(\mathbf{r}_1|\mathbf{r}'_1)\phi_\alpha(\mathbf{r}_1)\phi_\alpha^*(\mathbf{r}'_1)\, \mathrm{d}^3\mathbf{r}_1\, \mathrm{d}^3\mathbf{r}'_1. \tag{9.3}$$

To see this, expand Ψ in terms of the complete set of normalized and symmetrized states $\{\Phi_{\{n_\gamma\}}\}$ in which n_1 particles are in the first single-particle state, with wave-function ϕ_1, n_γ in the γth state, etc.:

$$\Psi(\mathbf{r}_1 \ldots \mathbf{r}_N) = \sum_{\{n_\gamma\}} A(\{n_\gamma\})\Phi_{\{n_\gamma\}}(\mathbf{r}_1 \ldots \mathbf{r}_N). \tag{9.4}$$

Then

$$\langle n_\alpha \rangle = \sum_{\{n_\gamma\}} n_\alpha |A(\{n_\gamma\})|^2 \Big/ \int |\Psi|^2\, \mathrm{d}^3\mathbf{r}_1 \ldots \mathrm{d}^3\mathbf{r}_N. \tag{9.5}$$

Also, combining eqn (9.1) and (9.3),

$$\langle n_\alpha \rangle = \frac{N \int \Psi^*(\mathbf{r}_1, \mathbf{r}_2 \dots \mathbf{r}_N)\Psi(\mathbf{r}'_1, \mathbf{r}_2 \dots \mathbf{r}_N)\phi_\alpha(\mathbf{r}_1)\phi^*_\alpha(\mathbf{r}'_1)\, \mathrm{d}^3\mathbf{r}_1 \dots \mathrm{d}^3\mathbf{r}_N\, \mathrm{d}^3\mathbf{r}'_1}{\int |\Psi|^2\, \mathrm{d}^3\mathbf{r}_1 \dots \mathrm{d}^3\mathbf{r}_N}$$

$$= N \sum_{\{n_\gamma\}} \sum_{\{n'_\gamma\}} A^*(\{n_\gamma\})A(\{n'_\gamma\}) \int \Phi^*_{\{n_\gamma\}}(\mathbf{r}_1 \dots \mathbf{r}_N)\phi^*_\alpha(\mathbf{r}'_1) \times$$

$$\times\, \Phi_{\{n'_\gamma\}}(\mathbf{r}'_1, \mathbf{r}_2 \dots \mathbf{r}_N)\phi_\alpha(\mathbf{r}_1)\mathrm{d}^3\mathbf{r}_1 \dots \mathrm{d}^3\mathbf{r}_N\mathrm{d}^3\mathbf{r}'_1/(\textstyle\int |\Psi|^2\, \mathrm{d}^3\mathbf{r}_1 \dots \mathrm{d}^3\mathbf{r}_N) \tag{9.5a}$$

The factors $\Phi^*_{\{n_\gamma\}}(\mathbf{r}_1 \dots \mathbf{r}_N)\phi^*_\alpha(\mathbf{r}'_1)$ and $\Phi_{\{n_\gamma\}}(\mathbf{r}'_1, \mathbf{r}_2 \dots \mathbf{r}_N)\phi_\alpha(\mathbf{r}_1)$ in the integrand in the numerator are wave-functions for an $N+1$ particle system which are neither symmetrized nor normalized with respect to the particle at $\mathbf{r}'_1$ or $\mathbf{r}_1$. Performing the integrals over the **r**s shows, after some combinatorics, that the two expressions (9.5 and 9.5a) for $\langle n_\alpha \rangle$ are identical.

If the system is spatially homogeneous and large, γ will be a function only of $|\mathbf{r}_1 - \mathbf{r}'_1|$ except near the boundaries of the system. Often, we may neglect boundary effects and use periodic boundary conditions; $\langle n_\mathbf{k} \rangle$ then reduces to the Fourier transform of γ:

$$\langle n_\mathbf{k} \rangle = \frac{1}{V} \int \exp[\mathrm{i}\mathbf{k} \,.\, (\mathbf{r}_1 - \mathbf{r}'_1)]\gamma(\mathbf{r}_1|\mathbf{r}'_1)\, \mathrm{d}^3\mathbf{r}_1\, \mathrm{d}^3\mathbf{r}'_1. \tag{9.6}$$

The kinetic energy of the system is then

$$\text{K.E.} = (2m)^{-1} \sum_\mathbf{k} k^2 \langle n_\mathbf{k} \rangle. \tag{9.7}$$

The two-particle correlation function is defined:

$$\rho_2(\mathbf{r}_1, \mathbf{r}_2) = \Gamma(\mathbf{r}_1, \mathbf{r}_2|\mathbf{r}_1, \mathbf{r}_2). \tag{9.8}$$

Again for spatially homogeneous systems, ρ_2 will be a function only of $|\mathbf{r}_1 - \mathbf{r}_2|$, and may then be written in terms of the radial distribution function, $g(r)$:

$$\rho_2(\mathbf{r}_1, \mathbf{r}_2) = \rho^2 g(r_{12}). \tag{9.9}$$

$g(r_{12})$ is proportional to the probability of finding any particle at $\mathbf{r}_2$, given that there is one at $\mathbf{r}_1$. It may be determined experimentally from the liquid structure factor (which is, apart from a constant, its Fourier transform) by scattering X-rays or neutrons from the system. (See, e.g., Henshaw (1960).)

The mutual potential energy of the particles is

$$\text{P.E.} = \tfrac{1}{2} \int v(|\mathbf{r}_1 - \mathbf{r}_2|)\rho_2(\mathbf{r}_1, \mathbf{r}_2)\, \mathrm{d}^3\mathbf{r}_1\, \mathrm{d}^3\mathbf{r}_2, \tag{9.10}$$

where $v(r)$ is the potential energy of two particles separated by a distance r and where use has been made of the symmetry of the wave function. For spatially homogeneous systems:

$$\text{P.E.} = \frac{N}{2}\rho \int v(r)g(r)\,\mathrm{d}^3\mathbf{r}. \tag{9.11}$$

Finally, the particle density is given by:

$$\rho(\mathbf{r}) = \gamma(\mathbf{r}|\mathbf{r}) \tag{9.12}$$

so that if the system is placed in an external potential $U(\mathbf{r})$, the resulting potential energy is

$$U = \int U(\mathbf{r})\rho(\mathbf{r})\,\mathrm{d}^3\mathbf{r}. \tag{9.13}$$

The total energy of the system is then:

$$E = \frac{1}{2m}\sum_{\mathbf{k}} k^2\langle n_{\mathbf{k}}\rangle + \tfrac{1}{2}\int v(\mathbf{r}_1-\mathbf{r}_2)\rho_2(\mathbf{r}_1,\mathbf{r}_2)\,\mathrm{d}^3\mathbf{r}_1\,\mathrm{d}^3\mathbf{r}_2 + \int U(\mathbf{r})\rho(\mathbf{r})\,\mathrm{d}^3\mathbf{r} \tag{9.14}$$

At first sight, it appears that one might use the 'density matrix', Γ, as a variational function, since each term in eqn (9.14) may be expressed in terms of $\Gamma(\mathbf{r}_1, \mathbf{r}_2|\mathbf{r}'_1, \mathbf{r}_2)$. This would save a lot of hard work, since the direct use of the wave-function involves the use of N coordinates, whereas $\Gamma(\mathbf{r}_1, \mathbf{r}_2|\mathbf{r}'_1, \mathbf{r}_2)$ is a function only of three. Unfortunately, such a procedure would be incorrect, because one does not know that to every Γ there corresponds a wave-function such that eqn (9.2) is satisfied. Indeed, one can at once see that there are many Γs which do not have a corresponding wave-function, since, if eqn (9.2) is to be satisfied, we must have

$$\Gamma(\mathbf{r}_1, \mathbf{r}_2|\mathbf{r}'_1, \mathbf{r}'_2) = \varepsilon\Gamma(\mathbf{r}_2, \mathbf{r}_1|\mathbf{r}'_1, \mathbf{r}'_2) \quad \text{etc.} \tag{9.15}$$

Before we can use the variational principle in terms of Γ, we need a set of conditions sufficient to ensure that it is derivable from a wave-function. The search for such a set has come to be known as the N-representability problem. Its solution has been determined in principle (see Garrod and Percus (1964)), but unfortunately the conditions are too complicated to be of use in a practical calculation.

Practical calculations have been done on atomic systems, using an idea associated with the variational theorem. If we were to use a sufficient set of conditions on Γ, as described above, and could do an exact variational calculation on E, as expressed in eqn (9.14), we should obtain the exact ground-state energy. If we relax one or more of the conditions on Γ, and then perform a variational calculation, the minimum in our energy functional, if it occurs, will be at an energy equal to, or lower than the true ground-state energy, as shown symbolically in Fig. 9.1. We shall thus achieve a lower bound on the ground-state energy. In this method, we thus write down a set of necessary conditions on Γ (such as 9.15) and vary E with respect to

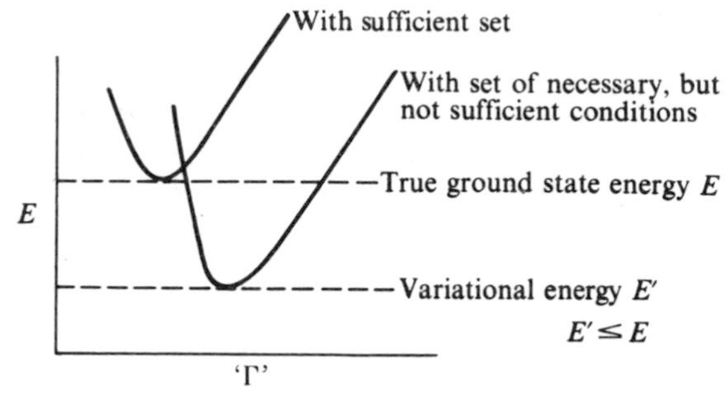

FIG. 9.1.

all Γs subject to these conditions. The more closely the set of necessary conditions approaches a set sufficient to ensure that Γ may be derived from a wave-function, the more closely will our variational energy approach the true ground-state energy. The search for sets of conditions which will lead to useful results has been to some extent successful for atomic systems (see, e.g., Bopp (1959), Coleman (1963), Kijewski and Percus (1967)). Very little work has been done on the application of this idea to solids and fluids.

9.2. The pair-wave approximation

As an alternative approach, we might assume some specific approximate form for the wave-function, and then use the variational theorem. The simplest form to assume would be a symmetrized, or antisymmetrized product of one-particle wave-functions. However, we have seen in previous chapters that the correlations induced by strong interparticle forces at small separations are often very important. Indeed, the expectation value of the potential energy for a system of particles with hard-core interactions is not finite unless the wave-function is zero for interparticle separations less than the diameter of the hard core. The simplest form, for bosons, which can achieve this is

$$\Psi = \prod_{i<j} f(r_{ij}) \tag{9.16}$$

where r_{ij} is the distance between the ith and jth particles and where the product is to be taken over all pairs in the system. For the ground state of a Bose system, f may be taken to be real. Such a form is sometimes referred to as a Bijl–Jastrow wave function (Bijl (1940), Jastrow (1955)). If the inter-atomic potential is very large for $r < a$, we expect $f(r)$ to be very small in that region. Further we should expect f to approach a constant† (which we can choose equal to unity, since it is convenient to work with a wave-function

† The constant must not be zero, since Ψ would then be zero for nearly all configurations, as most particles are well separated.

which is not normalized) for large values of r, provided there are no long-range correlations in our system. For the atoms of a crystalline solid, for example, f would not have this property, but would depend on the vector displacement between pairs of particles, of which it would be a periodic function, with the periodicity of the lattice. The expectation value of the energy for the wave-function of eqn (9.16) is, in the absence of an external potential,

$$E = \frac{\int \prod_{i<j} f(r_{ij})[-N(2m)^{-1}\nabla_1^2 + \frac{1}{2}N(N-1)v(r_{12})] \prod_{i<j} f(r_{ij})\, d^3\mathbf{r}_1 \dots d^3\mathbf{r}_N}{\int |\Psi^2|\, d^3\mathbf{r}_1 \dots d^3\mathbf{r}_N}$$

$$= [(2m)^{-1}[-\nabla_1^2 \gamma(\mathbf{r}_1|\mathbf{r}_2)]_{\mathbf{r}_1=\mathbf{r}_2} + \tfrac{1}{2}\int v(r)g(r)\, d^3\mathbf{r}]V \tag{9.17}$$

where $g(r)$ is defined by eqn (9.9), which, in the present case reduces to

$$g(r_{12}) = V^2 \int \prod_{i<j} |f(r_{ij})|^2\, d^3\mathbf{r}_3 \dots d^3\mathbf{r}_N \Big/ \int \prod_{i<j} |f(r_{ij})|^2\, d^3\mathbf{r}_1 \dots d^3\mathbf{r}_N. \tag{9.18}$$

Since the energy is now expressed entirely in terms of f, a variational calculation may be performed with f as the variational function. Unfortunately, an integro-differential equation of great complexity results. We may, however, obtain an expansion of the energy in powers of the density by following the cluster development of Ursell and Mayer (see Mayer and Mayer (1940)) in classical statistical mechanics. To do this, one writes

$$h(r) = f^2(r) - 1 \tag{9.19}$$

and notes that h differs from zero only in the region (of volume v_c, say) of interparticle correlations. Then, we may write for g, for example:

$$\begin{aligned} g(r_{12}) = f^2(r_{12})\Bigg[1 + \frac{N}{V}\int h(r_{13})h(r_{32})\, d^3\mathbf{r}_3 + \\ + \frac{N^2}{V^2}\Bigg\{\iint [2h(r_{14})h(r_{43})h(r_{32}) + 4h(r_{13})h(r_{24})h(r_{43})h(r_{23}) + \\ + h(r_{41})h(r_{13})h(r_{32})h(r_{24})h(r_{43})]\, d^3\mathbf{r}_3\, d^3\mathbf{r}_4 + \\ + \tfrac{1}{2}\left(\int h(r_{13})h(r_{32})\, d^3\mathbf{r}_3\right)^2\Bigg\} + \cdots\Bigg]. \end{aligned} \tag{9.20}$$

The second term in the square bracket will be of order ρv_c times the first one, provided $h(r)$ dies away sufficiently quickly for large r. Similarly, the third term is of order $(\rho v_c)^2$ times the first. Thus, provided the system is sufficiently dilute, or the interaction of sufficiently short range, for ρv_c to be small, we have an expansion for g which is useful, however strong the interparticle forces. We shall not describe the use of this expansion here.† Instead, we write

$$f(r) = \exp\{\tfrac{1}{2}u(r)\}. \tag{9.21}$$

The kinetic energy of the system may then be written:

$$\text{K.E.} = N(2m)^{-1}\int \nabla_1\Psi^*.\nabla_1\Psi\,\mathrm{d}^3\mathbf{r}_1\ldots\mathrm{d}^3\mathbf{r}_N \bigg/ \int |\Psi|^2\,\mathrm{d}^3\mathbf{r}_1\ldots\mathrm{d}^3\mathbf{r}_N \tag{9.22}$$

$$= (8m)^{-1}\int [\nabla_1 u(r_{12})]^2\rho(\mathbf{r}_1,\mathbf{r}_2)\,\mathrm{d}^3\mathbf{r}_1\,\mathrm{d}^3\mathbf{r}_2 +$$

$$+(8m)^{-1}\int \nabla_1 u(r_{12}).\nabla_1 u(r_{13})\rho_3(\mathbf{r}_1,\mathbf{r}_2,\mathbf{r}_3)\,\mathrm{d}^3\mathbf{r}_1\,\mathrm{d}^3\mathbf{r}_2\,\mathrm{d}^3\mathbf{r}_3. \tag{9.23}$$

ρ_2 is defined in eqn (9.8) and

$$\rho_3(\mathbf{r}_1,\mathbf{r}_2,\mathbf{r}_3) = \frac{N(N-1)(N-2)\int \Psi^*(\mathbf{r}_1,\mathbf{r}_2\ldots\mathbf{r}_N)\Psi(\mathbf{r}_1\ldots\mathbf{r}_N)\,\mathrm{d}^3\mathbf{r}_4\ldots\mathrm{d}^3\mathbf{r}_N}{\int |\Psi|^2\,\mathrm{d}^3\mathbf{r}_1\ldots\mathrm{d}^3\mathbf{r}_N}. \tag{9.24}$$

But

$$\nabla_1\rho_2(\mathbf{r}_1,\mathbf{r}_2) = \nabla_1 u(\mathbf{r}_{12})\rho_2(\mathbf{r}_1,\mathbf{r}_2) + \int \nabla_1 u(r_{13})\rho_3(\mathbf{r}_1,\mathbf{r}_2,\mathbf{r}_3)\,\mathrm{d}^3\mathbf{r}_3. \tag{9.25}$$

Putting this into eqn (9.23), we obtain

$$\text{K.E.} = (8m)^{-1}\int \nabla_1 u(r_{12}).\nabla_1\rho_2(\mathbf{r}_1,\mathbf{r}_2)\,\mathrm{d}^3\mathbf{r}_1\,\mathrm{d}^3\mathbf{r}_2 \tag{9.26}$$

$$= (8m)^{-1}\Bigg[\int \frac{[\nabla_1\rho_2(\mathbf{r}_1,\mathbf{r}_2)]^2}{\rho_2(\mathbf{r}_1,\mathbf{r}_2)}\,\mathrm{d}^3\mathbf{r}_1\,\mathrm{d}^3\mathbf{r}_2 -$$

$$-\int \frac{\nabla_1\rho_2(\mathbf{r}_1,\mathbf{r}_2)\nabla_1 u(r_{13})\rho_3(\mathbf{r}_1,\mathbf{r}_2,\mathbf{r}_3)\,\mathrm{d}^3\mathbf{r}_1\,\mathrm{d}^3\mathbf{r}_2\,\mathrm{d}^3\mathbf{r}_3}{\rho_2(\mathbf{r}_1,\mathbf{r}_2)}\Bigg]. \tag{9.27}$$

The final step seems to be in the direction of increasing complexity, and is useful only in the case of low density. In this case the second term on the right-hand side of eqn (9.27) may be seen to be of order ρv_c times the first, since $\nabla u(r)$ will be appreciable only for r within v_c. Thus, for low densities, we may neglect this term and:

$$\frac{E}{N} = (8m)^{-1}\rho\int \left[\left(\frac{\mathrm{d}g}{\mathrm{d}r}\right)^2 \bigg/ g\right] 4\pi r^2\,\mathrm{d}r + \tfrac{1}{2}\rho\int v(r)g(r)\,4\pi r^2\,\mathrm{d}r. \qquad \ddagger(9.28)$$

† See, for example, Jastrow (1955), Gersch and Smith (1960).

‡ This equation may also be obtained using expansions of the type of eqn (9.20).

In this approximation, eqn (9.25) becomes

$$\nabla_1 \rho_2(\mathbf{r}_1, \mathbf{r}_2) = \nabla_1 u(r_{12}) \rho_2(\mathbf{r}_1, \mathbf{r}_2), \tag{9.29}$$

which, for every ρ_2, may be solved for u. Thus in this approximation there is no N-representability problem, and ρ_2, or g, may be used as a variational function. Thus, applying Euler's equations† to the expression for the energy in eqn (9.27),

$$(4m)^{-1} \frac{\mathrm{d}}{\mathrm{d}r} \left(r^2 \frac{\mathrm{d}g}{\mathrm{d}r} \bigg/ g \right) = -(8m)^{-1} r^2 \left(\frac{\mathrm{d}g}{\mathrm{d}r} \right)^2 \bigg/ g^2 + \tfrac{1}{2} r^2 v(r),$$

which may be rewritten:

$$-(mr)^{-1} \frac{\mathrm{d}^2}{\mathrm{d}r^2}(r\sqrt{g}) + v(r)\sqrt{g} = 0 \tag{9.30}$$

This equation is rather similar to the Schrödinger equation for two particles interacting through a potential $v(r)$. If higher-order terms in the expansion are included,‡ the correlation function g is found to satisfy an integro-differential equation again resembling the Schrödinger equation for two particles, but with an effective potential consisting of the actual two-body potential with an additional potential produced by neighbouring particles. This then provides an extension of the Hartree equation to strongly-interacting systems.

Let us consider eqn (9.30) in the simple case of a hard-sphere potential:

$$\begin{aligned} v(r) &= \infty, \qquad r < a \\ &= 0, \qquad r \geqslant a. \end{aligned}$$

In this case, $g = 0$ for $r < a$, and for $r \geqslant a$:

$$\frac{\mathrm{d}^2}{\mathrm{d}r^2}(r\sqrt{g}) = 0. \tag{9.31}$$

But, from eqns (9.2, 8 and 9),

$$\int g(r)\, \mathrm{d}^3\mathbf{r} = V$$

and thus $g(r)$ must be equal to unity almost everywhere. Thus $g(r) = 1$ for large r, and in the present case, from eqn (9.31),

$$g = \left(1 - \frac{a}{r}\right)^2.$$

† See, e.g. Courant and Hilbert (1953).
‡ See, e.g. Aviles (1958).

From eqn (9.28) we then have

$$\frac{E}{N} = 2\pi\rho a/m \tag{9.32}$$

a result first obtained by Lenz (1929).

In order to improve on this approximation, we must return to eqn (9.25). An approximation which has been much used for ρ_3 is called the superposition approximation, introduced by Kirkwood (1942, 1950) in classical statistical mechanics:

$$\rho^3\rho_3(\mathbf{r}_1, \mathbf{r}_2, \mathbf{r}_3) = \rho_2(\mathbf{r}_1, \mathbf{r}_2)\rho_2(\mathbf{r}_2, \mathbf{r}_3)\rho_2(\mathbf{r}_3, \mathbf{r}_1). \tag{9.33}$$

This form is physically reasonable, and satisfies the requirement that ρ_3 should vanish when any two particles are close to each other in the case of an infinitely repulsive potential; further, ρ_3 is independent of the separation of any two particles if they are far apart. A similar approximation has been used for gases in classical statistical mechanics, and various authors have tested it. They find that the second and third virial coefficients are given correctly, but that the fourth is in error by twenty or thirty per cent. In the theory of classical liquids, Green (1960) has used the approximation in different ways, and tested the consistency of the results, and concludes that the error introduced by the approximation is probably small, but not completely negligible.

If we use the superposition approximation in eqn (9.25), we obtain

$$\nabla_1\rho_2(\mathbf{r}_1, \mathbf{r}_2) = \nabla_1 u(r_{12})\rho_2(\mathbf{r}_1, \mathbf{r}_2) + \rho^{-3}\rho_2(\mathbf{r}_1, \mathbf{r}_2)\int \nabla_1 u(r_{13})\rho_2(\mathbf{r}_2, \mathbf{r}_3)\rho_2(\mathbf{r}_1, \mathbf{r}_3). \tag{9.34}$$

∇u may be expressed in terms of ρ_2, or g (which can be determined experimentally), by an iterative solution of this equation, which then gives an expansion of the energy in powers of the density (of which, in the above example, we used only the first term). Abe (1958*a*) considers the case in which the first two terms are taken. In terms of g, eqn (9.34) may be written:

$$g'(r_{12}) = u'(r_{12})g(r_{12}) + \rho g(r_{12})\int u'(r_{31})g(r_{23})g(r_{31})\cos(12, 13)\,\mathrm{d}^3\mathbf{r}_3, \tag{9.35}$$

where primes denote differentiation, and cos(12, 13) is the cosine of the angle between $\mathbf{r}_{12}$ and $\mathbf{r}_{13}$. Wu and Feenberg (1961) took the experimentally determined g for liquid helium, and solved eqn (9.35) for $u'(r)$ numerically. The resulting u' was then integrated with the boundary condition $u(\infty) = 0$, which follows from eqn (9.21) and the choice $f(\infty) = 1$. The function $u(r)$ so obtained can be used to estimate the kinetic energy from eqn (9.26); Wu and Feenberg obtained $2{\cdot}9\times10^{-15}$ erg/atom. The experimental figure is a little uncertain, but probably it is about $2{\cdot}5 \times 10^{-15}$ erg/atom. This is very good agreement for this type of calculation, and, although we should not take it to indicate that all is well with our approximations in eqns (9.16) and (9.33), at least we may place a little more confidence in them.

Similar work using the pair wave-function for the Bose gas has been done by, for example, Lieb (1963), Lieb and Sakakura (1964), Lieb and Liniger (1964), Penrose (1960), Karp (1959), Hiroike (1962) and Abe (1958*b*). McMillan (1965) has used eqn (9.16), together with Monte-Carlo techniques for 32 and for 108 atoms in a cube to discuss many properties of the ground state of liquid helium, and Parry and Rathbone (1967) have used it to compute the single-particle density matrix, γ, and hence the momentum distribution function. Bowley (1970) has discussed the surface tension by this method. Gersch and Smith (1960) have studied the connection between this method and the approximations one uses in diagrammatic perturbation theory. Bogoliubov and Zubarev (1955) showed that the pair wave-function form was consistent with a collective coordinate description of the system to the extent that phonon–phonon interactions were neglected. Reatto and Chester (1967) used the pair-wave function taking into account both short- and long-range correlations. They showed that the latter must be present if the system is to support low-lying phonon modes. They further showed that the infinite-range correlations were decisive in determining whether or not the interacting system exhibits Bose–Einstein condensation.

In the case of fermions, eqn (9.16) may no longer be used as an approximate wave-function, since it is symmetrical, rather than antisymmetrical under the interchange of the coordinates of two particles. A suitable antisymmetric generalization of eqn (9.16) is

$$\Psi = \prod_{i>j} f(r_{ij})\Delta$$

where Δ is the Slater determinant of a set of single-particle wave-functions:

$$\Delta = \sum_{P} (-1)^P \prod_{i} \phi_{\alpha_i}(\mathbf{r}_j)/(N!)^{\frac{1}{2}},$$

where the sum is taken over all permutations of the N particle coordinates amongst the N states, and $(-1)^P$ is $+1$ for even permutations and -1 for odd. In the case of particles in a box, for example, of a large number of helium three atoms, the ϕ_αs are taken to be plane-wave states, but for smaller systems, for example, the atomic nucleus, other single-particle states may be appropriate.

To express the expectation value of the energy in terms of f is a more complicated problem in this case. Iwamoto and Yamada (1957*a*, *b*) have developed an expansion for the energy in terms of f, similar to the expansions described above. The dependence of the Fermi momentum on the density results in an expansion in mixed powers of ρ and $\rho^{\frac{2}{3}}$. It has been discussed and applied to the case of liquid helium three by Wu and Feenberg (1962). Emery (1958) used the lowest-order terms to derive an approximation for the nuclear binding energy. An extensive review of this approach to the fermion problem was given by Clark and Westhaus (1966), and this contains many references to previously published work. More recently, several authors have used the method to analyse high-energy electron scattering from nuclei (see, for example, Ciofi degli Atti and Kabachnik (1970), Ripka and Gillespie (1970)).

One advantage of the pair-wave approximation is the ease with which the low-order terms in an expansion in powers of the density may be found. It is the simplest generalization of the Hartree–Fock approximation which can be used for potentials with hard cores. It is also useful for describing finite systems, or surface effects in large systems. The relative simplicity of the method may be judged by comparing the work leading to eqn (9.32), and the solution of the t-matrix equation which we should have to write down if we were to do the same problem in terms of diagrammatic perturbation theory. On the other hand, higher-order terms lead to greatly increased complexity, and the method cannot be extended to non-zero temperatures in any obvious way. When one is discussing systems for which ρv_c is not small, expansions in powers of the density have to be abandoned, and an approximation such as the superposition approximation has to be used; the effects of this approximation are usually difficult to estimate.

The chief disadvantage of the method is that an approximation is made right at the start, in that the correct wave-function is not of the form given in eqn (9.16). Again, the effects of this approximation are difficult to estimate. Schiff and Verlet (1967) used this wave-function approximation for liquid helium three and four, with molecular dynamical methods, for a system of 864 atoms. The discrepancies between their results for the energy and equilibrium density and the experimental values are of the order of those due to the uncertainties in the interatomic potential. But in assessing what this most impressive result can tell us about eqn (9.16) one should remember that, in variational calculations, good results for the energy may sometimes be obtained for rather poor wave-functions.

A comparison of the results of the pair-wave approximation with a perturbation expansion is given by Sim, Woo and Buchler (1970), and Sim and Woo (1970). They find that the three leading orders in the interaction strength are given correctly by the pair-wave approximation; in the fourth order, only certain terms are included. The use of this wave-function corresponds to summations of selected diagrams to all orders. Similar work is reported by Lee (1971*a*, *b*) and Day (1971).

APPENDIX A

ONE AND TWO-BODY OPERATORS IN SECOND QUANTIZED FORM

WE wish to show that the representation of an operator $F = \sum_{s=1}^{N} f(\mathbf{r}_s, \mathbf{p}_s)$ for an N-body system is, in terms of creation and annihilation operators, $\sum_{ij} f_{ij} a_i^+ a_j$, where f_{ij} is the matrix element of $f(\mathbf{r}, \mathbf{p})$ between the states ϕ_i and ϕ_j:

$$f_{ij} = \int \phi_i^* f(\mathbf{r}, \mathbf{p}) \phi_j \, \mathrm{d}^3\mathbf{r}.$$

The proof is straightforward but tedious, and we present it partly to show how tiresome it would be to use configuration space all the time. All the essential ideas of the proof are contained in exercises 1.1.1, 1.2.1, and 1.3.1, and we suggest that the reader attempts these exercises before reading this appendix.

We compare the matrix elements of the two forms of F. Consider the boson case first, for the second-quantized form:

$$\left\langle n_1', \dots n_i', \dots n_j', \dots \left| \sum_{ij}' f_{ij} a_i^+ a_j \right| n_1, \dots n_i, \dots n_j, \dots \right\rangle$$
$$= \sum_{ij}' \delta_{n_1, n_1'} \dots \delta_{n_i+1, n_i'} \dots \delta_{n_j-1, n_j'} \dots f_{ij}[(n_i+1)n_j]^{\frac{1}{2}}.$$

We have excluded the term $i = j$, and indicated this by putting a prime on the summation symbol, since this term needs special consideration.

In configuration space, let Φ $(\mathbf{r}_1, \dots \mathbf{r}_N)$ be the wave-function for which n_1 particles are in the first single-particle state, ... n_i in the ith, and so on, and $\Psi(\mathbf{r}_1 \dots \mathbf{r}_N)$, for which n_i' are in the ith state, etc. As a result of the symmetrization of the wave-functions, Φ will be a sum of $N!/(n_1! \dots n_i! \dots n_j! \dots)$ terms (the numbers of ways of arranging the N particles so that n_1 are in the first state, n_2 in the second, and so on). There will thus be a normalization factor of $[n_1! \dots n_i! \dots n_j! \dots/N!]^{\frac{1}{2}}$.

Consider a typical term in the operator $\sum_s f(\mathbf{r}_s, \mathbf{p}_s)$, say $f(\mathbf{r}_t, \mathbf{p}_t)$, and let us evaluate the contribution to the matrix element $\int \dots \int \Psi^* F \Phi \, \mathrm{d}^3\mathbf{r}_1 \dots \mathrm{d}^3\mathbf{r}_N$ from all terms in Φ in which the tth particle is in the jth single-particle state, and all terms in Ψ in which the tth particle is in the ith state. There will be $(N-1)!/[n_1! \dots n_i! \dots (n_j-1)! \dots]$ such terms in Φ. When the integration over all the $\mathbf{r}$s apart from $\mathbf{r}_t$ is performed, we shall have a non-zero result only if $n_1' = n_1 \dots n_i' - 1 = n_i \dots n_j' = n_j - 1$. In this case there will be only one term in Ψ giving a non-zero result for each of the special terms we are considering in Φ. The integration over $\mathbf{r}_t$ then gives f_{ij}. Summing over t will give a factor N, as the wave-functions are symmetrized, and allowing i and j to vary will give the sum over i and j. Hence, collecting together all the factors:

$$\int \dots \int \mathrm{d}^3\mathbf{r}_1 \dots \mathrm{d}^3\mathbf{r}_N \Psi^*(\mathbf{r}_1 \dots \mathbf{r}_N) \sum_{s=1}^{N} f(\mathbf{r}_s, \mathbf{p}_s)\Phi(\mathbf{r}_1 \dots \mathbf{r}_N)$$

$$= \sum_{ij}{}' \delta_{n_1,n_1'} \dots \delta_{n_i+1,n_i'} \dots \delta_{n_j-1,n_j'} \times$$

$$\times N \times [n_1'! \dots n_i'! \dots n_j'! \dots /N!]^{\frac{1}{2}} \times [n_1! \dots n_i \dots n_j! \dots /N!]^{\frac{1}{2}} \times$$

$$\times (N-1)!/[n_1! \dots n_i! \dots (n_j-1)! \dots] +$$

$$+ \text{similar term for } i = j$$

$$= \sum_{ij}{}' \delta_{n_1,n_1'} \dots \delta_{n_i+1,n_i'} \dots \delta_{n_j-1,n_j'} \dots [(n_i+1)n_j]^{\frac{1}{2}} +$$

$$+ \text{similar term for } i = j.$$

We thus see that the two matrix elements are the same. The terms for which $i = j$ are left as an exercise.

For fermions, in the second-quantized form,

$$\left\langle n_1' \dots, n_i' \dots, n_j' \dots \left| \sum_{ij}{}' f_{ij} a_i^+ a_j \right| n_1 \dots, n_i, \dots n_j \dots \right\rangle$$

$$= \sum_{ij}{}' (-1)^{\gamma} [(1-n_i)n_j]^{\frac{1}{2}} \delta_{n_1,n_1'} \dots \delta_{n_i+1,n_1'} \dots \delta_{n_j-1,n_j'} f_{ij},$$

where

$$\gamma = \sum_{k=1}^{k=j-1} n_k + \sum_{k'=1}^{k'=i-1} n_{k'}'.$$

In configuration space, we shall have a factor of $(N!)^{-1}$ from the two normalization factors and $(N-1)! \times n_j$ for the number of terms in Φ with the tth particle in the jth single-particle state. Integration over the $\mathbf{r}$'s will give the δ-functions and f_{ij} as before, together with a factor $(1-n_i)$ to allow for the fact that we must have one, and only one, particle in the ith single-particle state in Ψ. Summation over t gives a factor of N. The sign of each contributing term must next be considered. We define the sign of our wave-functions so that the Slater determinant is written with the states going downwards in our chosen order, and the particle-labels across in ascending order: for example, for three particles in the three states labelled 3, 5, and 8:

$$\Delta = \begin{vmatrix} \phi_3(\mathbf{r}_1) & \phi_3(\mathbf{r}_2) & \phi_3(\mathbf{r}_3) \\ \phi_5(\mathbf{r}_1) & \phi_5(\mathbf{r}_2) & \phi_5(\mathbf{r}_3) \\ \phi_8(\mathbf{r}_1) & \phi_8(\mathbf{r}_2) & \phi_8(\mathbf{r}_3) \end{vmatrix}.$$

From the general properties of determinants, the sign of the term in $\Psi^*\Phi$ of the form $\phi_\nu^*(\mathbf{r}_1)\phi_\nu(\mathbf{r}_1)\phi_\mu^*(\mathbf{r}_2)\phi_\mu(\mathbf{r}_2)\ldots\phi_i^*(\mathbf{r}_t)\phi_j(\mathbf{r}_t)\ldots$ is $(-1)^\gamma$, where γ is the number of states appearing in Φ with labels lying between i and j. Thus, as before

$$\gamma = \sum_{k=1}^{k=j-1} n_k + \sum_{k'=1}^{k'=i-1} n'_{k'},$$

for each term. Thus we have:

$$\int \ldots \int \Psi^* F \Phi \, \mathrm{d}^3\mathbf{r}_1 \ldots \mathrm{d}^3\mathbf{r}_N = \sum_{ij}{}' \delta_{n_1 n'_1} \ldots \delta_{n_i+1, n'_i} \ldots \delta_{n_j-1 n'_j} \ldots (-1)^\gamma f_{ij} \times$$

$$\times [(1-n_i)n_j]^{\frac{1}{2}} + \text{similar term for } i = j,$$

since $\sqrt{n_j} = n_j$ and $\sqrt{(1-n_j)} = 1-n_j$ when $n_j = 0$ or 1.

The proofs of the equality of the matrix elements for the two-body operators follow in very much the same way. It is hoped that by now the reader is convinced of the convenience of the creation- and annihilation-operator formalism.

APPENDIX B

To prove:

(i) $\exp(A)B\exp(-A) = B+[A, B]+\frac{1}{2}[A,[A, B]]+\cdots$

(ii) $a_s^+\rho_0^{(\beta)} = \rho_0^{(\beta)}a_s^+\exp[\beta(\varepsilon_s-\mu)]$

(iii) $a_s\rho_0^{(\beta)} = \rho_0^{(\beta)}a_s\exp[-\beta(\varepsilon_s-\mu)]$

(iv) $\bar{a}_s(\tau) = a_s^+\exp[\tau(\varepsilon_s-\mu)]$

(v) $a_s(\tau) = a_s\exp[-\tau(\varepsilon_s-\mu)]$

(vi) $S^+\exp(-\beta\omega_0 S^z) = \exp(-\beta\omega_0 S^z)S^+\exp(\beta\omega_0)$.

Proofs:

(i) Define $B(t) = \exp(At)B\exp(-At)$

Then

$$\mathrm{d}B/\mathrm{d}t = A\exp(At)B\exp(-At) - \exp(At)B\exp(-At)A$$

$$= [A, B(t)]$$

Similarly,

$$\mathrm{d}^2B/\mathrm{d}t^2 = [A,[A, B(t)]]$$

and $\mathrm{d}^nB/\mathrm{d}t^n$ is equal to n successive commutations of $B(t)$ with A. But $\exp(A)B\exp(-A) = B(1)$, which, using Taylor's theorem in the form

$$B(t) = B(0)+t[\mathrm{d}B/\mathrm{d}t]_{t=0}+t^2/2![\mathrm{d}^2B/\mathrm{d}t^2]_{t=0}+\cdots,$$

may be written

$$\exp(A)B\exp(-A) = B+[A, B]+\tfrac{1}{2}[A,[A, B]]+\cdots+ \qquad \text{Q.E.D.}$$

(ii) From (i)

$$\exp\left[\beta\sum_{s'}(\varepsilon_{s'}-\mu)a_{s'}^+a_{s'}\right]a_s^+\exp\left[-\beta\sum_{s'}(\varepsilon_{s'}-\mu)a_{s'}^+a_{s'}\right]$$

$$= a_s^+ + \beta(\varepsilon_s-\mu)a_s^+ + \frac{1}{2!}\beta^2(\varepsilon_s-\mu)^2a_s^+ + \cdots$$

$$= a_s^+\exp[\beta(\varepsilon_s-\mu)].$$

Operating on the left with $\exp[-\beta\sum_{s'}(\varepsilon_{s'}-\mu)a_{s'}^+a_{s'}]$ we have

$$a_s^+\rho_0^{(\beta)} = \rho_0^{(\beta)}a_s^+\exp[\beta(\varepsilon_s-\mu)] \qquad \text{Q.E.D.}$$

(iii) follows in a similar way.

(iv) From (i) we have

$$\bar{a}_s(\tau) = a_s^+ + \tau(\varepsilon_s-\mu)a_s^+ + (\tau^2/2)(\varepsilon_s-\mu)^2a_s^+ + \cdots$$

$$= a_s^+\exp[\tau(\varepsilon_s-\mu)].$$

(v) and (vi) follow in a similar way.

APPENDIX C

THE LINKED CLUSTER THEOREM

WE first consider the two examples shown in Fig. C.1. These are fourth-order diagrams with two disconnected parts, and, if the linked-cluster theorem is true, we pick up their contributions in $[\mathscr{C}(\beta)]^2/2!$ where $\mathscr{C}(\beta)$ is the sum of the contributions of all connected diagrams. We have shown in eqn (2.43) that the same integrals, matrix elements, and Green functions appear in $[\mathscr{C}(\beta)]^2/2!$ and in the contribution of the disconnected diagrams. Here we wish to show that the numerical factors are the same in both cases.

We work in the simplest diagrammatic representation, without the complications introduced in sections 2.4 and 2.6 to reduce the number of diagrams. We must, however, consider all the diagrams of each type. Thus, there are four diagrams with identical contributions to that of Fig. C.1, with a different τ-label for the upper vertex, and if we are to include all diagrams of the type indicated in (i), we must include a factor of 4 in our contribution. We could also take a factor (of 2) for the number of ways of interchanging the τ-labels of the lower part of the diagram which lead to the same contribution from different contractions. An identical factor appears in the corresponding third-order term in $\mathscr{C}(\beta)$, however, and so we omit it for simplicity. We also have a factor of $1/4!$ since this is a fourth-order term. In $[\mathscr{C}(\beta)]^2$, this term will appear with a factor of 2 from the binomial theorem, because it is a cross term, and with factors of $1/3!$ and $1/1!$, because it is a product of a third-order and a first-order term. The linked-cluster theorem tells us to divide $[\mathscr{C}(\beta)]^2$ by a factor of $2!$. The two numerical factors are thus the same.

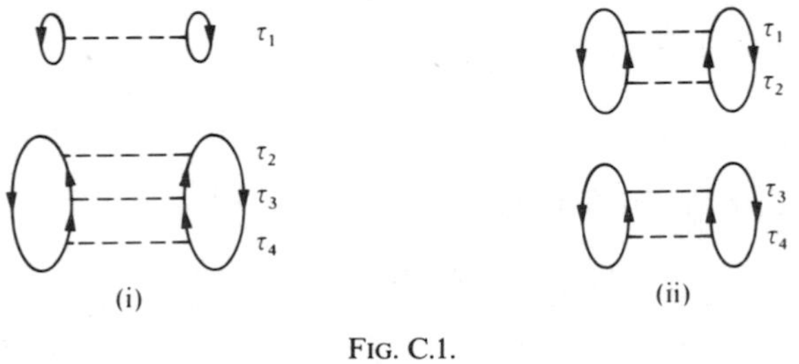

FIG. C.1.

Fig. C.1(ii) is of a slightly different type, in that the two disconnected parts are identical. We thus have a factor $4!/(2!)^3$ for the number of ways of choosing the four τ-labels so that there are two for the upper part and two for the lower. We also have a factor of $1/4!$ since it is a fourth-order term. In

$[\mathscr{C}(\beta)]^2/2!$ this term appears with a factor $(1/2)^3$, since it is the product of two identical second-order terms.

In general, consider a qth-order diagram with r disconnected parts, n_i identical parts of order m_i, n_j identical parts of order m_j, etc., where

$$\sum_i n_i m_i = q \quad \text{and} \quad \sum_i n_i = r.$$

In the expansion for Q this will have a factor $q!/[(m_i!)^{n_i}(m_j!)^{n_j} \ldots n_i!n_j! \ldots]$ for the number of ways of arranging the q τ-labels so that there are n_i identical sets with m_i in them, etc. We shall also have a factor $1/q!$ since this is a qth-order diagram. The corresponding term from the expression given by the linked-cluster theorem has a factor $1/r!$ from the theorem, a factor $[(m_i!)^{n_i}(m_j!)^{n_j} \ldots]^{-1}$ from the products of n_i diagrams of order m_i, etc., and a factor $r!(n_i!n_j! \ldots)$ from the multinomial theorem. The two factors are thus identical and the theorem follows.

APPENDIX D

WE present an alternative proof of eqn (2.48) as an example of a technique which is often useful in performing the sums over l which occur in the ω-representation of perturbation theory. We wish to prove that

$$g_s(\tau) = -\beta^{-1} \sum_l \exp(-i\omega_l \tau)/[i\omega_l - \varepsilon_s + \mu], \tag{D.1}$$

where $\omega_l = (2l+1)\pi/\beta$ for fermions, and $2l\pi/\beta$ for bosons. We shall first consider the fermion case for $-\beta < \tau < 0$. Consider the integral:

$$\mathscr{I} = \frac{1}{2\pi i} \int_C d\zeta \frac{\exp(-\zeta\tau)}{[\zeta - \varepsilon_s + \mu][\exp(\beta\zeta) + 1]}$$

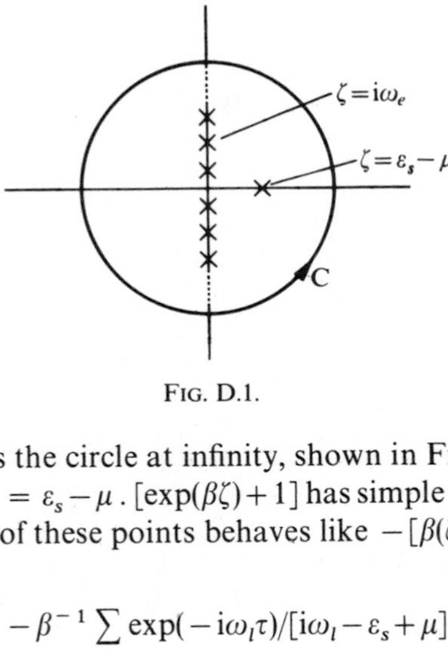

FIG. D.1.

where the contour C is the circle at infinity, shown in Fig. D.1. $1/[\zeta - \varepsilon_s + \mu]$ gives a simple pole at $\zeta = \varepsilon_s - \mu$. $[\exp(\beta\zeta) + 1]$ has simple poles at $\zeta = i\omega_l$ and in the neighbourhood of these points behaves like $-[\beta(\zeta - i\omega_l)]^{-1}$. Thus, by Cauchy's theorem,

$$\mathscr{I} = -\beta^{-1} \sum_l \exp(-i\omega_l\tau)/[i\omega_l - \varepsilon_s + \mu] +$$
$$+ \exp[-(\varepsilon_s - \mu)\tau]/\{\exp[\beta(\varepsilon_s - \mu)] + 1\}.$$

But the integral over the circle at infinity is zero, since $-\beta < \tau < 0$. Thus $\mathscr{I}$ is zero and eqn (D.1) follows at once.

A similar proof may be given for $\tau > 0$ by considering the integral

$$\frac{1}{2\pi i} \int_C \frac{\exp[(\beta - \tau)\zeta]}{(\zeta - \varepsilon_s + \mu)[\exp(\beta\zeta) + 1]}.$$

For the case of bosons, we replace the factor $[\exp(\beta\zeta) + 1]^{-1}$ in the integrals by $[\exp(\beta\zeta) - 1]^{-1}$ as might be expected.

In general, it is possible to rewrite in terms of integrals all the sums which appear in the theory by multiplying either by f^+ or by f^-, where:

$$f^+(\zeta) = \exp(\beta\zeta)[\exp(\beta\zeta) - \varepsilon]^{-1}$$

$$f^-(\zeta) = [\exp(\beta\zeta) - \varepsilon]^{-1}.$$

The choice is made so that the integral round the circle at infinity is zero, so that the sum over l is replaced by a sum of residues at the other non-analytic points of the function we are integrating. Often, one is summing functions which have a series of poles, or discontinuities, along the real axis. In this case, for fermions, one changes the summation first to an integration around the imaginary axis, as above, and then, if the function is analytic everywhere except on the real axis, to an integration along the real axis (the contour C_F in Fig. D.2). Some care is necessary in the case of bosons, as one term in the sum

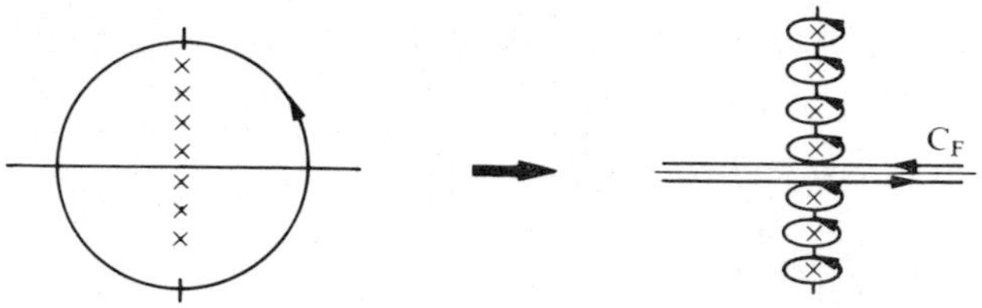

FIG. D.2.

to be performed has $i\omega_n = 0$. The contour C'_B must then include the whole of the real axis apart from the origin (Fig. D.3). For a fairly complicated example in the use of this technique, see section 7.4.

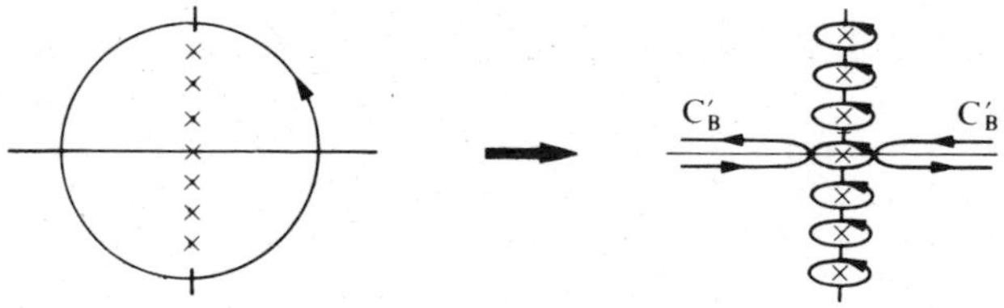

FIG. D.3.

APPENDIX E

THE DIRAC DELTA-FUNCTION

WE wish to prove:

$$\operatorname*{Lt}_{\gamma \to 0} \frac{1}{x - \mathrm{i}\gamma} = \frac{P}{x} + \mathrm{i}\pi\delta(x)$$

where P indicates that the principal value is to be taken of any integral in which $(x-\mathrm{i}\gamma)^{-1}$ appears; i.e.

$$\mathscr{I} \equiv \operatorname*{Lt}_{\gamma \to 0} \int_{-\infty}^{\infty} f(x)(x-\mathrm{i}\gamma)^{-1}\,\mathrm{d}x = P\int_{-\infty}^{\infty} f(x)x^{-1}\,\mathrm{d}x + \mathrm{i}\pi f(0),$$

provided that $f(x)$ is sufficiently well-behaved. We may rewrite $\mathscr{I}$:

$$\mathscr{I} = \operatorname*{Lt}_{\gamma,\gamma' \to 0}\left(\int_{-\infty}^{-\gamma'} + \int_{-\gamma'}^{\gamma'} + \int_{\gamma'}^{\infty}\right) f(x)(x-\mathrm{i}\gamma)^{-1}\,\mathrm{d}x$$

where $\gamma' \gg \gamma$

$$= P\int_{-\infty}^{\infty} f(x)x^{-1}\,\mathrm{d}x + \operatorname*{Lt}_{\gamma,\gamma' \to 0}\int_{\gamma'}^{\gamma'} [f(0) + xf'(0) - x^2 f''(0)](x-\mathrm{i}\gamma)^{-1}\,\mathrm{d}x$$

provided $f(x)$ is sufficiently well-behaved. Thus, performing the integrals and taking the limit,

$$\mathscr{I} = P\int_{-\infty}^{\infty} f(x)x^{-1}\,\mathrm{d}x + \mathrm{i}\pi f(0). \qquad \text{Q.E.D.}$$

Similarly,

$$(x-\mathrm{i}\gamma)^{-1} - (x+\mathrm{i}\gamma)^{-1} = 2\pi\mathrm{i}\delta(x).$$

APPENDIX F

THE EQUATION OF MOTION FOR THE DENSITY MATRIX

THE normalized density matrix may be written in the form

$$\rho = \sum_n |n\rangle p_n \langle n| \qquad (\alpha)$$

where p_n is the probability that the system is in the state described by the state-vector $|n\rangle$. We then have, for any operator A, and any complete set of states $\{|s\rangle\}$

$$\begin{aligned}\mathrm{Tr}\{\rho A\} &= \sum_s \langle s| \sum_n |n\rangle p_n \langle n|A|s\rangle \\ &= \sum_n p_n \sum_s \langle n|A|s\rangle\langle s|n\rangle \\ &= \sum_n p_n \langle n|A|n\rangle\end{aligned}$$

as required, using the completeness of the set $\{|s\rangle\}$.

Using the form (α), we may write down the equation of motion for ρ:

$$\begin{aligned}\mathrm{i}\frac{\mathrm{d}\rho}{\mathrm{d}t} &= \sum_n \mathrm{i}\left(\frac{\mathrm{d}|n\rangle}{\mathrm{d}t} p_n\langle n| + |n\rangle p_n \frac{\mathrm{d}\langle n|}{\mathrm{d}t}\right) \\ &= \sum_n (H|n\rangle p_n\langle n| - |n\rangle p_n \langle n|H)\end{aligned}$$

from Schrödinger's time-dependent equation. Thus

$$\mathrm{i}\frac{\mathrm{d}\rho}{\mathrm{d}t} = [H, \rho].$$

APPENDIX G

WICK'S THEOREM AT $T = 0$

WE consider the case of fermions and first divide our creation and annihilation operators into two sets, A and B. A contains the annihilation operators for s such that $\varepsilon_s > \mu^0$ and the creation operators for $\varepsilon_s < \mu^0$, and B contains the remaining creation and annihilation operators. We define the 'normal product' of a set of operators, written $N(...)$, as the product with the operators arranged in such a way that the members of set A are all to the right of the members of set B: the product is multiplied by $(-1)^s$, where s is the number of interchanges of Fermi operators needed to achieve the desired order. The order of the operators of set A amongst themselves, or of set B amongst themselves, is immaterial. We note that if a is any member of set A, and $|\Phi_0\rangle$ is the ground-state vector for the non-interacting system, then:

$$a|\Phi_0\rangle = 0$$

Similarly,

$$\langle\Phi_0|b = 0. \tag{G.1}$$

The difference between the τ-ordered product and the normal product of two creation or annihilation operators is called the contraction of these operators, written, for example, $\{a_{s_1}(\tau_1)\bar{a}_{s_2}(\tau_2)\}''$. It is a number, since:

$$\{a_{s_1}(\tau_1)a_{s_2}(\tau_2)\}'' = 0 = \{\bar{a}_{s_1}(\tau_1)\bar{a}_{s_2}(\tau_2)\}''$$

$$\begin{aligned}\{a_{s_1}(\tau_1)\bar{a}_{s_2}(\tau_2)\}'' &= \delta_{s_1s_2}\exp\{(\tau_2-\tau_1)\varepsilon_{s_1}], & \tau_1 > \tau_2 \quad \varepsilon_{s_1} > \mu^0 \\ &= 0 & \tau_1 \leqslant \tau_2 \quad \varepsilon_{s_1} > \mu^0 \\ &= 0 & \tau_1 > \tau_2 \quad \varepsilon_{s_1} < \mu^0 \\ &= -\delta_{s_1s_2}\exp[(\tau_2-\tau_1)\varepsilon_{s_1}] & \tau_1 \leqslant \tau_2 \quad \varepsilon_{s_1} < \mu^0.\end{aligned}$$

We may write

$$\begin{aligned}\{a_{s_1}(\tau_1)\bar{a}_{s_2}(\tau_2)\}'' &= \delta_{s_1s_2}\exp[(\tau_2-\tau_1)\varepsilon_{s_1}][1-\langle n_{s_1}\rangle_0] & \tau_1 > \tau_2 \\ &= -\delta_{s_1s_2}\exp[(\tau_2-\tau_1)\varepsilon_{s_1}]\langle n_{s_1}\rangle_0 & \tau_1 \leqslant \tau_2,\end{aligned} \tag{G.2}$$

where:

$$\begin{aligned}\langle n_s\rangle_0 &= 1, & \varepsilon_s < \mu^0 \\ &= 0, & \varepsilon_s > \mu^0\end{aligned}$$

which is very similar to the definition of the contraction in eqn (2.36). Notice, however, that the Fermi energy, μ^0, referred to here is that for the non-interacting system.

Wick's theorem at zero temperature states that the τ-ordered product of a set of operators is equal to the normal product, plus the sum of the normal products with one pair contracted, all pairs being taken in turn, plus the sum of the normal products with two pairs contracted, and so on. The sign of each term is $(-1)^t$, where t is the number of interchanges of Fermi operators needed to bring together the contracted operators. Thus, for example, for the four Fermi operators, C, D, E, F,

$$\begin{aligned}\mathscr{T}\{CDEF\} = N(CDEF)+\{CD\}''N(EF)-\{CE\}''N(DF)+\{CF\}''N(DE)+\\ +\{DE\}''N(CF)-\{DF\}''N(CE)+\{EF\}''N(CD)+\\ +\{CD\}''\{EF\}''-\{CE\}''\{DF\}''+\{CF\}''\{DE\}'',\end{aligned} \tag{G.3}$$

and, in general,

$$\begin{aligned}\mathscr{T}\{C_1C_2C_3C_4 \dots C_n\} = N(C_1C_2C_3C_4 \dots C_n)+\{C_1C_2\}''N(C_3 \dots C_n)-\\ -\{C_1C_3\}''N(C_2C_4 \dots C_n)+ \cdots +\\ +\{C_1C_2\}''\{C_3C_4\}''N(\dots C_n)+ \cdots .\end{aligned} \tag{G.4}$$

The final terms will be of the form $\{C_1C_2\}''\{C_3C_4\}'' \dots \{C_{n-1}C_n\}''$ if n is even and $\{C_1C_2\}''\{C_3C_4\}'' \dots \{\dots\}''C_n$ if n is odd. Note that when one takes the expectation value in the non-interacting ground state of this equation, all the normal product terms vanish, and we are left with only products of paired contractions, as in the form of Wick's theorem in Chapter 2.

We prove the theorem by induction. For two operators, it corresponds to the definition of the contraction. Suppose that it is true for the ordered product of n operators. Then,

$$\mathscr{T}\{C_1C_2 \dots C_{n+1}) = (-1)^s\mathscr{T}\{C_1C_2 \dots C_{p-1}C_{p+1} \dots C_{n+1}\}C_p, \tag{G.5}$$

where C_p is the operator with the smallest τ-label, and s is the number of interchanges of operators needed to place it at the end of the product. Suppose C_p belongs to set A. If we multiply eqn (G.4) on the left by C_p, it can be taken inside the normal products without further ado, since it is already in the correct position. Further, all contractions with C_p are zero, as it belongs to set A and has the smallest τ-label of all the operators. Similarly, if C_p belongs to set B, it can be commuted through all the operators on the right-hand side of eqn (G.4) and then put inside the normal product. The anticommutator of C_p with each operator is just the contraction of C_p with that operator, since C_p belongs to set B and has the lowest τ-label. The theorem thus holds for $n+1$ operators whether C_p belongs to set A or B, and the theorem follows by induction.

REFERENCES

ABE, R. (1958*a*). *Prog. theor. Phys.* **19**, 57. (1958*b*). *Prog. theor. Phys.* **19**, 713.

ABRIKOSOV, A. A. (1965). *Physics* **2**, 5.

ABRIKOSOV, A. A., GOR'KOV, L. P., *and* DZYALOSHINSKII, I. Ye (1959). *Zh. éksp. teor. Fiz.* **36**, 900, translated in *Soviet Phys. JETP* **9**, 636. (1965). *Quantum field theoretical methods in statistical physics.* Pergamon.

ATKINS, K. R. (1959). *Liquid Helium.* Cambridge University Press.

AVILES, J. B. (1958). *Ann. Phys.* **5**, 251.

BALIAN, R. *and* DE DOMINICIS, C. (1971). *Ann. Phys.* **62**, 229.

BARDEEN, J., COOPER, L. N., *and* SCHRIEFFER, J. R. (1957). *Phys. Rev.* **108**, 1175.

BAUMANN, K. *and* RANNINGER, J. (1962). *Ann. Phys.* **20**, 157.

BAYM, G. *and* MERMIN, N. D. (1961). *J. math. Phys.* **2**, 232.

BETHE, H. A. *and* GOLDSTONE, J. (1957). *Proc. R. Soc.* A **238**, 551.

BIJL, A. (1940). *Physica* **7**, 869.

BLOCH, C. *and* DE DOMINICIS, C. (1958). *Nucl. Phys.* **1**, 459. (1959). *Nucl. Phys.* **10**, 181.

BLOOMFIELD, P. E. *and* NAFARI, N. (1972). *Phys. Rev.* A **5**, 806.

BOGOLIUBOV, N. N. (1947). *J. Phys. U.S.S.R.* **11**, 23.

BOGOLIUBOV, N. N. *and* ZUBAREV, D. N. (1955). *Zh. éksp. teor. Fiz.* **28**, 129, translated in *Soviet Phys. JETP* **1**, 83.

BONCH-BRUEVICH, V. L. *and* TYABLIKOV, S. V. (1962). *The Green function method in statistical mechanics.* North Holland.

BOPP, F. (1959). *Z. Phys.* **156**, 348.

BOWLEY, R. M. (1970). *J. phys. Chem.* **3**, 2012.

BRANDOW, B. H. (1971). *Ann. Phys.* **64**, 21.

BRENIG, W. *and* PARRY, W. E. (1963). *Z. Phys.* **175**, 40.

BROWN, G. E. (1971). *Rev. mod. Phys.* **43**, 1.

BRUECKNER, K. A. *and* WADA, W. (1956). *Phys. Rev.* **103**, 1008.

BRUECKNER, K. A. (1959). In *The many-body problem.* Methuen.

CHAPMAN, S. *and* COWLING, T. G. (1952). *The mathematical theory of non-uniform gases.* Cambridge University Press.

CHESTER, G. V. (1963). *Rep. Prog. Phys.* **XXVI**, 411.

CHESTER, G. V. *and* THELLUNG, A. (1961). *Proc. Phys. Soc.* **77**, 1005.

CIOFI DEGLI ATTI, C. *and* KABACHNIK, N. M. (1970). *Phys. Rev.* C **1**, 809.

CLARK, J. W. *and* WESTHAUS, P. (1966). *Phys. Rev.* **141**, 833.

COLEMAN, A. J. (1963). *Rev. mod. Phys.* **35**, 668.

COURANT, R. *and* HILBERT, D. (1953). *Methods of mathematical physics*, Vol. 1. Interscience.

DAVIS, H. L. (1960). *Phys. Rev.* **120**, 789.

DAY, B. D. (1971). *Phys. Rev.* A **4**, 681.

DE DOMINICIS, C. T. (1957). University of Birmingham. Ph.D. thesis.

DEVLIN, J. F. *and* VERTOGEN, G. (1972). *Physica* **57**, 455.

DONIACH, S. (1966). *Phys. Rev.* **144**, 382.

DYSON, F. J. (1956). *Phys. Rev.* **102**, 1230.

EDWARDS, S. F. (1958). *Phil. Mag.* **3**, 1020.

EMERY, V. J. (1958). *Nucl. Phys.* **6**, 585.

FEENBERG, E. (1969). *Theory of quantum fluids.* Academic Press.

FRADKIN, E. (1959). *Nucl. Phys.* **12**, 465.

GARROD, C. *and* PERCUS, J. K. (1964). *J. math. Phys.* **5**, 1756.

GAUDIN, M. (1960). *Nucl. Phys.* **15**, 89.

GAVORET, J. *and* NOZIÈRES, P. (1964). *Ann. Phys.* **28**, 349.

GERSCH, H. A. *and* SMITH, V. H. (1960). *Phys. Rev.* **119**, 886.
GILLIS, N. S. (1968). *J. math. Phys.* **9**, 2007.
GIOVANNINI, B., PETER, M., *and* KOIDÉ, S. (1966). *Phys. Rev.* **149**, 251.
GOLDSTONE, J. (1957). *Proc. R. Soc.* A **239**, 267.
GOTTFRIED, K. (1966). *Quantum mechanics*, Vol. 1. Benjamin.
GÖTZE, W. *and* WAGNER, H. (1965). *Physica* **31**, 475.
GREEN, H. S. (1960). *Handb. Phys.* **X**, 1.
TER HAAR, D. (1966). *Elements of thermostatics*. Holt, Rinehart, and Winston, Inc.
HAAS, C. W. *and* JARRETT, H. S. (1964). *Phys. Rev.* **135** A, 1089.
HENSHAW, D. G. (1960). *Phys. Rev.* **119**, 9, 14.
HENSHAW, D. G. *and* WOODS, A. D. B. (1961). *Phys. Rev.* **121**, 1266.
HIROIKE, K. (1962). *Prog. theor. Phys.* **27**, 342.
HOLSTEIN, T. *and* PRIMAKOFF, H. (1940). *Phys. Rev.* **58**, 1098.
HUBER, A. (1969). In *Mathematical methods in solid-state and superfluid theory*, edited by R. C. Clark and G. H. Derrick. Oliver and Boyd.
HUGENHOLTZ, N. M. (1965). *Rep. Prog. Phys.* **XXVIII**, 201.
HUGENHOLTZ, N. M. *and* PINES, D. (1959). *Phys. Rev.* **116**, 489.
IWAMOTO, F. *and* YAMADA, M. (1957*a*). *Prog. theor. Phys.* **17**, 543. (1957*b*). *Prog. theor. Phys.* **18**, 345.
JASTROW, R. (1955). *Phys. Rev.* **98**, 1479.
JOHNSTON, J. R. (1970). *Am. J. Phys.* **38**, 516.
KADANOFF, L. P. *and* BAYM, G. (1962). *Quantum statistical mechanics*. Benjamin.
KARP, I. L. (1959). *Phys. Rev.* **115**, 223.
KEHR, K. (1967). *Physica* **33**, 620.
KELLY, H. P. (1963). *Phys. Rev.* **131**, 684. (1969). *Adv. chem. Phys.* **XIV**, 129.
KENAN, R. P. (1966). *J. appl. Phys.* **37**, 1453.
KIJEWSKI, L. *and* PERCUS, J. K. (1967). *J. math. Phys.* **8**, 2184.
KIRKWOOD, J. G. *and* BOGGS, E. M. (1942). *J. chem. Phys.* **10**, 394.
KIRKWOOD, J. G., MAUN, E. K., *and* ALDER, B. J. (1950). *J. chem. Phys.* **18**, 1040.
KITTEL, C. (1958). *Phys. Rev.* **110**, 1295. (1963). *Quantum theory of solids*. Wiley.
KOBE, D. H. (1966). *J. math. Phys.* **7**, 1806.
KOHN, W. *and* LUTTINGER, J. M. (1960). *Phys. Rev.* **118**, 41.
KRAMERS, H. A. (1927). *Atti Congr. int. Fiz. Como.* **2**, 545.
KRONIG, R. (1926). *J. opt. Soc. Am.* **12**, 547.
KUBO, R. (1956). *Can. J. Phys.* **34**, 1274. (1957). *J. phys. Soc. Japan* **12**, 570. (1958). *Lectures in theoretical physics* **1**, 120. Boulder.
KUPER, C. G. (1968). *An introduction to the theory of superconductivity*. Clarendon Press, Oxford.
KWOK, P. C. *and* WOO, J. W. F. (1971). *Phys. Rev.* A **3**, 437.
LANDAU, L. D. (1941). *Zh. éksp. teor. Fiz.* **11**, 592. Translation *J. Phys. U.S.S.R.* **5**, 71. (1944). *Zh. éksp. teor. Fiz.* **14**, 112. Translation *J. Phys. U.S.S.R.* **8**, 1. (1947). *J. Phys. U.S.S.R.* **11**, 91. (1956). *Zh. éksp. teor. Fiz.* **30**, 1058. Translated in *Soviet Phys. JETP* **3**, 920. (1957). *Zh. éksp. teor. Fiz.* **32**, 59. Translated in *Soviet Phys. JETP* **5**, 101. (1958*a*). *Zh. éksp. teor. Fiz.* **34**, 262. Translated in *Soviet Phys. JETP* **7**, 182. (1958*b*). *Zh. éksp. teor. Fiz.* **35**, 97. Translated in *Soviet Phys. JETP* **8**, 70 (1959).
LANDAU, L. D. *and* KHALATINKOV, I. M. (1949) *Zh. éksp. teor. Fiz.* **19**, 637, 709.
LANGER, J. S. (1960). *Phys. Rev.* **120**, 714. (1961). *Phys. Rev.* **124**, 1003. (1962*a*). *Phys. Rev.* **127**, 5. (1962*b*). *Phys. Rev.* **128**, 110.
LEE, D. K. (1971*a*). *Phys. Rev.* A **3**, 345. (1971*b*). *Phys. Rev.* A **4**, 1670.
LENZ, W. (1929). *Z. Phys.* **56**, 778.
LEWIS, W. W. *and* STINCHCOMBE, R. B. (1967). *Proc. phys. Soc.* (*London*) **92**, 1002, 1010.
LIEB, E. H. (1963). *Phys. Rev.* **130**, 2518.
LIEB, E. H. *and* LINIGER, W. (1964). *Phys. Rev.* **134**A, 312.
LIEB, E. H. *and* SAKAKURA, A. Y. (1964). *Phys. Rev.* **133** A, 899.
LUTTINGER, J. M. (1960). *Phys. Rev.* **119**, 1153. (1961). *Phys. Rev.* **121**, 942. (1968). *Phys. Rev.* **174**, 263.
LUTTINGER, J. M. *and* NOZIÈRES, P. (1962). *Phys. Rev.* **127**, 1431.
LUTTINGER, J. M. *and* WARD, J. C. (1960). *Phys. Rev.* **118**, 1417.

McMillan, W. L. (1965). *Phys. Rev.* **138**A, 442.
Mahan, G. D. (1966). *Phys. Rev.* **142**, 366.
Martin, P. C. *and* Schwinger, J. (1959). *Phys. Rev.* **115**, 1342.
Matsubara, T. (1955). *Prog. theor. Phys.* **14**, 351.
Mattis, D. C. (1965). *Theory of magnetism*. Harper & Row.
Mayer, J. E. *and* Mayer, M. G. (1940). *Statistical mechanics*. Wiley and Sons.
Miller, A., Pines, D., *and* Nozières, P. (1962). *Phys. Rev.* **127**, 1452.
Miller, J. H. *and* Kelly, H. P. (1971). *Phys. Rev.* A, **4**, 480.
Mills, R. (1971). *Phys. Rev.* A **4**, 394.
Mills, R. L., Kenan, R. P., *and* Korringa, J. (1960). *Physica* **26**, S 204.
Montroll, E. W. (1959). *Rendiconti della Scuola Istituto di Fisica 'Enrico Fermi' Varenna.* (1960). *Lectures in theoretical physics* **3**, 221. Boulder.
Montroll, E. W. *and* Ward, J. (1958). *Physics of Fluids* **1**, 55.
Nozières, P. (1963). *Theory of interacting Fermi systems*. Benjamin.
Nozières, P. *and* De Dominicis, C. (1969). *Phys. Rev.* **178**, 1097.
Nozières, P. *and* Luttinger, J. M. (1962). *Phys. Rev.* **127**, 1423.
Nozières, P. *and* Pines, D. (1958). *Phys. Rev.* **109**, 741.
Okubo, S. (1971). *J. math. Phys.* **12**, 1123.
Parry, W. E. *and* Turner R. E. (1962*a*). *Ann. Phys.* **17**, 301. (1962*b*). *Phys. Rev.* **128**, 929. (1963). *J. math. Phys.* **4**, 530.
Parry, W. E. *and* Rathbone, C. R. (1967). *Proc. phys. Soc.* **91**, 273.
Peierls, R. E. (1934). *Helv. phys. Act. Suppl.* **24**, 7. (1955). *Quantum theory of solids*. Oxford University Press.
Penrose, O. (1960). *Proc. R. Soc.* A **256**, 106.
Peterson, R. L. (1967). *Rev. mod. Phys.* **39**, 69.
Phillips, T. G. *and* Rosenberg, H. M. (1966). *Rep. Prog. Phys.* **XXIV**, 1, 285.
Pines, D. (1961). *The many-body problem*. Benjamin.
Pines, D. *and* Nozières, P. (1966). *The theory of quantum liquids*. Benjamin.
Pitaevski, L. P. (1959). *Zh. éksp. teor. Fiz.* **37**, 1794. Translated in *Soviet Phys. JETP* **10**, 1267 (1960).
Puff, R. D. (1965). *Phys. Rev.* **137**A, 406.
Quinn, J. J. *and* Ferrel, R. A. (1958). *Phys. Rev.* **112**, 812. *J. Nucl. Energy* Part C **2**, 18.
Reatto, L. *and* Chester, G. V. (1967). *Phys. Rev.* **155**, 88.
Rickayzen, G. (1965). *Theory of superconductivity*. Benjamin.
Ripka, G. *and* Gillespie, J. (1970). *Phys. Rev. Letters* **25**, 1624.
Roth, L. M. (1968). *Phys. Rev. Letters* **20**, 1431.
Schiff, D. *and* Verlet, L. (1967). *Phys. Rev.* **160**, 208.
Schweber, S. S. (1961). *An introduction to relativistic quantum field theory*. Row, Petersen & Co.
Schwinger, J. (1952). *A.E.C. report* NYO-3071.
Seavey, M. H. *and* Tannenwald, P. E. (1958). *Phys. Rev. Letters* **1**, 168.
Sim, H.-K., Woo, C.-W., *and* Buchler, J. R. (1970). *Phys. Rev.* A **2**, 2024.
Sim, H.-K. *and* Woo, C.-W. (1970). *Phys. Rev.* A **2**, 2032.
Sinclair, R. N. *and* Brockhouse, B. N. (1960). *Phys. Rev.* **120**, 1638.
Sommerfeld, A. (1928). *Z. Phys.* **47**, 1.
Soven, P. (1969). *Phys. Rev.* **178**, 1136.
Spencer, H. J. (1968*a*). *Phys. Rev.* **167**, 430, 434. (1968*b*). *Phys. Rev.* **171**, 515.
Stevens, K. W. H. *and* Toombs, G. A. (1965). *Proc. phys. Soc.* **85**, 1307.
Stinchcombe, R. B., Horwitz, G., Englert, F., *and* Brout, R. (1963). *Phys. Rev.* **130**, 155.
Suhl, H. *and* Werthamer, N. R. (1961). *Phys. Rev.* **122**, 359.
Tahir-Kheli, R. A. *and* ter Haar, D. (1962). *Phys. Rev.* **127**, 88, 95.
Thouless, D. (1957). *Phys. Rev.* **107**, 1162. (1960). *Ann. Phys.* **10**, 533. (1961). *The quantum mechanics of many-body systems*. Academic Press.
Thouless, D. J. (1964). *Rep. Prog. Phys.* **XXVII**, 53.
Titchmarsh, E. C. (1939). *The theory of functions*. Oxford University Press.
Toigo, F. *and* Woodruff, T. O. (1970). *Phys. Rev.* B **2**, 3958. (1971) *Phys. Rev.* B **4**, 371.
Tyablikov, S. V. *and* Bonch-Bruevich, V. L. (1962). *Advances in Physics*, **11**, 317.

VAKS, V. G., LARKIN, A. I., *and* PIKIN, S. A. (1967). *Zh. éksp. teor. Fiz.* **53**, 281. Translated in *Soviet Phys. JETP* **26**, 188 (1968).
VALATIN, J. G. (1961). In *Lectures in field theory and the many-body problem*, edited by E. R. Caianiello. Academic Press.
VELICKY, B. (1969). *Phys. Rev.* **184**, 614.
WANG, Y.-L., SHTRIKMAN, S., *and* CALLEN, H. (1966). *Phys. Rev.* **148**, 419.
WARD, J. C. (1950*a*). *Phys. Rev.* **77**, 293. (1950*b*). *Phys. Rev.* **78**, 182.
WILKS, J. (1967). *The properties of liquid and solid helium.* Clarendon Press, Oxford.
WOO, J. W. F. *and* JHA, S. S. (1971). *Phys. Rev.* B **3**, 87.
WOOD, J. W. F. *and* JHA, S. S. (1971). *Phys. Rev.* B **3**, 87.
WU, F. Y. *and* FEENBERG, E. (1961). *Phys. Rev.* **122**, 739. (1962). *Phys. Rev.* **128**, 943.
YOLIN, E. M. (1965). *Proc. phys. Soc.* **85**, 759.
ZUBAREV, D. N. (1960). *Usp. fiz. Nauk.* **71,** 71. Translated in *Soviet Phys. Usp.* **3**, 320.

INDEX